ENTHUSIAST'S RESTORATIC

HOW TO RESTO

TRIUMP TR5/250 & TR6

Also from Veloce Publishing

First published in 2001 by Veloce Publishing Limited, Veloce House, Parkway Farm Business Park, Middle Farm Way, Poundbury, Dorchester, Dorset, DT1 3AR, England. Telephone 01305 260068/Fax 01305 268864/e-mail info@veloce.co.uk/web www.veloce.co.uk or www.velocebooks.com
Reprinted 2002, 2003, 2004 and March 2008. This new, revised edition first published October 2010, reprinted February 2014 and April 2018.
ISBN: 978-1-787113-43-5/UPC: 6-36847-01343-1
 British Library Cataloguing in Publication Data – A catalogue record for this book is available from the British Library. Typesetting, design and page make-up all by Veloce Publishing Ltd on Apple Mac. Printed and bound by CPI Group (UK) Ltd, Croydon, CR0 4YY.

ENTHUSIAST'S RESTORATION MANUAL™

HOW TO RESTORE TRIUMPH TR5/250 & TR6

ROGER WILLIAMS

Contents

Acknowledgements & about the author

ACKNOWLEDGEMENTS

This book would never have been written without the help of a great number of people. Help in the sense of encouragement, but more particularly practical help by the unstinting provision of information, photographs and diagrams. Whilst I have appreciated every contribution, the full list is too extensive to mention everyone, but I hope all will accept my grateful thanks.

I wish to particularly single out a small number of absolutely crucial contributors, starting with John Sykes of TR Bitz. John suggested the book, got me started with a huge initial contribution, provided much by way of technical help along the way, and finally read the manuscript and made many invaluable suggestions. Alex, my wife, has provided moral support and a very practical contribution through hours spent in front of her laptop. (As an aside, I really wonder how authors wrote books before the days of computers, spell checks and auto-correction!) Gary Bates of TRGB and Malcolm Jones of PDI made significant contributions through the provision of technical information and boxes of photographs. Neil Revington of Revington TR, Alan Wadley of the TR Workshop and Steve Hall of TR Enterprises were other major sources of information, while Howard Vesey and Adam Bell of Faversham Restorations provided countless photographic opportunities. Books like this would remain stillborn without the collective help of these professionals, for which I am most grateful.

This particular book has benefited from the very large photographic and technical contribution of Mark Price of Hartford Michigan.

Last but not least, I must record my thanks to Steve Redway (of the TR Register), Jon Korbin (VTR Vehicle Registrar) and Bill Piggott for their invaluable help with my assessment of the numbers of TRs remaining. I also extend my grateful thanks to Bill Piggott for his Foreword.

ABOUT THE AUTHOR

Roger Williams was born in 1940 in Cardiff, brought up in Guildford and attended Guildford Royal Grammar School.

Aircraft became Roger's first love and he joined the de Havilland Aircraft Company in 1957 as a production engineering apprentice, and very quickly added motor cars to his list of prime interests. During the ensuing six years he not only completed his apprenticeship and studies, but built two Ford-based "specials" and started on a career in the manufacturing engineering industry as production engineer. Works managerial and directorial posts followed, and these responsibilities, together with his family commitments, reduced his time for motoring interests to exiting the company car park at the fastest possible speed!

Roger's business interest moved on to company doctoring, which he enjoyed for some ten years, specialising in turning round ailing engineering businesses. In 1986 he started his own consultancy business and renewed his motoring interests. His company specialised in helping improve client profitability by interim management or consulting assignments, whilst his spare time was – and continues to be – devoted to motor cars or writing.

Roger has owned numerous MGBs, all of which he rebuilt over a period of some seven years. He still has two of his all-time favourites – the V8 powered variants – and has two MGB books in print. More recently Roger has become involved with the Triumph marque and has restored a TR6 and, currently, a Stag.

Roger is married and lives in south west France in semi-retirement. He has two married daughters and is a Fellow of the Institution of Mechanical Engineers and a Fellow of the Institution of Engineering and Technology.

Foreword

I feel honoured to have been asked to write the foreword to this new series of books relating to the much-loved TR sports cars built between 1953 and 1981. As an author of TR subjects myself, I know only too well how difficult is the incorporation of highly technical subject matter within a readable framework, but in this awkward task I feel that Roger Williams has succeeded admirably.

We now have, for the first time, a comprehensive guide to the purchase, maintenance and restoration of the TR sports car, written in an easily understandable form which is both accurate and practical. Such a series of books on the various TR models was much needed; we have had books on TR history, books on TR originality, books on TR competition success, but not previously any definitive title on TR restoration and purchase, despite several earlier attempts.

What I should stress is that this new series is an addition to the TR enthusiast's library, not a replacement for existing literature such as workshop manuals and parts catalogues. Roger Williams' books are designed to be used in conjunction with official Triumph literature and to disseminate the collective and edited experience and expertise of owners and professional restorers as an overlay to the purely descriptive "by-the-book" methods of the TR manuals. Many aspects of TR restoration covered by Roger are absent from the manuals, perhaps for no other reason than that those who compiled the factory literature decades ago could not have envisaged that "restoration" of such cars would ever take place.

Roger's perception – with which I agree – is that the average TR owner is now older, as is indeed the case with most classic car marques. There are several reasons for this, but the principal one must be the cost factor; the cars are quite simply too expensive for a younger person to acquire and restore. These books aim to make home restoration easier for the amateur TR enthusiast, and enable him to save the considerable cost of a professional restoration. If they succeed in that aim, and in my opinion they do, then they will have made TR ownership more affordable to young and old alike, which can only be a good thing.

Roger's style is easy to read, yet informative; he writes from a practical "hands-on" point of view, encapsulating the opinions and hard-won experience of many TR restorers, both professional and amateur. This series of books will pay for themselves many times over – they have achieved exactly what was intended, and I commend them to you.

Bill Piggott
North Yorkshire, England

Introduction & using this book

INTRODUCTION

This is the first of a series of books written to help would-be and existing owners appreciate how to select, buy and restore a basically standard TR. This book will primarily focus on the TR models that used the superb six-cylinder engine and Independent Rear Suspension; in other words the TR250, the TR5 and the TR6.

Virtually every reader will already know that the Triumph TR2 to TR6 cars are regarded as the "classic" sports cars from Triumph's extensive model range, and, as such, have a special place in the hearts of all motoring enthusiasts. The TR7 and TR8 are still not regarded in quite the same light for some reason. There again, there are a few who do not consider the Michelotti cars to be in the same classic tradition as the earlier sidescreened cars. Since there is no logical reason for excluding them, my overview of the range of Triumph Roadsters includes the TR7 and 8 which recieve detailed attention in their own volume in this series of restoration manuals.

So, looking across the whole TR range, the cars were produced from 1953 to 1981, and all were extensively exported to the USA (earning many much needed dollars for the UK economy) and were always at the forefront amongst popular priced high performance sports cars. So why select the six-cylinder models for our first detailed exploration? Simply because the "six" – coupled to the technically advanced (for its day) IRS and fuel injection – saw, to my mind, the widest range of technical innovation of the whole TR era. The fact that more of these cars still exist may also have something to do with it, as might the fact that our family has a delightful TR6!

I also mentioned that I planned to focus upon the restoration of a standard, rather than a highly modified TR. That decision was made partly for reasons of available space, but also because I appreciate that some readers may not want to change the original factory specification of their TRs. For those who would like to read about upgrading their TR, this omission will be rectified in due course by a volume that deals solely with "Improvements".

However, I should tell you that I have had some difficulty deciding what is an improvement and what is not! It's easy when it comes to some issues, like, for example, installing a V8 engine in a TR, which is certainly such a major change that it's excluded from this volume. However, how would you view fitting an overdrive to a TR model for which it was only available as an optional extra, or what about welding additional strengthening gussets to known weak spots in a TR chassis? I have included both these last two modifications and a number of other, strictly speaking, non-original details, so this book is not solely about restoring a completely standard car. However, I have tried to highlight the fact when I'm suggesting a variation from standard and my reasons – which you will find are mostly safety or reliability inspired.

The Triumph TR marque was a success story. The Beach Boys sang of "Fun Fun Fun" round about the middle of the TR production run. Their song actually refers to the T-Bird (Ford Thunderbird), but could so easily have been about the TR range – for that was what these cars were all about and why they have remained so popular.

However, the cars are all getting older as, sadly, are their owners. I would guess that the majority of current owners first saw a TR when they were rather younger, thought how much they'd like one but couldn't possibly afford it, and only realised this long dormant ambition in later life. Simultaneously, the costs of a professional restoration have inevitably soared as inflation has risen inexorably, whilst the marque's numbers have shrunk. This combination of circumstances could mean that a younger generation of potential owners rarely sees the remaining TRs in use, so are neither fired with the same ambition to own a TR as those of us who are more senile once were, or are understandably daunted by the cost of purchasing or restoring a TR. I cannot instil the ambition of ownership by mere words, you have to see (and hear) the real thing! So it is up to current owners, motoring clubs and spares suppliers to so publicise these superb classics that those starting out on their careers learn to love the marque in preference to the growing range of (often Japanese) competition.

The prime objective of this book is to help owners and would-be owners carry out themselves as much of the repair and restoration work as possible at home. Indeed, I believe that many can contemplate participating in the TR experience only if they can carry out much of the work themselves, and know when and why to subcontract certain key tasks. This book will help, encourage and guide them, for only by encouraging a constant flow of new – and hopefully younger – enthusiasts to know, love and afford these classics will we ensure the long-term health of the TR marque. I sincerely hope that the information within these pages will help many new and existing owners select, restore and enjoy any Triumph Roadster.

USING THIS BOOK

The author, editors, publisher and retailer cannot accept any responsibility for personal injury, mechanical damage or financial loss, which results from errors or omission in the information given. If this disclaimer is not acceptable to you, please immediately return your unused pristine book and receipt to your retailer who will refund the purchase price paid.

Safety! – During work of any type on your car, **your personal safety MUST always be your prime consideration.** You must not undertake any of the work described in this book yourself unless you have sufficient experience, aptitude and good enough workshop facilities and equipment to ensure your personal safety at ALL times.

As stated in the Introduction, the primary purpose of this book is to guide the reader through the selection, purchase, repair and home restoration of a classic TR sports car. The book is not intended to be a workshop, operations or spares manual, but is meant to supplement and complement these invaluable sources of information. Consequently, you would be well advised to purchase the manual(s) relevant to your particular model before embarking upon a significant repair, and certainly before starting a complete restoration.

All of the component/service (approximate) prices given in the text were those prevailing in the UK at the time of publication. These prices will be subject to normal market forces and will, of course, tend to rise with inflation. You would be well advised to allow for these factors when calculating your budget. Bear in mind that it's possible that the goods and services mentioned will become unavailable or altered with the passage of time.

Note that dimensions given in the illustrations are in millimetres (unless otherwise stated) and that line illustrations are not to scale.

References to 'right side' and 'left side' are from the point of view of standing behind the car looking forward.

You may find references to non-TR5, 250 and 6 models in some text and some pictures. This is because some material within this book is common to other TR restoration books within the series. Use the information relevant to your model.

Chapter 1

Selecting your first TR, and buying tips

WHICH TR TO CHOOSE?

In all probability you have already decided which TR is your heart's desire. A decision probably made quite subconsciously on the basis of a very brief glimpse of some lucky so-and-so flashing past you, a drop-dead-gorgeous blonde's hair streaming behind, or some equally illogical reason. Me, well I LOVED the sound of the six-cylinder engine and lines of the TR6. Anyway I am (almost) past noticing blondes; on the very rare occasion I might spot one it takes all my concentration to focus on the lady in question, consequently, I rarely notice the car she is in!

However, if you have not quite made up your mind, there are some very logical and unemotional points we should discuss as part of your TR selection process. Sadly, the first consideration must be money, so before deciding anything decide where you think you fit into the various price bands for each model: clearly, it is pointless looking at £10K cars with £5K available. Take care, however, for you need to cross-check prices carefully.

Along with the basic price issue is the related question of whether you are buying from a dealer or privately. What is your Safety Net Position? Put another way, how much redress do you want? Only you can decide this; many will be happy to pay, perhaps, £1000 more to a reputable dealer, comfortable in the knowledge that if anything goes wrong – even, say, on the M25 on a Bank Holiday Monday – they can pick the up 'phone and get help. Buying privately may save some money but there's little comeback, so you are best advised to take this route only if you have good technical knowledge or the close support of someone who "knows his onions".

Furthermore, buying privately does not guarantee the lowest price. There are occasions when buyers take their privately purchased car to a dealer for work or restoration only to hear that they could have bought a similar car for less from that dealer! Bear in mind, too, that it is very difficult to value a classic car. The magazine *Classic Cars* carries a monthly valuation review under the headings of Mint, Average and Rough, whilst *Practical Classics* also offers valuation information under Excellent, Regularly-Used and Rebuild-Required categories. While you would be wise to take into account these valuations, the price guide in *Classic and Sports Cars* magazine is probably your best cost reference. All price guides need to be used very carefully, however, for, at best, they give an average figure for each category of car. You may note that a particular model's 'show' (or, as I interpret the heading, 'mint') price is quoted as £10,000 when, in fact, several cars have an asking price of £15,000 – and may well be good value for money even at that price. At the other end of the scales you could find a restoration project price of £4500 but could buy from several dealers for £2500! Nothing, therefore, beats seeing the car, asking questions and getting the feel of the market.

It is essential that you view a number of cars that fall into your target of model, price and condition before actually getting your money out of the bank! If you really have not made up your mind which TR to shortlist, it will resolve several uncertainties at once if we go through the range of TRs open to you. Not only will this give you a feel for each car's appearance (via a photograph, of course), but I will try and outline each car's major features, too. We should also evaluate its scarcity and value at the time we went to press by recording the price spread for each car as shown in a recent issue of *Classic and Sports Cars* magazine. I hope you will accept in a spirit of fun the personal views I record

1-1. No, there's no 'hidden' TR, but this seemed so in keeping with the early days of TR motoring – at least in the UK – that I felt it set the scene for times past. Not that I expect all readers to remember those days, but this was how things were when the first TRs hit the road.

1-2. Interested in classic TRs but can't tell one from the other? This picture could help you, from left to right: TR2, TR3A, TR4A, TR6 and TR7.

for each model, even if they do not completely align with your own opinion. Lastly, I hope you find it interesting and relevant if I sketch in the major technical developments of the TR range and the main competitors in the popular sports car market, for you will notice a definite correlation between each company's technical improvements and what their competitor does next!

A REVIEW OF THE TR RANGE

Origins

Spurred on by MG's success, particularly in the USA, Triumph set about its own two-seater sports car design early in the 1950s. The contemporary MG that Triumph must have used as an initial benchmark was the highly successful MG TD model. This had a 57bhp, 1250cc engine and "traditional" upright body styling comprising "humped" scuttle, flat-folding windscreen and cut-away doors.

The TR prototype for what turned out to be an extensive range of Triumph TRs, was an amalgam of a Standard Nine chassis and Triumph Mayflower suspension. It had a re-linered (to 1991cc) version of a 2088cc (85mm bore), wet-linered, ex-Standard Vanguard/Triumph Roadster engine, a Triumph Roadster 2000 gearbox and 3.7 :1 ex-Mayflower rear axle. The front suspension for all TRs up to 1976 was almost the same throughout the whole period, and was, in fact, based on the Triumph Mayflower Saloon of the early 1950s! The car was shown at the 1952 Motor Show as "20TS", and visitors to the show must have thought the TR's smooth aerodynamic body shape both a revelation and really far-sighted, compared to the traditional MG TD.

However, the TR was subsequently tested by Ken Richardson and the chassis in particular declared a death trap! With Ken Richardson's close supervision the car was redesigned. The characteristic TR2 faired headlamps and flat windscreen were retained, but the rear body shape was squared-up slightly from the original prototype. It was the chassis that received the most radical change in order to do away with the original's flexing.

The prototypes that followed were called TR (for Triumph Roadster) 1 but there were no TR1 production units made, and, to the best of my knowledge, all prototypes were scrapped. They were a vital step in realising the Triumph company's ambition to offer a choice to those seeking a low cost, fun sports car who, to date, had had to look no further than the nearest MG showroom. The next link in the TR chain occurred in 1953 with the introduction of the TR2.

Sidescreen TRs

My personal opinion is that the TR2, TR3 and TR3A offer the appeal of rarity, and will turn heads wherever they go. I think few would argue, however, that they are not best suited as daily transportation except in the most pleasant of climates. The space available within the cockpit is limited, as is luggage capacity. They leak. The steering is heavy and, in standard form, there's a couple of inches (roughly 50mm) of "play" at the steering wheel. They are best bought as a second or third car for occasional use, and the price of them needs to be seen in that context.

You should be aware that these cars do not have wind-up side windows. The TR2 and TR3 have (removable) sidescreens. Obviously, if you enjoy a Californian-like climate this is, as they say, no problem. If you are resident in less balmy climes, you need to decide whether this is acceptable. It depends upon the use you expect to put the car to and your own fortitude. If it's your first and only mode of transport, well, I would not recommend a sidescreened TR in anything but the warmest of climates! There again, many TR owners would totally disagree. I can still remember the admiration I felt when we saw off a lovely couple I'd just met in their 2, top-down, from a hotel "do". Nothing remarkable about that, I hear you mutter, but it was a very frosty, mid-January night with a temperature of about -10C (about 20F). Mind you, they did don leather helmets, and, I guess, sneaked the heater on just a little after they had left the hotel grounds! As a general rule-of-thumb, though, I think it prudent to steer you towards the wind-up windows of the TR4 (and onwards) if you envisage the car providing long-distance, all-weather transportation. That said, there is a major focus these days on the sidescreened cars and they are generally selling well, with good examples finding new homes quite quickly. Lets look at each of the sidescreen cars in a little more detail.

1-3. The TR2, first of the TR range of sports cars. Introduced in the early 1950s, the TR2 retains its classic style today.

TR2 (1953- 1955)

(Photograph 1-3)

By the time production of the TR2 commenced, the chassis had been dramatically stiffened, but the ex-Vanguard, wet-linered, four-cylinder engine was retained in 1991cc/83mm bore format. The original gearbox and rear axle were also retained. The formula proved a success, for 8636 cars were built, of which 5182 were exported.

The MG of the day was now the "TF" model with its 1250cc engine and identical performance to the preceding MG model. MG was forced to uprate the engine to 1466cc but, not surprisingly, this did little to stimulate sales and the TR2's streamlined body shape must have appeared a significant improvement to the buying public and motor manufacturers alike. Furthermore, the TR's 90bhp gave it a top speed of 108mph. This was faster than the TF and no doubt contributed to the TR2 breaking into a sports car market long dominated by MG.

This first TR used drum brakes on all four wheels. The front suspension was by unequal length wishbones and telescopic shock absorbers, whilst steering was by worm and peg. The rear suspension was totally conventional for the era, using a pair of leaf springs to provide the suspension and locate the rigid rear axle. Rear shocks were lever-arm type. Overdrive was available as an optional extra on top gear, which was an innovative development.

One development took place during the production run – the original doors stopped at the bottom of the sill, effectively hiding the sill, and cars with these deeper doors subsequently became known as 'long door' TR2s. Some found the original doors struck the kerb, which made exiting the car difficult, so a shallower door with a visible sill was introduced in 1954. The later cars became known as 'short door' TR2s.

It is estimated that a total of only some 2500 TR2s remain in existence today worldwide, the majority of which – nearly 1800 – in the USA. I estimate that nearly 600 remain in the UK, however. *Classic Car* valuations range from £7500, and £9000 to £12000 at the time of going to press.

TR3 (1955 to 1957)

(Photograph 1-4-1)

10,032 of the 13,377 TR3s produced were exported, maintaining Triumph's steady penetration into this (to Triumph) new market. This was in spite of the fact that the competition was not standing still, and the improved handling (due to a lower centre of gravity) streamlined MGA was launched in 1955 with rack and pinion steering, and a new 1489cc, 72bhp engine.

TR innovation continued, too, with the introduction of 11 inch front disc brakes, several years ahead of Triumph's arch rivals! These were initially an optional extra but were standardised in 1956. The TR3 retained the 1991cc, wet-liner Vanguard engine, which, as an aside, was a development of a Ferguson Tractor engine, although power was increased to 95bhp. The TR3 used the more robust Vanguard Girling rear axle, but otherwise had few significant differences.

Approximately 1700 TR3s are thought to remain throughout the world, with close to 900 in the US and approaching 600 in the UK.

Classic Cars values are £7000, and £8500 right up to £12,500.

TR3A (1957 to 1962) and TR3B (1962)

(photograph 1-4-2)

MG upgraded the MGA in 1958 with a 1588cc engine and disc brakes, and again in 1961 with a 1622cc/93bhp engine, and the car did much to revive MG's image.

Triumph introduced the "3A" with its optional 2138cc 100bhp engine. The 3A's appearance was revised slightly by a new front (panel and grill), and exterior door handles that were lockable! The Triumph had yet to match the MG's handling, however, and weatherproofing was still by sidescreens. Nevertheless, the 3A was an outstanding success, as illustrated by the 58,309 cars produced – of which 52,478 were exported – or an average of over 10,000 cars per year. This production rate is something in the order of three times the average annual output of the previous model.

TR development continued and, in 1962, the TR3s standardised on the 2138cc engine and front disc brakes. To Triumph this was just a continuation of the 3A's production run but, to the majority of Triumph enthusiasts, this amounted to the introduction of the "TR3B". This model still does not officially exist but 3334 were manufactured – solely for export markets! Whatever the model was called, this production run was the last of the sidescreened TRs since the TR4 design showed the way forward by introducing TR drivers to the comfort of wind-up windows.

There are thought to be a total of about 9500 surviving TR3As but only perhaps 250/300 TR3Bs are available for us to enjoy. Of the sidescreened TRs, the 3A is not only the most numerous, but probably the best loved. This is probably why the *Classic and Sports Car* values for the 'A' are £7000, and £8500 to a high of £12,500, whilst the 'B' has the slightly lower but still impressive price profile of £7500, £9000 and £12,000.

1-4-1. The second of what became known as the 'sidescreen' TRs, this is the TR3. A relatively small proportion of TRs left the factory fitted with wire wheels, but this car is non-standard by virtue of its retro-fitted, heavy duty wire wheels, and broader section tyres. It still looks very attractive, though.

1-4-2. The TR3A model, which was built in greater numbers than the 3. It would be difficult to tell the two apart at this distance were it not for the wide pressed aluminium radiator grille, although the 3A's door handles are just visible. The 3A has a huge and committed following still.

Non-sidescreen TRs

The TR4, TR4A, TR5 (and TR250) and the TR6 provide much more in the way of creature comforts in almost every respect. To be fair, I do not believe there are many TRs made before the TR7 which are in true daily use. Many are available for daily use but when snow, ice or salt are around their owners mostly take the bus. Not, mind you, because they're concerned about their car's performance or reliability, but are understandably anxious to keep it in pristine order.

These later TRs were made in (increasingly) larger numbers than the earlier models, which means two things: there were more to survive (and more did survive); and more spares are more readily available to help keep the survivors on the road. Furthermore, the numerous common parts within the TR4A, 5, 250 and 6 range of models means that, generally, the volume of spares used is higher and the cost is consequently more affordable.

TR4 (1961 to 1965)
(photograph 1-5).
For the TR4 Triumph revealed its clever "one major change at a time"

1-5. The TR4 is quite different in appearance with its very pretty Michelotti-designed body. Underneath, it was not so different, however, and took advantage of the best features of the sidescreen model's engine and running gear. The TR4 offered greater comfort in the form of higher door lines and wind-up windows, and provided the basis for two subsequent models.

development policy retaining an almost unchanged chassis, suspension and engine design, but introducing an Italian redesigned body style – the Michelotti shape, that was, in fact, retained for the subsequent TR4A and TR5. There were some other improvements – notably rack and pinion steering and a wider track (no doubt intended to improve handling), and synchromesh on all four forward gears. Manufacturing volume was retained at the 10,000 units per annum level with a total of 40,253 cars produced, of which a very creditable 36,803 (over 90%) were exported.

However, the competition leapt ahead by introducing the MGB in 1963. This was significant by virtue of its unitary construction that integrated a now redundant chassis into a stress-carrying bodyshell with significant weight and rigidity benefits.

There are thought to be about 4000 TR4s worldwide, with the majority (2250) Stateside. The model has a UK valuation spread of £4000, £7000 and, in show condition, £13,000.

The TR4A (1964 to 1967)
(photograph 1-6)
Whilst the competition fundamentally took a development "time-out", Triumph TRs took their own leap forward with the introduction of the TR4A, with its new chassis and IRS (Independent Rear Suspension). Critics of the day complained that the engine had not been updated, and that the car was under-powered. They may have been right, but it is my view that Triumph probably got its development program about right, and the introduction of a new chassis and IRS was enough to swallow in one step. Besides, the TR4A was more different than it appeared at first sight as the body had numerous under-the-surface changes, although the outer panels remained the same. I wonder if the critics of the day realised or cared about such detail?

However, that is hardly material to our current day review of the offerings available to you, the prospective TR-er, so let us conclude with the current numerical assessment. Of the 28,465 TR4As produced, 22,826 were exported and about 4000 remain worldwide. Some 1400 to 1500 can be found in the UK, but the US can boast something approaching 2500. Valuations range from a low of £4500, excellent examples are valued at £7000, and up to £12,500 is asked for show standard examples.

The TR5/TR250 (1967 to 1968)
In 1967, MG made what appeared to be, and should have been, a significant development, with the introduction of the 150bhp, straight-6, 2912cc MGC. This MG appeared initially to have the major advantage of a unitary constructed bodyshell.

Fortunately, Triumph TRs also developed – at least in the engine department, and at least in some markets – with the introduction of the TR5 (photograph 1-7) and its US version, the TR250 (photograph 1-8). Both variants were fitted with a 2498cc, straight 6-cylinder engine developed from Triumph's saloons, but the induction systems, compression ratios and rear axle ratios were dramatically

1-6. The TR4A had the same beautiful Michelotti body and basically the same four cylinder engine as earlier TRs. Underneath, however, Triumph introduced the exciting concept of independent rear suspension, and a number of detailed changes, too. I doubt you will need me to tell you that the 4A is on the left, and we get another chance to enjoy the TR2 shown in photo 1-3.

different, depending on the destination of the finished car. Vehicles bound for the USA had twin Stromberg carburettors, low compression (and correspondingly low performance), and were designated TR250. Some 8480 were built but only 600 or so survive.

The UK and many other export markets did rather better in that not only was the TR5's engine petrol injected (hence the PI), but the compression ratio, camshaft and performance of the car were rather more compatible with a sports car. In fact, the 2947 TR5s produced were quite brisk as a result of their reputed (but probably optimistic) 150bhp. Top speed was in excess of 115mph but, more importantly, acceleration was very satisfying.

Why such a short production run? Retention of the basic TR4/4A Michelotti shape was only an interim "bridge" until the restyled TR6 shape was ready. The 5 acted as the perfect development stepping stone for the petrol injected, 6-cylinder engine, and was completely compatible with Triumph's development policy.

It is very satisfying to report that about 800 TR5s are thought to remain, which is a very high number, considering the small initial production run and the years that have passed. For once the UK can claim the lion's share, but I did establish to my surprise that at least 9 examples reside in the US.

TR5 valuations reflect its desirability, spreading from £5800 through £10,000 to £14,000. The TR250 comes, of course, solely with left-hand-drive with about 1500 remaining, mostly in the USA. I guess the *Classic and Sports Car* valuations of £3500, £5000 and £8000 in the UK reflects these facts.

While many TR owners, and many non-owners, for that matter, would not agree, it is my belief that the TR5 is the most desirable of the TR range; its performance, rarity, spares availability and superb lines make it the best. There must be those who do agree, however, as it is currently amongst the most expensive TRs available.

1-7. Arguably the best TR of the lot – the TR5, which is certainly one of the rarest of the marque. The Michelotti body styling augments this technically superb vehicle, with its fuel injected, six-cylinder engine and independent rear suspension.

1-8. The USA had a carburettor inducted version of the TR5 called the TR250. The carburettors are the giveaway under the bonnet whilst, outside, it is easily distinguished by the bonnet cross stripes.

The TR6 (1969 to 1976)

The TR6 enjoyed a 7 year life – the longest production run of all the TRs – but not the highest sales volume, as we will see shortly.

Like the earlier 5, the 6 was also produced in two versions – the de-tuned US model and the sportier UK/rest-of-the-world offering. The situation is slightly complicated by the fact that the tune of both versions was adjusted in 1972: basically, further de-tuning both versions and (on US models) introducing the first of an ever-escalating amount of de-tox, anti-pollution equipment. The chassis of these cars was fundamentally the same as the TR5 (some variations will be explored in the appropriate chapters), and, as we have already established, the engine from the 5 was put to further use.

What was different – but, cleverly, not as different as it at first appeared – was the TR6 body. The Karmann Company redesigned it to utilise many of the existing TR4A/TR5 body pressings. The external boot/trunk and bonnet shapes were changed significantly, and I cannot help but point out that Triumph TR6s still used a separate chassis and body construction, which does not bear comparison with the 6-13 year headstart MG enjoyed with the far superior unitary construction roadster bodyshell. I suppose it makes little difference to our perception and valuation today of how up-to-date a car's design was at the time of manufacture. To prove the point, I confess to owning the pictured (photograph 1-9) TR6, and loving it! Just imagine, however, what a car the TR6 would have been with a unitary bodyshell, it would have knocked spots off the competition ...

In fact, perhaps fortunately for Triumph, it did not have to, for MG was being knocked about by the press and the buying public. All hated the MGC's poor handling, and as a consequence the MGC was withdrawn in 1969, leaving the TR6 with no competition in the popular 6-cylinder sports car market. Sales responded and production volumes were respectively 77,938 (carburettor US) and 13,912 (UK), with average units sold reaching 13,000 per annum.

The numbers of TR6s remaining today are also encouraging: there are nearly 6500 examples of the US carburettor model still around, and it is thought that some 4500 petrol injection cars are still in existence worldwide. The *Classic Car* values span £4800, £7800 with £12,000 for show quality examples of the latter model. You can probably think in terms of a 10% reduction for the carburettor version in the UK.

1-9. The TR6 was a very clever updating exercise by Karmann (of Karmann Ghia fame), using all but the front and rear skin panels of the TR5. This 1972 car has a particular place in my affections, and is shown beside my 1970 3500cc MGB. This might not be the right place to say which is the quicker of the two, but the TR does have a rather nicer registration plate ...

The "Wedgies"

These TRs had a surprisingly shallow (front to back) boot/trunk area which is considerably smaller than it's predecessor's – the TR6. The 6, however, was unusually well equipped as far as luggage capacity was concerned. Not worried about luggage you cry. Fair enough, but do note the 7 and 8 are slightly more difficult to enter by virtue of the very deep sills (*a la* E-type Jaguars).

Once seated, however, two major 7/8 advantages became clear: the cockpit width is wider than its non-sidescreened predecessors, and the top of the TR's windscreen seems higher than the earlier TRs – which not only helps forward visibility but makes driving the TR7 quite comfortable, even with the hood raised.

The TR7 is known in the USA as the "flying doorstop", I suspect not always affectionately, but even if you are not entirely sold on its wedge shape, do note that it definitely reduces wind noise at high speed, doubtless aided by the lower frontal area and sharply swept screen.

The TR7 and TR8 (1975 to 1981)
(photograph 1-10).
I mentioned earlier that the TR6 had not achieved the distinction of being the highest number model TR produced. In fact, the TR7 achieved 112,368 units and the 8 a further 2743, making a total of 115,111 in 6 years of production. Exports were a creditable 88,000 units, while the average annual sales figure was now slightly over 19,000.

The car did finally enjoy a unitary-construction bodyshell, and Triumph took the opportunity to make other changes, too. In fact, Triumph swept aside its previously conservative "one major change at a time" approach and changed ... everything ... radically! The buying public of the day clearly liked the "clean sheet of paper" approach, as sales figures show.

I believe sales could have been

1-10. The TR7 was a radical change of shape and technology, and is none the worse for that. The model attracted more buyers than any other TR. The same aerodynamic shape was also used for the 3500cc TR8 variant.

higher still had build quality and continuity of production allowed the car's full potential to be realised. The sleek, wedge-shaped bodyshell was wider than the previous TRs, wind noise was much reduced, and the rear suspension reverted to a well-located live axle with a consequentially improved ride. All factors that I applaud. However, do not go looking for a Roadster (open-topped) version of the TR7 before 1979; there were only Coupes up to this date.

This oversight may have something to do with the fact that the number of cars thought to remain in existence today (Roadsters and Coupes) is in the order of 15000, possibly 20000. The TR8 has stayed the course very well indeed, but I find the numbers of TR7's still in existence slightly disappointing considering this was the most recently produced model , yet it appears to have suffered almost the highest attrition rate of all the TRs.

The company may not have engendered long-term love for the model either by reverting to a 105bhp, four-cylinder, 1998cc engine for the TR7 and, at least initially, a 4-speed, no-overdrive gearbox. The gearbox decision was corrected with a good 5-speed 'box but the V8 engine the car needed was only fitted late in the model's life, creating the superb TR8. Too late, though, to undo the suicidal policies, actions and quality of previous years.

So, after a basically sad story, is this a TR you should contemplate? Yes, I believe the 7 has many fine qualities and the 8 (or a well-converted TR7-V8) is a great car. The attrition rate I spoke of earlier has the benefit of ensuring there are lots of spares available, which could work to your advantage. Indeed, the coupe versions are so available that their price almost enables Roadster owners to have one in the back yard just for spares! Even more important is the fact that the unitary bodyshell build allows for the car to be used as everyday transport, and the coupe versions should allow even those without garaging to consider a TR7 as everyday use. It is still a TR, after all ...

These TRs offer a more comfortable ride than earlier cars, yet with probably slightly superior cornering capability. Track is approximately 10% wider than the non-sidescreened cars. If I had to drive 500 miles in a day and could choose any of the TRs, it would be the TR7 (V8, of course!) Roadster.

This leads me to conclude on a cheerful note. There seem to be about 1700/1750 remaining; about 100 or so in Europe and the rest in the US. A very low attrition rate, from which, I conclude (as the advertisement says), size really does matter!

Conclusion

So you think you're narrowing down the field by virtue of the price you are prepared to pay; that's not the end of it, however. The useful life of the car is an equally important factor.

Obviously, we all want our classic to last indefinitely, but even a well restored classic has a finite life span. You should reckon that an average, well restored car will probably require further major attention in about 8-12 years. Therefore, do you want a car that requires further restoration in 1 or 10 years? If it's the former, then clearly you are looking for a car in the bottom half of the "Excellent" range, or possibly top of the "Good" range. If you're not into restoration, and want years of (hopefully) trouble-free enjoyment, then you are into the "Show" range of cars on offer, so make your shortlist of TR models accordingly.

If you are buying a donor car for its components then you'll be more interested in the parts than the price. The completeness or availability of parts, rather than, say, condition of the chassis, will be most important, as you could have decided, even before seeing the car, that you will fit a new chassis. If you are buying a CP 150bhp fuel injected 6, is the correct distributor and fuel injection equipment in place, or at least supplied loose with the car. If loose, why? The distributor is rare in that it has a mechanical tachometer drive, which you must ensure is there.

If you are after a car to bring up to concours standard, then you are looking for a host of ORIGINAL detail – far better to have an original part in any condition than a brand new, shiny, non-original replacement.

Hopefully, by now you've settled on a shortlist of models. Let's hope it's a model you can afford, in which case it's time to refine your options still further by spending a few seconds on the matter of colour. If you are buying a donor vehicle for immediate restoration, then colour is not too important. If, on the other hand, you are buying a car to drive and enjoy for the next 10 years, then getting it in a colour you like moves up the priority list somewhat! There's no point in even looking at a car, particularly one at the top of your price bracket, if you hate the colour. Further, although it sounds obvious, do establish what colours you do wish to pass over; some colours enjoy an impressive official name that we may not initially recognise for what it is. I will upset too many if I actually mention the official colour by name, but I recall travelling many miles to view a Triumph that I thought would be a very acceptable shade of blue, but which turned out to be (to me) a quite unacceptable colour. So, before you leave home, establish what the base colour really is!

I have taken you through the practical steps, which I hope has helped. At the end of the day, you will have to go and see for yourself, in which case be prepared for the heart taking over! Everyone has a view and/or opinion of which is the best model, what colours are beautiful and what body shape is the most pleasing. It might be as well to look at some of the problems and pitfalls that await the first-time TR buyer, starting with a few more golden rules ...

BUYING – SOME GOLDEN RULES

Take particular note of the first rule: Do NOT buy the first car you see – shop around.

Strangely enough, the second rule applies to personal chemistry and requires you ignore the car for starters! Talk to the man who is selling the car and establish whether he is the sort of guy you would want to buy a car from. Are you comfortable with what he is saying, how he says it and the general interaction between you? Of course, many vendors – particularly professional car salesmen – have a natural ability to sell anything; nevertheless, this is an excellent starting point.

Thirdly, establish that the vendor has the right to sell this car. Is it his car, does he have the log book? If so, ask to see it. Is he the registered owner, and if, not why not? Do all numbers in the log book tally with the car? Is the engine number (in particular) right for the car: a 150bhp CP engine in a 125bhp CR series TR6 may be no bad thing, but the reverse would devalue the car. TR2 engines appear in TR3s. Some six-cylinder cars have an engine with an MG prefix. This could be an ex-Triumph Saloon engine with 135bhp power output, but it could also be a factory replacement engine, in which case you should see the original receipt. Remember, if you do not check this detail you can be sure that your buyer will be sure to note the discrepancy when you come to sell the car! Do all of this before you agree a price.

Throughout the review, resist the temptation to pick fault with the car. Try and establish a mutual rapport with the vendor so that he will find it harder to reject your offer. Do not go into the reasons why you cannot pay his asking price; they are not his problem. You can say that the car is basically what you are looking for, and that it is sound apart from a couple of remediable items. However, you've set your heart on a car with, say, overdrive, an unleaded engine conversion and chrome wire wheels, and if you were to pay his asking price and fix the problems, then you wouldn't be able to afford to bring the car up to the desired specification. NEVER offer what you think the car is worth. For example, if the asking price is £7500, but you believe it will cost around £1100 to bring up to condition and with the facilities you require, you may be tempted to offer the vendor £6400. DON'T! Offer £5300, by all means, which, of course, will be refused, but hopefully this will start the haggling that will result in an amicable mid-point handshake.

If you simply cannot reach agreement be prepared to walk away. This will give the seller time to think again about your offer and you the opportunity to check out the cost of rectifying the faults found. Don't make the mistake here of presuming that if a rebuilt differential is priced at £250, then this is the total cost. Add on the necessary mounting rubbers, and perhaps some new brackets, and the total cost is likely to rise to £400. So take notes, go away and find out what the total costs are in each case. Armed with this irrefutable information you are in a better position to argue your case.

It's also a good idea during this time to do some more homework, particularly if the seller is a dealer. Phone the area secretary of your local TR Register and ask if anyone knows the car, dealer or owner. Most area secretaries will be pleased to give you all the advice you need if you explain that you are seriously considering buying the car. Be on your guard against non-specialist dealers, especially those that do not have workshops; anyone that sells from home and does not have his own repair facilities should be treated with caution. Even if you get a warranty from this type of trader it may be very difficult trying to enforce it should you ever need to. If you buy your car from a one-make specialist who has his own workshop you will get more detailed and expert attention, and if things should go wrong, the cost to the dealer with his own workshop of putting things right will not be as high as it would be to a non-workshop concern (which would have to pay another company the full workshop rate).

Chapter 2

What to check when buying

Much of the inspection work – particularly the extensive chassis checks – will require your crawling underneath an elevated car. Dress accordingly, but, above all, ensure the car is safely jacked up and securely positioned on axle stands. If in doubt, take your own axle stands, trolley jack and a powerful torch when you leave home to view each prospective purchase.

BODY, PAINTWORK AND TRIM

Bodywork – An overall impression of the car's bodywork is best gained from the rear of the body. All TRs are longitudinally fish-shaped – by which I mean that there is a gradual but steady side curve from front to back.

Go to one rear corner and kneel down to sight along the coachline. The car should curve steadily. Check the other side is the same. What you should particularly look out for is one very curvaceous side and one straight or flat side, which means the car has had a knock, poor rebuild or even a front half/rear half marriage from two different cars! The latter problem doesn't mean that the car should be rejected out of hand, but its correction will need to be carefully, and therefore invisibly, carried out.

Sadly, it is an inescapable fact that all TRs have a number of corrosion weak spots. Some of the illustrations shown are certainly on the dire side of what you may be offered, but do serve to show the sort of rust holes that can occur, and should alert you to watch for fillered cover-ups in the areas listed below. Most vulnerable spots are common to all models. Each should be examined carefully and a magnet used to test for body filler when in doubt. Rear inner wing repairs are possible but involve a great deal of skilled work, and cars with rear inner wing rot are best avoided. Establish whether this is a problem from inside the boot with the boot trim panels removed, and by looking and feeling each side of the fuel tank (photograph 2-1). External bubbling on the rear deck panel and forward deck section will mean trouble in the near future.

2-1. Rear inner wing problems, which signal that this is a car to be particularly cautious about buying. In fact, in this picture the wheelarch looks sound enough, and it is the inner wing panel that shows most corrosion. However, if you were to poke the wheelarch with a small screwdriver you would certainly find a similar amount of corrosion there, too. It's repairable, but even an experienced body repair specialist will find this tricky and time-consuming, and you'll find it expensive. Subsequently, the back half of this TR5 was replaced completely.

Note that top repairs can be done quite professionally as photograph 2-2 shows, and will hide, for a short while, the true seriousness of the situation – which is why an examination from inside the boot is particularly valuable. Feel where the rear wing is bolted to deck sections alongside the fuel tank. There will be cars where you will see daylight because the panel has completely rusted away.

2-2. The rear deck on this car would not have looked too bad at first glance since the body filler, seen here sideways on, will have covered much corrosion. In fact, the true extent of the problem would only have been revealed by looking at these panels from the inside of the boot/trunk, or when the wing/fender was removed.

2-3. A good example of the care needed when examining a potential purchase, as this TR6 looked great from the outside. However, a peek from within the engine compartment would have revealed impending trouble!

2-4. Not as smart-looking, this TR4 had a rusting door, but you would never have thought that the rear baffle would be completely rusted away. This, too, should have been obvious from a look from within the engine compartment. This is not untypical of what you must expect when you remove the front wings. Devoid of a worthwhile amount of original paint, the joint with the wing is corroded, and about one third of the splash plate has rotted away at the top, allowing corrosion to attack and hole the top inner wing. Note, too, that the bottom inner wing is corroded and also holed.
This car has still got its original sills as you can see the indentation (arrowed) pressed in the sill to allow water to escape from the original plenum chamber drain hose, also arrowed. Less obvious, this picture also demonstrates the typical upturn that occurs at both rear corners of the bonnet with years of use. It is a good idea to reinforce the inner lip of the bonnet for about 12ins (300mm) in front of both rear corners to stop it kinking in future. A special repair panel is available.

2-5.

2-6. This example demonstrates the extra vulnerability of the pre-TR6 cars. The stainless strip looks very attractive, but the reaction between dissimilar metals undoubtedly accelerates the corrosion.

2-7.

2-7 & 2-8 (above). The first photograph seems to be a duplication of 2-6, but in fact serves to emphasise the body filler problem. Although this looks bad, picture 2-8 shows the full extent of the corrosion with a lump of filler removed. You could not have failed to see the extent of this corrosion had you looked from the boot/trunk side of these panels.

Above and around the headlamps the corrosion works back to inner wings where the bonnet hinges mount, so look there, too (from inside the engine bay). From here you will be able to see the corrosion shown in photograph 2-3. Baffles behind front wheels: the corrosion often occurs by the fuse-box on the inner wing. You will also see this from inside the engine bay, although the real problem may be obvious from a very quick look under the front wings. The baffle has completely gone in photograph 2-4, and looks pretty corroded in photograph 2-5.

As photographs 2-6, 2-7 and 2-8 illustrate, the pre-TR6 wing-to-body joints generate corrosion all along their length due to an electrolysis reaction of dissimilar metals. The stainless steel strips react with the body's mild steel material, and the problem is exacerbated by the sad fact that the factory did not apply paint or sealer to these joints. Do apply both paint and sealer to all joints upon reassembly, and, in the case of cars with a stainless strip, ensure the sealer is applied to both sides of the stainless strip.

Bottoms of doors: run your hand

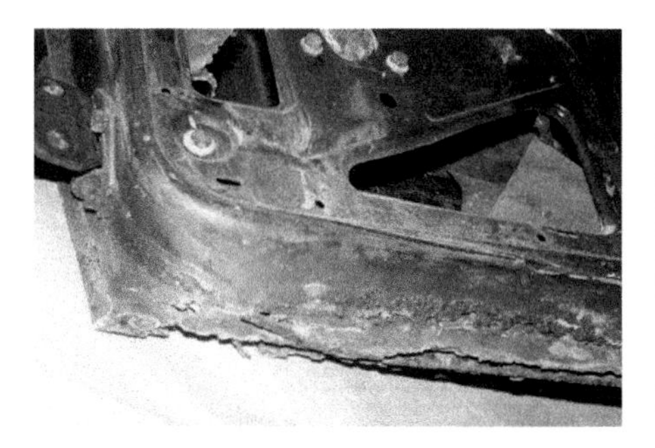

2-9. The doors seem somewhat vulnerable in the TRs we are exploring; here, the classic door bottom and edge of the door skin needs adding to your 'to-do' list. The second picture is heartening in that, although the door skin looks beyond help, the frame looks sound enough to warrant re-skinning. Fortunately, ex-Californian doors are still available, or new skins can be applied to any reasonable doorframe.

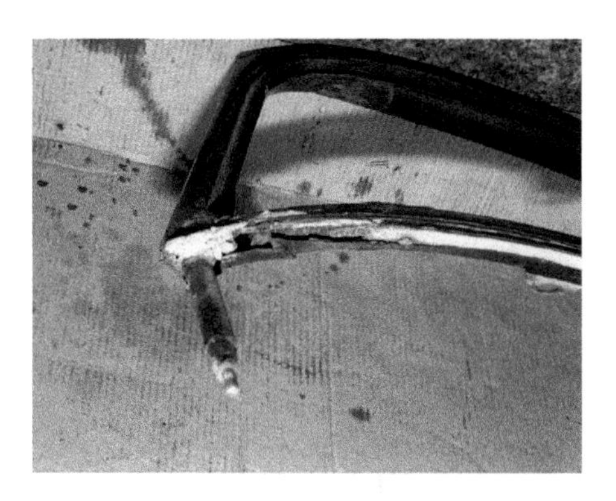

2-10. A rather worse-than-average windscreen frame. This one is actually in such bad shape (and has been filled) that you will have spotted it without much difficulty. It can be repaired, as we will see later, but there are, in fact, some good, ex-USA frames available which probably makes repairing examples like this not worth the time and trouble, not to mention the high level of skill needed!

(carefully) along the bottom where the door shell joins the outer skin. As can be seen in photograph 2-9, the rust can sometimes be obvious without the need to feel along the underside of the door. The windscreen frame is bolted on. It is a lightweight pressing and can rust from the inside or from cracks (due to load imposed by hood). The early stages of corrosion can only be seen when the screen is off the car, but take a close look anyway, mindful of what you see in photograph 2-10. The front top of doors (where they sit adjacent to the windscreen panel) is, for some reason,

2-11. The top of this door has clearly been filled and then has cracked again. A not uncommon problem which has the merit of making it clear to potential buyers that a replacement door skin will be essential. What you will need to watch for is the appearance of the door(s) in this area, for they could have been skilfully filled and painted, making the underlying fault much more difficult to detect.

2-12. For some reason the window channels can cause cracking around the front top of the door frame.

prone to the sort of corrosion shown in photograph 2-11. Door glass movement can also crack the tops of either or both doors, particularly if they are weakened by corrosion, as photograph 2-12 demonstrates.

There is a weakness where the

2-13. You are unlikely to come across as blatant an example as this, but be on your guard for corrosion problems at the base of the rear inner wing where it attaches to the 'B' post. The damage will be worse than you can see but should stop short of this example!

2-14. The condition of the boot/trunk floor should put you on your guard were you looking at this car but, again, the severe corrosion in this rear valance is masked by body filler. It would be more obvious if viewed from the underside of the car.

rear wing attaches to the B-post. This usually rots from the bottom up, as photograph 2-13 illustrates perhaps a little too vividly!

The rear valance is particularly vulnerable where it joints with the rear wings, but run your hand (carefully) along the whole bottom edge of the rear valance (particularly where the exhaust pipe exits) to check for corrosion. What's shown in photograph 2-14 does not seem too bad at first sight, but it is, in fact, full of body filler. You would get a hint of this from the condition of the adjacent boot/trunk floor, which would make you look closely at not only the rear valance, but the adjacent panels, too.

The boot lids on TR250s and 5s are susceptible to rust along the bottom

2-15. The bottom edge of this Michelotti boot/trunk lid appears only slightly corroded along the susceptible bottom 2 inches. No doubt, were you to gently sandblast (only) this edge, you would be surprised at the extent of the rust. However, you can still see the turned-in lip of the skin (arrowed) which shows that there have been no previous attempts to fill this area of the car – which is very good news.

2-17. The area in front of the splash-plate on this TR6 appears to have corroded particularly badly, and would have been very evident upon even the swiftest of examinations. There is evidence of corrosion (arrowed) further back than the splash-plate, too. Nice hardtop, though!

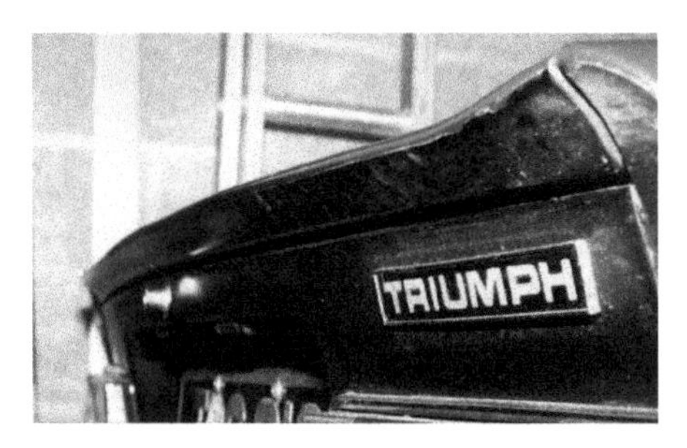

2-16. While this one looks pretty solid, the rear lip on the TR6's boot/trunk is vulnerable to the combined affects of water and corrosive exhaust gases, since the design of the rear generates a vacuum which sucks back onto the rear of the car and boot lip. You can possibly reduce the resulting corrosion by injecting lots of Waxoyl inside the boot lid, and by ensuring that your exhaust system pushes the gases away from the rear of the car.

2-18. This Michelotti rear wing, actually from a TR5, is almost as good as can be found on a car that is 30-plus years old. Nevertheless, hours of work will still be necessary to remove and replace the rusted rear lip/light surround (nearest the camera).

2-19. This hole was covered by the 6's front wing, but a good look at the four corners of the inner sills should have given a clue, although not perhaps quite to the extent of the problem shown in photos 2-20 and 2-21.

2 inches, where the panel is doubled-skinned. If corrosion is not evident, do check that the (internal) seam is visible, for often corrosion has been filled and the seam disappears! TR6 boot lids are most vulnerable along the rear lip overhang where the double skin traps moisture and the exhaust gases vortex up, causing double corrosion problems. The TR6 boot lid is not practically repairable, and a new one must be purchased, but you will need to repair the Michelotti item (photos 2-15 and 16).

Corrosion or damage (photo 2-17) to a TR6 wing is best handled by fitting a new, replacement wing (at about £175 each). The wings fit well and make the most economical repair. TR250 and 5 wings (photo 2-18) are, however, quite different in that the new wings are expensive at £300 apiece, and even a skilled and experienced panel beater can take two days to get each to fit; consequently you could be looking at £500/600 a wing – before painting! Fortunately, repair panels are available and, overall, it is more cost-effective to have a skilled and experienced specialist repair a damaged or corroded TR250/5 wing.

TR sills go in two places – the outside is usually pretty obvious and normally visible and replacement is straightforward. Consider the TR6 shown in photograph 2-19, where the outer sill is disintegrating behind the front wing.

Closely check the inner sill, too. The inner sill forms part of the floor panel, and outer sill corrosion of this severity is likely to manifest itself in at least the inner front corner as shown in photos 2-20 and 2-21. If you find the sill is suspect, also check the floor at the four corners nearest the wheels,

2-20.

2-20 & 2-21 (above). No-one could miss the advanced corrosion in this TR250 floor, nor fail to appreciate that new floors are essential. However, note the advanced corrosion in the inner sills, too. Furthermore, the lower steering column support bracket is missing from just below the grommet hole on the left side of the picture (arrowed). I made one from thicker material than standard for my 6. The photograph of the right-hand side of the car shows the plenum chamber coming right across the car, with its small stub-tube drain arrowed. This is an area that corrodes when both left and right corners of the plenum get blocked and water has no escape. A pair of rubber drain tubes exit the cockpit via one hole each side, one of which is visible situated about 3ins (75mm) below the bottom of the plenum chamber. A modification is recommended and explained in Chapter 4.

2-22. The corrosion here is obvious and clearly extensive, and signals a two-part body rebuild without looking any further!

2-23

2-23 & 2-24 (above). Bonnets are vulnerable, too.

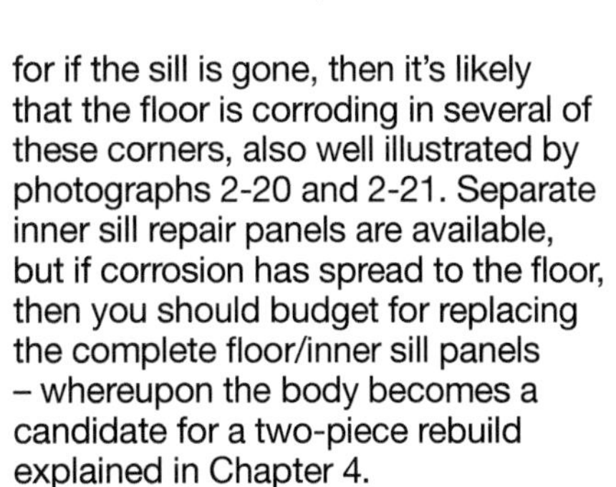

for if the sill is gone, then it's likely that the floor is corroding in several of these corners, also well illustrated by photographs 2-20 and 2-21. Separate inner sill repair panels are available, but if corrosion has spread to the floor, then you should budget for replacing the complete floor/inner sill panels – whereupon the body becomes a candidate for a two-piece rebuild explained in Chapter 4.

There used to be a tendency to fill the gaps between all four wings and the sills in an effort to make the car appear smoother than it really was. If you are offered a car where there is no clear and obvious gap between each wing and its mating sill, view that area with some suspicion: it may have started out a few weeks before looking like photo 2-22.

Photograph 2-23 is very typical and shows that the bonnet tends to corrode at the front. It is double-skinned in the 'nose' area where corrosion sets in.

2-25. A view upwards onto the back of the right-side plenum. The oblong hole allows access to the wiper wheelbox and is intended to be there, but the corrosion in the right corner is not shown on any Triumph drawing! Can you spot the drain tube that was intended to take water from the plenum down the outside of the passenger compartment? I doubt it had any work to do for many years, and that the floor on this car will be very badly corroded, too ...

2-26. ... Yes, I thought so!

The other area to look at is both rear corners, where owners have lifted the bonnet too enthusiastically, forgotten about the safety catch and, over the years, slightly raised the rear corner of the bonnet. You will find other, less dramatic examples as we progress through the detail but, in the case of the TR250 and 5, you can even find a split of about 12 inches in front of the bonnet rear edge, as illustrated by photograph 2-24. In this case, the corners need to be bent back down, welded, and, of course, repainted. This problem is so common that Revington TR has produced a strengthening insert, which is pictured later in the book.

The plenum chamber is a double-skinned cavity that stretches the full

2-27. It is almost inevitable that the battery tray will be corroded. Replacement panels are available, and once the shell has been repainted you can, today, buy a plastic moulding (highly recommended) that at least contains the acid until you can mop it up.

2-28. Clearly not a happy picture of the boot/trunk of this TR4A, looking forwards into the passenger compartment with the fuel tank removed. The photograph, in fact, shows that more than the boot/trunk floor needs to be repaired, for there should be two triangular strengthening gussets at the base of each inner wing, just in front of where the fuel tank normally sits. These dramatically increase the sideways strength of the rear of all TRs and must be left in place. Some owners weld a whole sheet of steel across the rear of their cars at this point to further increase structural rigidity of the rear of the car, as well as providing a firewall between them and the fuel tank. However, in this example a previous owner has hacked away most of both gussets, which, though it may have enabled him to install a couple of speakers for his hi-fi, did absolutely nothing for the back-end strength of the bodyshell. Not all bad news – the wheelarches look fine!

width of the car under the scuttle/windscreen. It catches water that enters the scuttle air vent and dissipates it through two outboard pipes down into each (front) inner wing area. You can see the hole for the drain tube in photograph 2-25. If the pipes become blocked, or the car stands outside for a prolonged period, water can gather in the plenum chamber, causing it to rot, as is shown all too clearly in photograph 2-25. There is a modification to duct the pipes away from the inner wing area, which we will explore in due course.

2-29 and 2-30 (right). The door gaps on IRS TRs widen as the chassis weakens: beware of a car where you see this. The excessive gap could be present if either/both chassis is rotting, or inexpertly repaired and 'hogging' has occurred.
2-30 shows the sort of gap. As a matter of interest, note the different door locking arrangement to the TR6 shown in 2-29.

To detect problems look first for signs of water ingress in the car under the windscreen. Wet carpets and rust on the floor panels (photograph 2-26) will give clues, but, with a good torch, also look up under the dashboard. Start by looking for the pipes in each corner (right under the windscreen corner) and work your way towards the centre of the car. This is a difficult repair which is expensive in that it involves removal of the steering column, dashboard and heater, so look carefully for it's better to find such problems sooner rather than later.

Battery trays go when battery acid corrodes them, as illustrated by photo 2-27. A replacement panel is available and is not too difficult or time consuming to fit. Nevertheless, if corrosion is present, add it to your (growing) list of faults.

As photograph 2-28 demonstrates, the boot/trunk floor is vulnerable, although – in this example – the rear inner wings look good, making this a restorable proposition. Contrary to most sports cars, the door gaps expand on IRS cars when the chassis structure weakens. Unsatisfactory and excellent examples can be seen in photographs 2-29 and 2-30.

Paintwork – Assess existing paintwork and decide whether it is satisfactory: like bodywork repairs, paintwork can be very expensive to properly correct. It takes time to remove all chrome work, wings and fittings, and carry out the chemical pickling necessary to bring the car back to bare metal. And note this work is required before pre-paint preparation and actual painting can take place. This means a cost of £2000 to £4500 to have the paintwork applied. Even new Heritage or rust free Californian shells need to be stripped (i.e. wings off) before painting in order to ensure the paint is applied to the mating surfaces and then reassembled with sealer if you are to enjoy a lasting impressive finish. Not to do the job properly repeats the mistakes of Triumph back in the '50s, '60s and '70s, and eventually wastes much of the remedial bodywork repairs you may have already carried out. Certainly, you can help the situation by doing the stripping yourself, and we will discuss that in some detail in a later chapter. Remember that good paintwork – though expensive – will pay for itself in due course.

Interior and external trim – You should not buy or reject a car on the basis of its hood, carpets and internal lining panels. A poor car beautifully trimmed is not a good buy, but a sound car with tatty trim is worth purchasing at the right price. Unless your intention is to have a potential concours car, the condition of external bright work and badges should not take up too much of your time and attention either. If the originality of your car to concours standard is important to you, note that, in this case, most cars need to have the original steering wheel, as genuine replacements are becoming progressively hard to find and expensive. If the original wheel is not in place, ask if it is available, or add another item to your 'to buy' list.

Purchasers of potential concours standard cars may need to look at the bright work with a more critical eye. Replacement front bumpers, in particular, are not up to the standard of the original item, and, consequently, a good original bumper will be important to potential concours entrants. Indeed, all bright work needs to be quite carefully viewed, for some badges, although available new, are not quite true replacements of the originals. In normal circumstances the variation will go unnoticed, but sharp-eyed concours judges will mark your car down for such a discrepancy. As an example: the original TR5 badge was in a buttermilk/cream colour but replacements are white. Most of us would not notice this, and neither would we detect a non-original jack, wheel brace, etc., but concours judges will!

MECHANICAL MATTERS

Engine condition details – Triumph six-cylinder engines are fundamentally tough, although, in spite of a long and respected life, do suffer weaknesses you should be aware of.

The position of the oil filter gives rise to initial moments of oil starvation on start-up, characterised by an initial rattle. A well-proven (spin-on) filter upgrade is available and is detailed in Chapter 7. The six-cylinder engines were fitted with a half-diameter crankshaft thrust washer, which requires you examine the engine for unacceptable crankshaft movement; more of this in a moment.

On start-up is there a rattling sound? Does it puff blue smoke, signifying general wear and tear? Do the tappets rattle even with the engine warm? Next, with the engine switched off, have a helper repeatedly depress the clutch while you hold the front crankshaft pulley/fan. Push the pulley backwards and feel for any front-to-back crankshaft movement (it should be a barely discernible 0.008ins). Too much needs investigation as the design weakness mentioned above can (in an extreme case) allow the thrust washer to drop out, which, in turn, allows the crank to move so far forward that the block and crank disintegrate. Any of the above need to be noted and included in your calculations as a new engine can cost £1000-£2000.

A saloon block is not unacceptable in principle, but it is prudent to keep (or ask for, as appropriate) documentary and photographic evidence that the original 150bhp balanced crank and camshafts are fitted.

Gearbox condition/assessment – When test-driving the car, try selecting each gear. It will require a firm, positive and deliberate action, and full depression of the clutch to do this. However, the gear lever should not fight, refuse to go in or make complaining noises as it goes (or does not go) into gear. Second gear, coming down the box, is the weakest aspect of the TR range of boxes. Try the gearbox in every gear on overrun (not in drive) to test for jumping out of gear. Watch the gear lever for clues during the overruns as it will inch forward slightly before popping out. If the clutch is stiff or heavy, then it's possible that the pin in the clutch fork has gone.

With the engine running and the gearbox in neutral, if there's a hissing noise when the clutch is depressed the release bearing is suspect. With the engine running and driving slowly forward in first gear, if the gearbox hisses, be warned that this is likely to mean that the lay-shaft's needle roller bearings are picking up on the hardened surface of the lay-shaft, and will require replacement in due course. This hiss will most likely disappear in top gear (i.e. direct drive). A broken clutch fork pin may cost pence but the lay-shaft will be a different matter, and you should budget for fitting a new clutch when your gearbox is out.

All of the foregoing problems will require gearbox-out repair, so will be correspondingly expensive. Obtain a professional estimate that can be taken into consideration.

Overdrive – Overdrive units were not fitted to every TR, although they were available as a highly desirable optional extra. In fact, even with the CC/CP versions of the latest car we are considering (the TR6) only about 5% of cars destined for the USA and 70% of UK cars were fitted with overdrive: strange but true. The final versions of the TR6 (CF/CR) did have overdrive fitted as standard, and, although not directly relevant to this book, you may be interested to hear that Triumph did not continue to supply overdrive as standard when it introduced the TR7.

Getting back to the six-cylinder cars, if you are faced with a potential purchase without overdrive, think in terms of a £500 difference in value compared to an otherwise equal car with overdrive. This is the going rate for a secondhand gearbox with overdrive. It will cost more than this if you take into account buying and installing it – but this difference goes most of the way to balancing the lack of overdrive. A reconditioned 'A' type box with overdrive will, in fact, set you back about £1000 if you have no compatible trade-in unit to offer. A straight non-overdrive gearbox has little trade-in value as everyone wants to trade-up to overdrive boxes.

Please note that the tail shaft within all overdrive boxes is actually shorter than that in non-overdrive boxes, consequently, even if you offer a non-overdrive box as a trade-in, your supplier will be short not only of the overdrive unit but of the correct tail shaft, too.

If the car you are considering has an overdrive unit fitted this is good news, but the overdrive can suffer as a consequence of the worn gearbox lay-shaft mentioned earlier. If the lay-shaft surface is breaking up, since gearbox and overdrive share the same oil, the swarf from the failing lay-shaft will have entered the overdrive unit and will be doing its best (usually successfully) to wear the overdrive pump and eccentric cam. So, not only are you faced with a gearbox rebuild but you must also budget and arrange for an overdrive strip, clean and inspection. Failure to do this will mean that swarf which has collected in the overdrive will get back into the gearbox and ruin the new lay-shaft and bearings once again.

When the overdrive control switch is flicked on, it should cut straight in, and with sufficient force to slightly jolt you. An overdrive that is slow to come in could be suffering from one of two faults, the least expensive of which is a faulty/sluggish solenoid that costs £50 and takes ten minutes to fix. However, it is prudent to assume that the previous owner has tried this remedy, and that the problem is due to a worn overdrive pump which is now taking time to build to the requisite 410psi operating pressure. Assume a worst-case

2-31. This is a picture of an IRS chassis – a beautifully restored one. Note that without the clutter of the rear suspension, you can see that all the stresses are focused just behind the middle of the chassis – at the narrowest/weakest point of the structure. Bear this in mind when examining a prospective purchase.

2-32. The dreaded swelling due to corrosion in the crucial central 'Tee-shirt' pressing signifies an expensive repair, and sooner rather than later. You should proceed with due caution as this area, along with the IRS mounting 'arm' sections, is the most prone to rust and most expensive to repair (in situ), since you either have to take the body off the chassis to get at the corrosion, or cut out the rear floors.

scenario and allow £500 to fix this, plus – nine times out of ten – a new clutch.

All of the foregoing problems necessitate removal of the gearbox before remedial work can begin, and should be costed accordingly.

The differential – Does it thud when the clutch is let in or out? If so, this is invariably because the right-hand front mounting pin has pulled out of its chassis mounting. The solution is to remove the differential, re-weld the pin from underneath and box the mountings to prevent re-occurrence. This is a professional job, so add £250 to your list of costs.

A 'click' suggests a worn sliding joint in the rear drive shafts which can only be remedied by replacement driveshafts at around £100 apiece. You may find that, collectively, the 6 U/Js in the propshaft and driveshaft power chain 'chunk' too and, whilst components are not expensive (6 will cost £50), fitting will take some time and you should get a professional estimate. The main propshaft in particular can only be removed with the differential (or gearbox) removed – so if either of these have to come out it is normally worthwhile replacing at least the two propshaft U/Js as a matter of course.

While in the area of the differential it is as well to alert you to an apparent problem with the rear wheel cambers that is, in fact, quite usual and normally need not concern you. For reasons that I am not sure were ever fully established, the left side rear wheel has a slight inward camber when the right side wheel stands upright. In other words, the left wheel leans in at the top slightly, and it does so on all IRS TRs – honest!

INSPECTING THE IRS TR CHASSIS

The TR4A, TR250, TR5 and TR6 all had Independent Rear Suspension, and it is most important that the chassis fits in with your game plan for, if not, a replacement chassis can be expensive to remedy. Even if you have already decided to buy a new chassis, examine the one in front of you with some care. The vendor does not know your plans and a poor chassis may be a bargaining point later in your discussions.

The first thing to check is the chassis 'leg' that connects to the central cruciform and carries the two trailing arms: see photograph 2-31. These chassis members are the first things to rot, and often you can crunch them with your finger, although they are the major load-bearing part of the chassis!

The second point to check is just as crucial and that is the central cruciform located just in

2-33. The tell-tale sign of a 'hogged' chassis. You'll need above average home restoration skills to correct this yourself. A professional TR restoration company may be able to, depending on the chassis detail, or you should budget for a replacement chassis that you can at least consider changing at home.

front of the differential, into which the aforementioned chassis legs are welded. The junction is 'plated' top and bottom with diamond-shaped pressings, in the trade called the 'Tee-shirt' pressings. When rusted metal is present underneath the pressings, the pressing swells, as shown in photograph 2-32. The metal can even de-laminate in some cases – and it's not unknown for a bodged repair to mean that the rusted parts of the pressing are pushed into the internal cavity and the whole mess plated over. So look (closely) for evidence of welding or added material on/in these diamond-shaped pressings.

Superficial and frequently unsafe 'bodges' are often carried out to get the car through an MOT. If corrosion is present above the bottom pressing, the only way the chassis can be repaired properly, safely and with long-term security is to remove the bottom plate completely and first replace the otherwise (probably considerable) hidden corroded metal in the junction and chassis members above the plates. Then, a complete new pressed plate should be re-fixed. A properly executed repair is likely to set you back £600-£700 (a little more if the differential mounting is simultaneously strengthened).

A badly repaired TR will show its chassis ends beneath the rear valance as per photograph 2-33. This is the infamous 'hogging' we will talk more about in the chassis restoration chapter. Budget for a body-off replacement chassis-type expense or walk away. As you go round the car, list the

2-34. A creased chassis leg tells you that this car has had a significant thump at some stage in its life. This is not necessarily terminal, since the IRS chassis can be shimmed to correct minor defects, but be on your guard from the moment you spot this. If you are not inclined at this point to lay on the ground in order to look for such damage, you can feel for creases by running your hand along the outside chassis rail, and taking a closer look only if you think you feel something untoward. Repairing the crease itself may not seem a major problem, but take note that it's a body-off task which may signal other, possibly more serious, accident damage, and consequential problems requiring close scrutiny and repair. The presence of a crease need not put an end to your interest in that particular car, as the car may have been well repaired and shimmed, and drive perfectly well. However, it may mean a more expert second opinion of the car – should everything else prove acceptable. In very bad cases the 12ins (300mm) or so of supposedly straight chassis in this area can actually be formed into a slight 'Z' shape – which is bad news since it is almost certain that the turret has been moved backwards by the impact. A repair is impossible on an assembled car, and even with a bare chassis, almost impossible for the amateur to properly repair at home without the necessary chassis jig and powerful hydraulic correction ram, not to mention the expertise to know where to 'doze' the problem out. Definitely a 'proceed with caution' car.

discrepancies you find and then take your list to a competent, experienced TR specialist. There's nothing on a TR that can't be fixed, it just costs money – you don't want it to be your money, though, do you?

Next, go to the front of the car, lift the bonnet and study closely the rear lower wishbone attachment brackets (visible through the inner wing apertures). It's acceptable and almost normal that these will have been welded, but they MUST have been welded competently.

2-35.

2-35 & 2-36 (above). Corrosion of this severity (once patched) in one spot should also set you looking, since there are likely to be similar rusted areas of chassis or body elsewhere. Photograph 2-36 may just show what you probably do not want to find; certainly once you have paid for the car and got it home!

Moving backwards, find where the chassis goes from two side members to one (roughly underneath the 'A' or front door pillar). If the car has suffered significant frontal impact it will show at the point where the chassis is down to a single 'leg' each side of the car. If you look down the side of these chassis members they should be perfectly straight. A crease will signify a significant impact – a typical example is shown in photograph 2-34 – and, if present, should trigger a much closer examination of the front of the chassis, as a replacement chassis could be needed. All is not lost if this is the case as reconditioned chassis for all independently sprung TRs are available for about £1000, and new replacement chassis can be purchased for about £2000.

If you are forced to consider such a task, remember that this will be the least amount you'll be likely to pay, as all suspension bushes, fuel and brake lines, and body mountings, etc., will have to be replaced. In truth, however, a poor chassis is unlikely to carry a super body and that translates into more remedial time and expense. Often, it works out at about double the time and treble the expense!

2-37. These creases in the front wheelarches look a problem, but relax, they are quite standard. The ones shown in this picture are to be found in every TR.

Potential TR-ers from Europe should note that spliced or patched chassis repairs (photograph 2-35) are not acceptable, particularly for the German TUV (MOT) test. All chassis repairs have to be 'invisible', which means that if a chassis member is to be repaired and the car is destined for the Continent, the chassis member must be replaced in its entirety.

Even if you do not find a single item of major concern, the chassis needs to be reviewed in the context of how long you expect the car to last. Is the chassis scaly and original, or is there any sign of the corrosion shown in photograph 2-36, in which case it may mean that your plan to get married in the car in ten years' time is somewhat ambitious ... If the car has been completely rebuilt and the original chassis was sand blasted, repaired and subsequently properly coated, and there is plenty of evidence to support this, then it could well last the ten years you may expect of it. Otherwise, consider the implications of having to carry out a body-off restoration in months, or possibly only a few years' time. Mark it up on your list accordingly!

If you have looked at several cars in your price bracket and found that all would seem to have relatively short life-spans before major restoration work is required, then you probably have an unrealistic mix of price and life expectancy. Nevertheless, have fun, take care and do look at potential purchases with a jaundiced eye!

Chapter 3

Preparing your restoration plan

You've taken the plunge and now have your beautiful (or maybe not yet beautiful) TR tucked up safely in your garage. If you paid "top-dollar" for the car and bought a near concours standard example, then most of the rest of this book should be of little more than academic interest in the short term. At the other extreme, for those who have bought a "restoration project", I strongly recommend more thought than action initially: you need to establish a restoration plan and a budget.

The purpose behind a restoration plan is to ensure that things are done in the most effective manner. However, there are some pitfalls to avoid, and a well thought-out restoration plan and associated budget can go a long way towards ensuring you avoid the three main reasons for those sad "abandoned project" advertisements. The reason most projects are abandoned are because they:

- went beyond the owner's technical knowledge/experience.
- cost more than was initially envisaged.
- took far longer than was expected.

These three issues are largely trade-offs: the more you can do yourself the less the project should cost, but the longer it will take.

This book is intended to help would-be restorers appreciate what they are getting into, and hopefully find a satisfactory trade-off for their particular car, skills, time and budget. There are policies you can adopt which will minimise the work involved and help you decide what to subcontract. I hope I have covered the vast majority in the previous chapter by helping you buy appropriately. Professionally restored cars are priced out of reach of the majority of enthusiasts, particularly younger enthusiasts. However, there are two things going for a home restored TR. Firstly, everything comes apart, which makes access much easier, and, secondly, if you do the majority of work yourself you can be pretty sure that you will recover most of what you've spent – which is not so with many an alternative. So, let's get started with preparing our restoration plan!

STRIPPING THE CAR

Plan for the extra space that a stripped motor car requires: as a rule of thumb you need three times the space. Secondly, do not throw anything away until the restoration is complete; however rusted, bent or rotten, keep everything for reference, and in case that odd clip, harness connection or bracket may no longer be available. You can certainly categorise and list the parts you strip from the car as "useless, requiring replacement" or "good probably reusable", and an intermediate category, too. The list of missing or useless parts will be very helpful when you get to visit some of the TR specialists. You would be wise to put all the parts for, say, the trim in one area for use later in the project. Other parts may be required earlier in the rebuild, and these should be stored in another – possibly more accessible – location – perhaps a rack in the main workshop?

I've suggested that some time should be spent thinking and planning, and there's much to be said for checking as many of the larger components as possible before removing them from the car. The engine, gearbox, radiator and rear axle are good examples of what is worthy of a moment's consideration.

Let's presume you have bought a non-running car that is more-or-less in one piece. Now, restoration plans can vary a great deal, but few restorers are keen to spend more than is necessary! For those on a very tight budget it could be particularly advantageous to postpone for a year or so a major expense like an engine rebuild. It would, in any event, be very helpful to learn what the gearbox is like and/or whether the overdrive works. So, before you rush too quickly into taking off some parts, consider whether it might be a good idea to find out in more detail what you have bought.

Do not forget to ensure that the car is safely and securely elevated off the ground, and to check fluid levels in all the components you are about to test. It may even be prudent to change some of the oils completely before you start. However, if after a few hours' work you can confirm that the engine runs reasonably well without any expensive noises, you may elect to postpone initially the expense of an engine rebuild. You may find the overdrive does not work, and establish (perhaps with the help of Chapter 9) that it is not a simple electrical problem. You may even hear some unpleasant gearbox noises; all of which may lead you to conclude that a rebuilt or replacement gearbox and overdrive are essential prior to reassembly of the car. These early tests may even completely reverse your original engine/gearbox restoration intentions. Such changes of plan are perfectly acceptable, for restoring these components can be – and, where practical, should be – postponed. There's absolutely no point in spending money before you need to!

There are, however, two major, absolutely vital parts of a TR that need to be put right first time round – the chassis and bodywork: you do not want to be postponing or part-restoring these.

It's a good idea to empty petrol/gas from a fuel tank that you are not going to use for a while and replace it with a gallon of diesel fuel before putting it somewhere safe. This does two things – ensures the solidarity of the tank and acts as a preservative for the period that the tank is in store.

Do not even think about taking the body off the chassis or, currently, the suspension off the chassis until you have established a TR specialist as your "partner" (discussed in the next section). You can remove the bolt-on panels (wings, doors, bonnet and boot/trunk-lid), exhaust, engine, gearbox, fuel tank, electrics, trim, seats, dashboard, etc., if you want to get a better look at the body and chassis, after which you had better set about establishing your contacts.

DO YOU HAVE THE CONTACTS?

If you have not done so already, I strongly advise that you join the TR Register (see Appendix 1), and call your local area secretary before attending the next area meeting. The least you will get from the local area TR group will be a list of local services and recommended suppliers, and you will establish contact with local members, some with very helpful experience who will be able to offer advice. Most amateurs do only one restoration and, consequently, the experience of others is absolutely indispensable if you are to avoid some (potentially expensive) mistakes.

Time spent planning is never wasted, nor is time spent finding and visiting local proven and trusted suppliers and establishing a rapport with all. Again, if you have not met any TR specialists, now would be a very good time to visit one or two. This book chronicles the combined expertise of the five TR restorers that I consider to be the premier specialists in the UK; (in alphabetical order) Revington TR, TR Bitz, TR Enterprises, TRGB and TR Workshop. However, I must not overlook the significant expertise of Faversham Restorations, which also gave a great deal of time and help. (Addresses can be found in Appendix 1.) It's very important to establish a rapport at the planning stage with the specialist you are going to buy parts from, probably use for such professional work as is required, and from whom you will seek advice.

Some parts are hard to obtain, and one of the benefits of buying your car, particularly an ex-US LHD TR, from one of the specialist TR businesses listed above is that you can specify that, as part of the overall deal, you must have all the essential components to restore the car (including conversion if applicable). For those who have bought their car privately, the next best thing is to approach your preferred specialist with the prospect of a mutually helpful partnership. This time the approach needs to be along the lines that you plan the restoration of a TR and, whilst you have the basic car, you will need at least the following spares (take as comprehensive a list of parts as you can prepare). Your partnership suggestion should be along the lines that you are prepared to buy all your spares from him and send any specialist work you cannot cope with to him. In return, you ask for help and advice when needed, introductions to suitable contractors (e.g. a sand-blaster) when the time comes, the prospect of a modest discount on purchases and a source for the hard-to-find parts.

Obviously, the specialist most local to your home is going to be the most convenient in many ways, but don't forget in these days of next-day delivery parcel post that personal chemistry, co-operation, experience and facilities are actually more important than geographical proximity in most cases. Take some photographs of your car and visit one or two.

The three US specialists that come highly recommended are (in alphabetical order) Moss Motors, The Roadster Factory and Victoria British Ltd. Their addresses also appear in Appendix 1.

CHASSIS OR BODY FIRST?

This important question requires some discussion. In fact, there are several important decisions to be made as your restoration plan starts to take shape, but I believe this to be the most crucial. If the chassis seems in good shape, and is to be used again (albeit after refurbishment), it's probably best to carry out body repairs BEFORE separating the bodyshell from its chassis. You can then use the chassis, in effect, as the jig to correctly locate those new body panels as are required. This route has the advantage of requiring only one body-off/body-on sequence, and is recommend for most non-professional restorers with a suitably good chassis.

The plan takes on a different sequence if you think it will be necessary to replace your chassis with an exchange/refurbished unit, or buy a new chassis. You are well advised to carry out the body repairs with the bodyshell on its permanent chassis. The home restorer will not have the facilities, particularly the bespoke jigs and measuring experience which any good professional TR body restorer has. So this restoration plan will require your sending the original chassis away for repairs, or acquiring the replacement before you start the bodyshell repairs. However, there's no point painting the new chassis and then doing your bodyshell repairs on that chassis as you will inevitably damage the chassis paint and, in any event, you will need to separate body from chassis to paint the underside of the repaired body! This type of plan, will therefore, mean two body-off/body-on exercises, including one before commencing body repairs.

Do not overlook the possibility that there could be an easier route to the initial removing/replacing of a bodyshell as part of this "two-lift" plan. Provided your plan calls for new floors and new sills, you may just as well take the body off in two pieces, remove the floor and sills, and then offer the body back up in two pieces to the (now satisfactory) chassis. It still has to come off once more for various painting operations, and the second time it must come off in one piece; however, the first off/on will at least be easier.

If you are uncertain about the true quality of your chassis, how do you decide which way round to effect body and chassis repairs? The first golden rule for cars with the IRS type of chassis is to establish whether the chassis is 'hogged' at the back. Hogging means that the rear of the chassis (roughly from the rear trailing arm chassis member) cants downwards. It is not uncommon and we will establish the cause in a moment, but clearly there is no point in using the chassis as a jig to align body panels if the 'jig' is flawed before you start on the body repairs! Hogging can be caused by the obvious – a shunt up the back. Inexpert chassis repairs can also bring it about, and probably this is the most common cause. Replacing the chassis members that carry the rear trailing arm would be one potential cause and replacing the diamond shaped 'tee-shirt' pressing at the centre of the chassis cruciform another. The main problem is often shrinkage brought about by cooling of the metal after welding. The components may be perfectly aligned before welding, and even properly tacked into position, but too much welding in one place without compensatory stitch welding on the opposite corner of the repair will 'pull' the chassis as the welding cools. Although the shrinkage on or close to the weld will be inconsequential, the cantilevered rear chassis member can easily droop an inch (25mm) at its tail end!

You do not actually need to look at the chassis itself to become suspicious, as the rear door gaps will probably tell much of the story. All TR door gaps open up as one progressively measures from the rear base of the door upwards. The bottom gap should be about 1/8in (3mm) and the top 1/4in (5/6mm) – but if the top has opened up to 1/2in (12/15mm) as in photograph 2-7, then you probably have a hogged chassis.

Viewing the car from 10 metres behind will also tell you, since you should not be able to see the ends of the chassis members below the rear valance. If you can see part of the chassis members as depicted in photograph 2-33, you certainly have a hogged chassis and repairing it becomes your first priority.

WHAT & WHERE TO PAINT?

Body and chassis painting are important elements of your plan. As the paintwork on a car is so highly visible, the first inclination is often to sub-contract the work to a professional restorer. Cost comes into play, and I have already made much about the cost of properly painting a car.

However, do bear in mind the difference that regularly spraying a car can make to your skills. If you last carried out a paint job three or four years ago, then at least have the outside sprayed professionally. Furthermore, there is considerable environmental pressure to reduce sales of cellulose paint; the carcinogenic nature of modern, 2-pack paints necessitates professional spray and oven facilities. There's no reason why you shouldn't paint the chassis at home (and there are other options, which will be explored in Chapter 6). For the sake of this planning review, however, let us presume you will sub-contract the body spraying and hand-paint the chassis at home.

When both chassis and body are finally finished you will wish to marry body to chassis once again. How, when and where? For those who choose to, or are forced to, repair the body on a different, but sorted chassis, there should ideally be a temporary marriage of the original body to its new, unpainted chassis. After the body repairs have been done this route necessitates the second 'body-off chassis' party we discussed earlier. This enables the chassis and body to be fully and properly painted before they are finally reunited.

So, in short, there's considerably more work involved if you have to do the body repairs after the chassis has been sorted. If you do the body repairs on the original chassis and then separate body from chassis, you are in a position (and many private restorations are carried out this way) to effect chassis repairs and painting while the body is away at a professional restorer to be painted. The same principle applies to those considering doing the body repairs after sorting the chassis, in that you will be refurbishing the various suspensions and mechanical components, and fitting out the chassis while the bodyshell is away for painting. All of the premier professional restorers are unanimous with respect to the foregoing advice.

Where opinion does differ, however, is whether or not it is important to carry out body restoration on a loaded IRS chassis. Unquestionably, loading the chassis with engine and gearbox (or representative weights) adds to the work involved, particularly as those that advocate this route also feel that the chassis needs to be sitting on its suspension. Several professional restorers regularly restore bodies on an unloaded IRS chassis, but you must ask around and form your own opinion. I think that the skills and experience of the best restoration companies may not be available to the first time restorer, and it may therefore be prudent to carry out your bodywork repairs on a loaded chassis.

A TYPICAL HOME RESTORATION PLAN

Every plan will vary in detail as the particular circumstances of each restoration are taken into account. The premises/tools/time/skills/experience available, distance from main subcontractors, and target restoration period will all influence the planning, along with the initial condition of the

3-1. Part-way into stripping a TR6 completely. You'll note that the doors, interior dash instruments, and electrics are completely out, with the front and rear wings (fenders) still to come off, and the engine to be lifted out. It would be prudent to remove the wings before taking the engine out. The floors on this example look to have no more than superficial rust, so this is unlikely to be the sort of car that warrants taking the body off in two parts.

3-2. However, this TR250 is already looking like a candidate for floor and sill replacement, and therefore cutting and removing the front half separately to the rear; so much so that I have marked the probable cut-line to adopt once the car is completely stripped.

3-3. If confirmation were needed, this should provide it. As a matter of interest, I understand that the metal was so thin that the floor of the TR250 would not support the weight of a spanner/wrench!

car and intended standard of the end result.

It's impossible to give a definitive plan, but the following thoughts may serve as prompts when shaping your own plan of attack. I have chosen to plan the IRS car on the assumption that the original chassis is too unsatisfactory to use as the body-jig. Not all IRS cars will be in this state.

- Strip the car completely, including engine, gearbox, differential, front and rear suspensions, electrics, trim (internal and external), dashboard and body panels. (Photograph 3-1).
- Decide whether the floors and sills are saveable. If the chassis is in as poor condition as this example shown

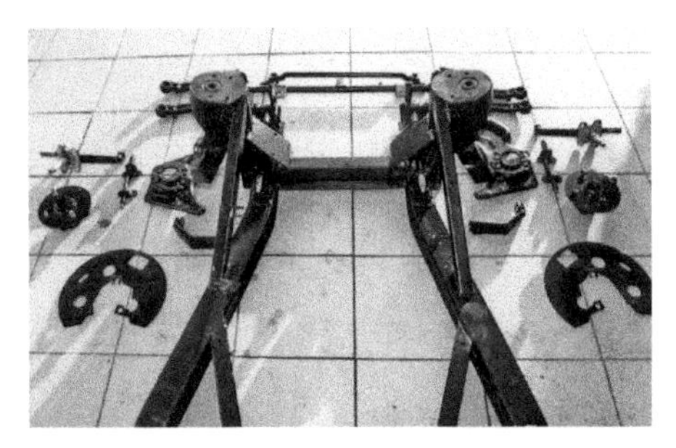

3-4. The various options for repairing, replacing or procuring a new chassis are explored in Chapter 5. While it does not matter which route you choose, it's important that the body is rebuilt on the chassis it will spend the rest of its life married to! Here is a new IRS chassis with some of the front suspension components laid out prior to re-assembly, although the major components will also have to be temporarily placed in position.

3-5. This car has been lifted to a more convenient working height by axle stands positioned under the chassis, which is certainly a safe method if the stands are positioned properly. The IRS structure is not stressed in the same way as if the stands, ramps or supports were positioned under the suspension components: more on this subject in Chapter 4. This chassis is at least stressed with its major components in position; with the dust-cover lifted we can see the engine and gearbox very clearly.

3-6. The pair of new floors have been securely mounted to the (new KTM) chassis – which, obviously, is going to act as both body assembly jig and permanent base for this Michelotti shell. The front and rear "halves" of the bodyshell will need to be carefully prepared around the respective mating faces. The old sill and any remnants of the old floors should be removed before you can contemplate offering either back up to the chassis/new floors.

3-7-1 (above) and 3-7-2. The front "half" is offered to the new floors. It sits comfortably in place as the result of the angled turret supports, which also have two body mounting points each. Note how the old floor and sill have been removed. These pictures were "posed" for my benefit and, in fact, there is some further making-good to carry out at the base of the 'A' posts before the front half will be properly positioned and tacked in place. Then we can try the doors on the 'A' posts and tentatively position the back half shell.

3-7-2.

3-8-1

3-8-1 & 3-8-2 (above). A general shot followed by a close-up of the rear half of this TR4's body being re-acquainted with its chassis. The floors are, of course, new but there's still some making-good to complete before the floors are welded to the rear half. Even then there will doubtless be much detailed renovation work to carry out – but this is a major step in any restoration project.

3-9. Complete the rest of the bodyshell repairs; in this example a rear wheelarch requires relatively minor repair.

3-10. Here, the repair has been completed. You could find you have dozens of similar repairs to complete, and many of these are studied in more detail in Chapter 4.

3-11. As the bodyshell repairs progress, you must fit crossbraces to the door apertures before attempting to remove the finished body from its chassis. Note that the crossbraces shown here are made from substantial angle-iron section, and follow the direct path across the top of the door aperture. Some form of crossbracing is indispensable, although we will discuss various alternative methods later. The braces shown here are very satisfactory, and will be welded to the inside of the front of the 'A' post, while the rear ends will be securely bolted to the inside of the 'B' post using two of the three hood mounting fastenings.

in photographs 3-2 and 3-3, the floors and sills are unlikely to be of further use. In this case cut the shell across the centre, just in front of the propshaft tunnel as shown in photographs 3-2 and remove the shell from its chassis in two halves as per picture 4-1.

- Repair, replace or buy a new chassis. An example is shown in picture 3-4.
- Remove the worst of the oil and dirt and (temporarily) replace the original major components and suspensions on the new chassis, noting that no refurbishing has taken place.
- Lift to a slightly more convenient working height and stand the car on its suspension using four 'ramps', or four stands that bolt to the hubs (photograph 3-5).
- Fit new floors to the chassis, refit

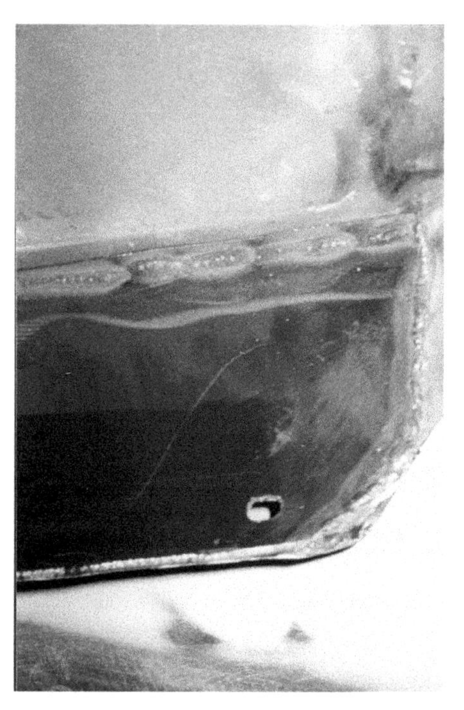

3-12. Just one example of the loose panel repairs you will have to carry out after you've finished the main body tub. As explained in more detail in Chapter 4, you'll probably use a complete replacement panel when repairing a TR6, but with TR250/5s it is usually better to repair the loose panels. Here, part of an old outer sill section has been used to provide the material and shape for a lower front wing repair on a TR5.

3-13. Remove the finished body from the chassis – usually best accomplished by four strong lifters (six, if you have the loose panels in place) – but in this case using a hoist, which has the added advantage of enabling the finished shell to be stored in the roof space until shot blasting is arranged. Very ingenious!

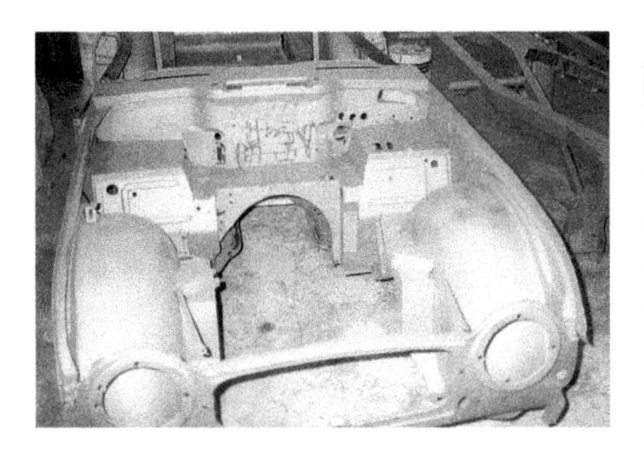

3-14. Noting that the door aperture crossbraces remain securely in place (top of the picture), shot blasting of the shell occurs only when the bodyshell repairs are substantially finished.

3-15. Certainly the odd minor spot of corrosion will be revealed by the blasting operation, and will require repair. I think I can spot one or two examples in this picture. However, the crossbraces are definitely still in place!

3-16. Note that other components await the painter's attention on the back wall, while the bare shell (no panels in place) can be wheeled around the shop with ease.

the original front and rear halves of the bodyshell onto the new floors, and cross-brace the door apertures as shown in photographs 3-6 to 3-11.

- Carry out body tub and panel repairs: photograph 3-12 shows one example. Remove all loose panels once the fit is to your satisfaction.

3-17. Prepare the chassis well and apply a good protective finish, such as this red oxide primer. For a variety of reasons, discussed in the main text, brush painting is not only the most cost-effective method, but also provides the most suitable long-term finish, provided you choose the paint carefully. If you damage the finish during subsequent restoration work it can be touched-in, and any damage that occurs later in the car's – hopefully – long life can also be rectified with the minimum of cost and fuss.

3-18. This photograph (actually of a TR4) obviously shows the engine and gearbox in place on this completed chassis, which is certainly an option. I prefer the simplicity of taking just a rolling chassis (possibly with diff and driveshafts in place, too) to the painter for fitting the painted body, since exhaust manifolds and/or radiator protrusions can add to the body-on task.

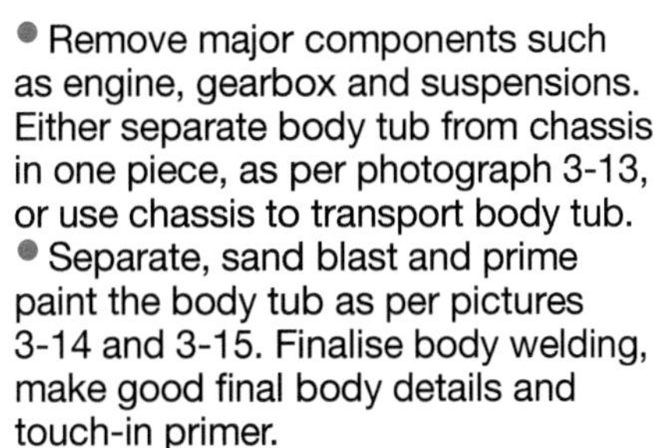

- Remove major components such as engine, gearbox and suspensions. Either separate body tub from chassis in one piece, as per photograph 3-13, or use chassis to transport body tub.
- Separate, sand blast and prime paint the body tub as per pictures 3-14 and 3-15. Finalise body welding, make good final body details and touch-in primer.
- Send body tub and loose panels to painter, noting importance of transporting the lot very carefully – possibly on the chassis (photograph 3-16).
- Paint chassis at home as shown in picture 3-17. Refurbish front and rear suspension components and affix to chassis.

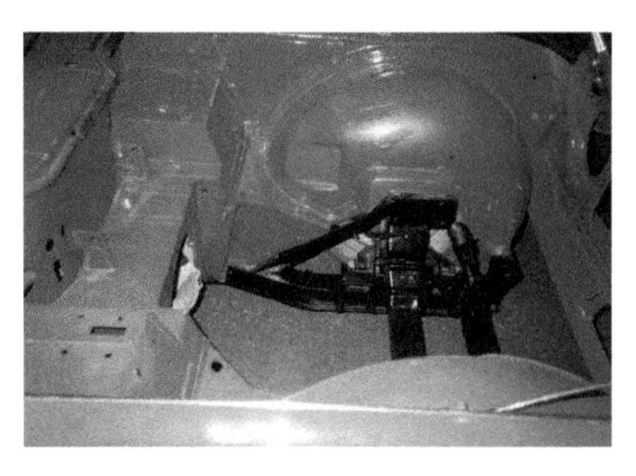

3-19. Body reunited to its chassis at the painters. We'll explore some of the preparatory steps you can take to ensure this task goes smoothly, and causes minimal interruption to usual paint-shop routine.

3-20. The engine going in – a major step forward! Some very useful detail is given in Chapter 7.

3-21. Here, the first of the separate panels is refitted. You'll note that the dash/fascia of this TR250 is, in fact, already in place, which just goes to show that there's more than one way to carry out a TR restoration. You may find several tasks easier to carry out before the doors are refitted.

3-22. Electrics in the course of refurbishment and reassembly to dash/fascia.

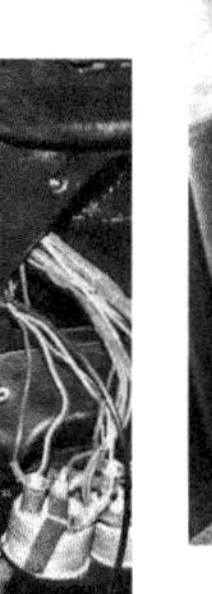

3-23. Refitting the loom is just one of dozens of detailed "final" tasks that, collectively, will take more time than you anticipate when preparing your restoration plan.

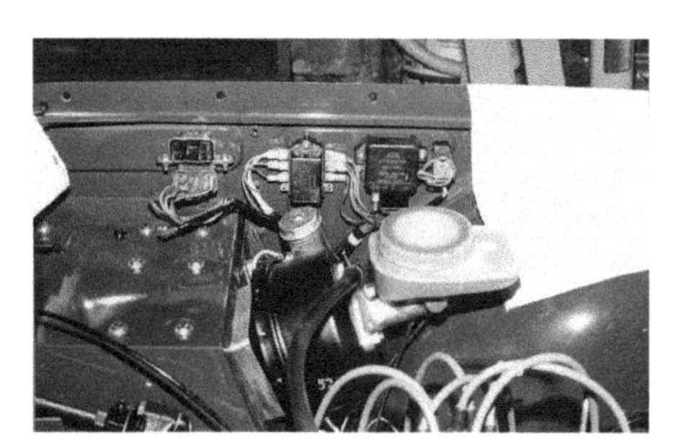

3-24. Then there are the hydraulics (and fuel piping) ...

- Take rolling chassis to painter, broadly as per photograph 3-18. Marry body to chassis (using appropriate body fixing kit) at the painters (photograph 3-19).
- Refurbish engine and gearbox and fit to car as per photo 3-20. Fit exhaust, induction, radiator, brakes, etc., to car.
- Fit body panels (boot, doors, bonnet) and external trim shown in photo 3-21.
- Refurbish as required and refit electrics, dashboard, internal trim,

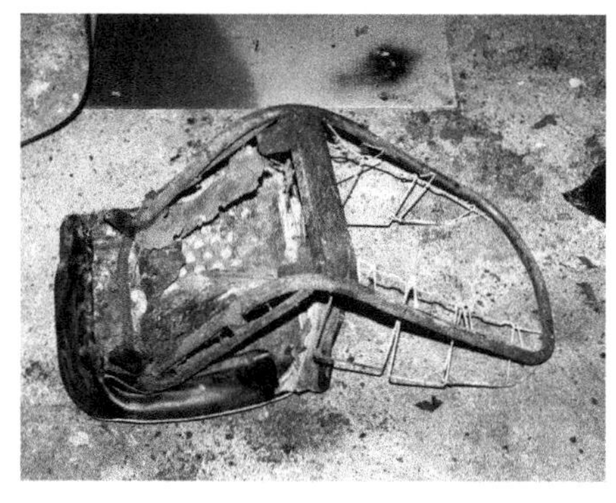

3-25. ... not to mention the seats ...

3-26. ... trim ...

3-27. ... and the hood.

hood, etc. A summary of the operations involved are shown in photographs 3-22 to 3-27.

The objective of the above is to help you plan your restoration to the best advantage. No point in painting, say, the chassis too early and then ruining your hard work by building the body on a painted chassis. No point in spending money on an engine rebuild and then leaving the engine standing in the corner of your garage for several years. This does neither the engine nor your bank balance any good and will almost certainly invalidate any warranty you initially had.

SPARE PARTS

It is vital that you order spares that are actually appropriate to your car rather than its registered year of manufacture or commission number. All the cars in question have lived long lives and very few will not have had secondhand (possibly reconditioned, but still used) parts fitted. While most will have been fitted without detriment to the car's operational effectiveness or reliability, not all of these parts will have exactly matched the original specification of the car.

I have spent some time during the preparation of these books detailing at least the various major design changes of the main parts. As two examples I can mention the brake calipers that progressed from imperial to metric specification, and the six-cylinder engine that, amongst other alterations, had plain and then recessed bores. Late examples of both these components can and have been fitted to early cars, whilst, conversely, early components have been fitted to cars manufactured long after the changeover was concluded. Often the owner at that time was unaware of the differences, while certainly you, as current owner, will not be expecting to find, say, a metric brake caliper or a recessed block on your 1969 TR6.

If you are not alert to these and similar possibilities, you could order the wrong brake hoses or cylinder head gasket, which, at best, will be inconvenient and could have more serious consequences. Clearly, the year, model and commission number of your car are important factors in identifying precisely what spares you need, but be aware that, at some earlier date, availability or expediency may have changed the basis on which you should be ordering spare parts.

Chapter 4

Body restoration

This chapter may use some component names, for example 'scuttle', that you may not be readily familiar with. Drawing D4-1 should help to clarify the locations of these panels.

THE MICHELOTTI – KARMANN DIFFERENCES

We are concerned with repairing either a Michelotti bodyshell in our TR250 or TR5, or a Karmann body that forms the basis of our TR6. Why mix these two different car shapes? In fact, they are based upon very similar body tubs once the bolt-on panels are removed, so there's greater similarity than many realise at first.

This demonstrates the skill of the Karmann designers in that they retained the majority of structural panels, and even made complete door assemblies interchangeable. On the other hand, you could be forgiven for not appreciating just how many differences there are between the TR4/TR4A shell and the TR5/TR250. They look the same, certainly at first glance, but, for example, the inner wings of the TR4/TR4A have more relation to the sidescreened cars than their, apparently, more alike Michelotti-designed sisters. 4/4A doorframes may not have much in common with the sidescreened TRs, but neither are they the same as TR5/250/6 doorframes – although they may appear interchangeable. The door skins can, however, be used on either car! Confused? So was I to start with, so let's examine some of the main panels to establish what is the same and what panel fits what car.

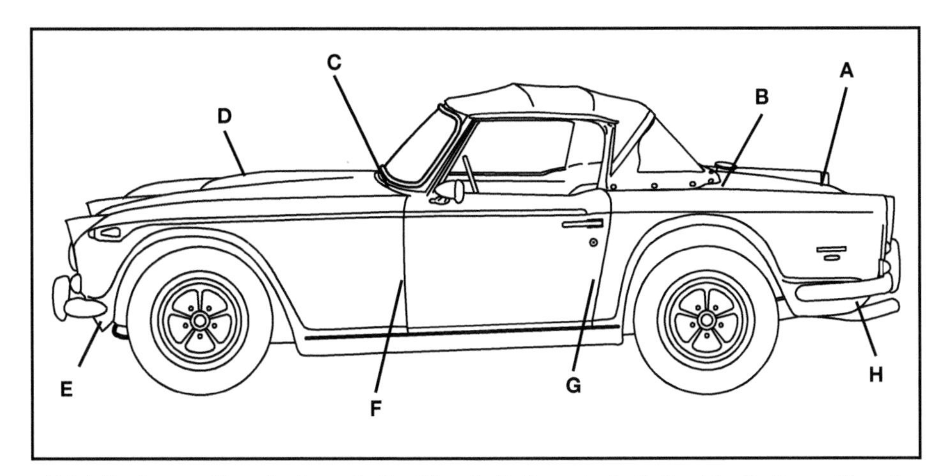

D4-1. Body panel terminology. A. Boot/trunk. B. Rear deck. C. Scuttle. D. Bonnet/hood. E. Front valance. F. 'A' post. G. 'B' post. H. Rear valance.

The centre of the TR4 bodyshell was also used on the TR4A, and consequently floors, inner and outer sills, front door posts, door skins, etc., are interchangeable and available new. The centre panels on the TR250/5/6 are all interchangeable within themselves, but different from the TR4/4A. The TR4, 4A, 5 use the same deck above the fuel tank, although the TR6 uses a different, flat deck.

The TR4 soft-top was stored in the boot, and so its frame needed minimal folding space at the rear of the cockpit: consequently, the 4 uses thinner forward deck extensions than the TR4A, 250, 5 and 6 – all of which had the same slightly thicker extensions.

Some of the Michelotti body panels are produced by original tooling, and fit very well, but you should be aware that many are made from replica/soft tools (notably the wings) and require considerable skill and patience to get to fit well. On the other hand, the vast majority of TR6/Karmann panels are still made from original tools and are a much better fit as a result. No-one is complaining, but it's important that those contemplating the purchase of a Michelotti-designed body are aware of the extra difficulties they may experience.

MINIMISING BODY REPAIR DIFFICULTIES

Not surprisingly, restoring the bodywork of any TR is not without difficulties. Identifying those difficulties and the solutions is one reason why you purchased this book. I'm not dodging the issue, but bodywork restoration on all but the TR6 is perhaps best avoided, or at least minimised, particularly if you have not restored a motor car (any car) before.

As mentioned a moment ago, the TR6's replacement panels come largely from original tooling, and the 'fit' is significantly better than with earlier TR panels. This not only makes panel replacement much easier for the amateur/first time restorer, but also the subsequent welding does not call for the super skills necessary when panels do not properly abut (fit) one another.

So what are the solutions if you have set your heart on a restoration case, Michelotti-bodied TR? You can avoid or at least minimise the body restoration problem by buying a car or bodyshell that has avoided the 'tin-worm'. An ex-Californian car is unquestionably best, but any car that has spent its life in a genuinely warm climate has to be an infinitely better restoration prospect for the amateur restorer than one with a body that is little more than a lattice-work of rust. You will, of course, pay more for a good ex-Californian car, £2000 (possibly even £3000) more than a comparable UK car, but it is certainly worth it. Naturally, the US car must be straight, and it is essential that you look closely at the chassis to ensure it has not had the 'cut-and-shut' (one car made from two) treatment.

Attracted by a newly rebuilt engine within a UK "unfinished restoration project" that has been advertised for sale? Don't be, at least not if the body is very rusty and funds dictate you will have to do the body restoration yourself. Mind you, there are still solutions for you to explore, but all carry a cost penalty.

Firstly, establish the availability and cost of buying a whole or even half (preferably the rear half) of a compatible ex-Californian body. They are becoming progressively harder to find, which is why you may be forced to seriously consider half-a-body like that shown in photograph 4-1. The rear is the more difficult to restore and, if you cannot find a good complete car, the rear half should be your priority as an amateur/first-time restorer. Don't worry about its reconstruction into a complete car, I will have explained that before this chapter is finished.

The back end of all TRs is constructed from lots of small panels, most of which are available, but the complexity and opportunity for error for the inexperienced restorer is correspondingly large. If you are inspecting a prospective purchase, be sure the panel at the rear of the propshaft tunnel is sound. If this panel is badly rusted and/or the rear wheelarches appear irreparable, there will likely be nothing left of the rear of the car; given the replacement panels available, repair will challenge even a very skilled and experienced professional restorer. In this circumstance you would be wise to proceed only if you can first identify an ex-Californian rear half or complete shell.

If you buy a complete Californian car or bodyshell, Chapter 18 tells you how to change the car to RHD. The TR6 panel fit is quite good and should provide the amateur with an interesting but achievable challenge. The later Michelotti cars (the TR5/250) have a much better panel fit than the 4 and 4A, but still present the amateur with alternative difficulties – and it is time to address those.

IRS CAR BODY REPAIRS

The live-axle/non-IRS/ladder chassis TR4 does not flex to the significant degree that the chassis of a TR4A/5/250/6 (IRS) does. Consequently, you can quite happily rebuild a TR4 body on an empty chassis supported on axle stands.

4-1. This is how you might expect to buy the rear 'half' of a TR body, in this case a TR4A. Note where the body has been cut in two, just in front of the propshaft tunnel. This means that much careful work is required to remove the old sills and floors before this back end can go for shot blasting and be offered up to the refurbished chassis with new floors pre-assembled. The same approach is used regardless of which model TR it is.

Many think that this is not possible with IRS cars, the chassis of which are so different that bodies must be rebuilt on the chassis, and the chassis supported in a very specific manner.

The IRS TRs have a major difference to almost any other classic sports car you can think of. As corrosion sets in and weakens the structure of most sports cars, including non-IRS Triumphs, the centre of the shell sags. Hopefully, this occurs very slowly, but nevertheless, as corrosion advances, most sports car door gaps close up, and eventually the doors become difficult to open.

Unique to IRS Triumphs (the 4A, 250, 5 and 6), it is the rear of the chassis that sags, opening the door aperture rather than closing it. The actual opening occurs between the rear door edge and the 'B' post. This difference occurs because the rear suspension mounting point of an IRS car is a long way further forward than on other sports cars that spring to mind. The consequence of this is that, in a home restoration, an IRS car MUST not only be rebuilt on its final (solid) chassis, but you would be well advised to fit your chassis with its suspension and major (weighty) components. If, for whatever reason, you are unable to, or do not wish to, use your actual (perhaps beautifully reconditioned) engine, gearbox and differential, you can simulate their weight with, say, paving stones.

However, the IRS chassis is so flexible that you would be wise to ensure that the laden chassis sits on its working suspension. By this I mean that the rolling chassis wheels sit on 4 ramps to ensure that weight distribution is near normal. If you are very careful from a safety viewpoint, you can use axle stands positioned under the spring-pans at the front, and towards the rear of the trailing arms at the back to simulate the working suspension. This, again, ensures the loads are through the chassis in as near usual position as possible

You may feel that these suggestions result in the chassis/body being below a convenient working height, in which case consider making up four stands that bolt to each of the four wheel hubs. This will give you maximum, almost unrestricted, access to the car; load the chassis correctly and position the car at a working height of your choosing. Personally, I would think it ideal if the sills stood a little over 40ins (about 1 metre) off the ground.

GENERAL BODY RESTORATION

The golden rules:

1 – ALWAYS complete one side of the body at a time. Perhaps the only exception to this advice is in respect to sorting out both doors, which we will cover below, and aligning both doors with the windscreen. Thereafter, it is definitely one side at a time.

2 – Whatever the car and whatever you do, DO NOT strip every panel off the car and then try and rebuild it from a giant jigsaw-like pile of parts. We will come to this in more detail later, but if you strip too much off the car you end up with few, if any, reference points to help you weld the car in something like the correct shape! Your removals should be restricted to both doors, lights, chrome, trim, seats, door handles, hood frame, dashboard, heater, wiring, fuel tank, etc. However, DO NOT take off the windscreen or any body panels, wings, etc., (whatever their condition) until you know exactly what your restoration plan is.

3 – DO NOT make the frequently made mistake of sending the body for sand blasting until much later in the restoration: I'll tell you when!

4 – DO NOT, on any account, imagine you can cover rusted base material with body filler. You will, sooner or later, be very disappointed. Cut out all rusted steel and replace it properly.

5 – NEVER entertain, even for a moment, patching a steel panel with a fibreglass panel base; you will totally devalue your car. I would not want to buy a road-going TR that had been fitted with fibreglass bolt-on panels (say, bonnet, boot or wings) but, again, those individuals who intend racing/rallying their cars may actually prefer lighter fibreglass panels. However, if you are restoring a road-going car, do consider very carefully whether or not to use fibreglass panels; you may well limit your car's appeal if you do.

6 – All cars should ALWAYS be rebuilt around their doors. Always!

7 – It is very important to use a fixed datum point when measuring the car before disassembly and during reassembly. Try not to use a corner or edge of a panel where possible as these can be easily bent, moved, even removed during the restoration. Hopefully, datum points such as the windscreen screw holes in the front bulkhead, the door hinge holes (even with a pointed threaded bolt inserted), the hinge holes on the rear deck and/or the hood mounting holes offer more consistent reference points. Obviously, you are going to take your measurements on both sides of the shell, but be sure to also take the diagonal measurements in order to avoid misalignments.

THE DOORS

So you couldn't find a good, ex-Californian body and need to rebuild your own. We had better face up to what needs to be done, then, and, crucially, the best order in which to tackle the work.

We have already established that TR restorations should always revolve around the doors. The point at which you should fit the doors will vary according to whether the body is to come off and will be remarried to the chassis in one or more pieces. To ease us into the concept of rebuilding the body around the doors, we will first explore the ways you can rebuild or acquire the doors you plan to use on the finished car – for this is your first priority.

This objective can be achieved, of course, by repairing the doors that came with the car. This solution will inevitably be your first choice and will involve reskinning the outside of the

4-2-1. Obviously, having taken all necessary safety precautions, the first step with restoring a TR door is to grind round the outside of the original door skin on the front, bottom, and rear edges.

4-2-2. You know you're making progress when most of the turned-over lip starts to hang loose from the door frame.

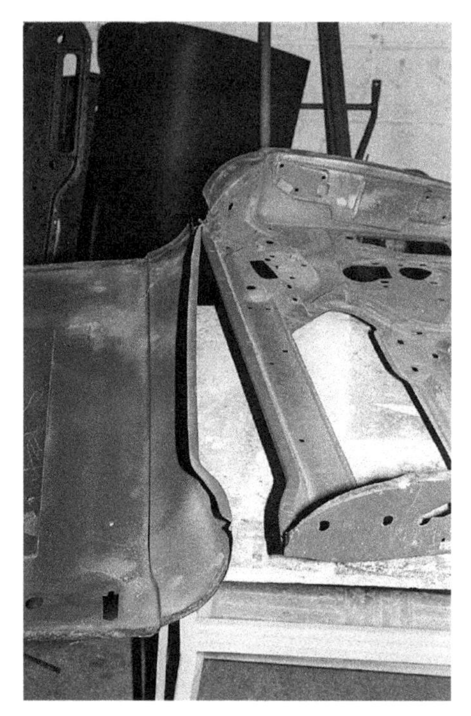

4-2-3. There will always be some spots where things are not quite as easy but, in due course, you should be able to fold back the old skin without any straining whatsoever.

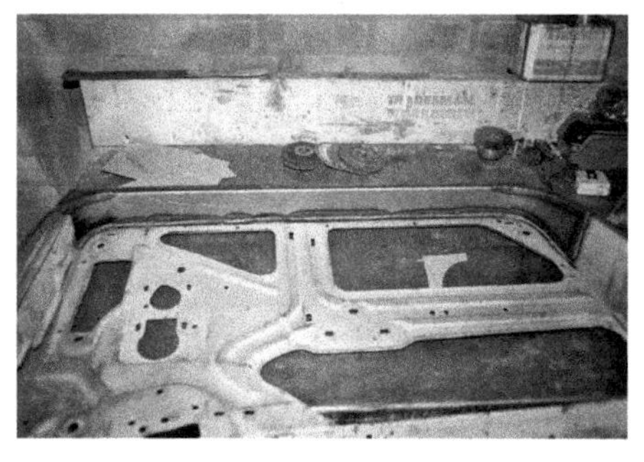

4-2-4. After the door skin, the base of a TR door is most vulnerable to corrosion. There is little option but to cut out and replace the pressing with a new panel – as shown here.

door. Unfortunately, it's also highly probable that the base of the door will require repair, even replacement with a pressed 'repair' section (see photographs 4-2-1 to 4-2-17). Amateurs are best using a door skinning tool, and only consider welding the skin to the doorframe when the door has been offered up to the car. Reskinning offers the amateur numerous opportunities to twist the door. The best plan initially is to just lightly tack the new door skin to its frame in about four or five places on the car with the 'A' and 'B' posts and sills in place. By this method you

4-2-5. It really is a good idea to keep every old item until you are sure it can serve no further useful purpose. Here, we see the old bottom pressing being compared to the newly repaired door bottom. It does look a significant improvement, although you may need to drill two important drain holes in the new bottom before skinning and painting.

4-2-6. As we discovered in Chapter 2, the door frames on our TR are also prone to corrosion, and cracking at the top, too. Here is an example with both top corners showing signs of rust; the front corner (left of the picture) is in need of replacement.

4-2-7. It always looks worse before it looks better. Here, note how the reference point of the hole almost central to the picture has been retained, along with the majority of edge references. A sizeable, right-angled piece has been cut away to remove corroded material.

4-2-8. With excellent results, although the repair piece was more complex than may at first have been realised when looking at picture 4-2-6!

4-2-9. With frame repairs complete, it's time to take the first steps towards actually re-skinning the door. The new skin will already have the protection of a factory-finished (black) primer, but it is important to provide the frame with as much protection as possible. First, prime the newly-repaired frame inside and out, with the exception of the inside lips, over which you will soon be folding your new door skin. Make quite sure that the frame edges are completely free of all corrosion: an angle grinder and abrasive pad are best for checking this.

4-2-10. The inside edges need protection too, of course, but you will eventually tack weld the skin to its frame, and therefore the paint needs to be of the 'weld-through' type shown here (like the sill lips we'll be tackling later in the chapter). Apply several coats and allow the primer to dry before offering the skin to the door.

will be able to twist (or reset) the door and avoid the all-too-common TR door problem of the bottom corner sticking out. When fully satisfied, apply some extra tacks across the skin and the doorframe, with the door still on the car. Remember, final welding should only be done towards the end of your body restoration with the doors on the car.

4-2-11. The top of the door is very visible, and must have enough room for the door glass and inner and outer seals to fit comfortably. The top corners of the skin therefore need to be (slightly) trimmed to fit, and provide for a parallel gap down the length of the door top. We were aiming for about a ¾in (20mm) gap.

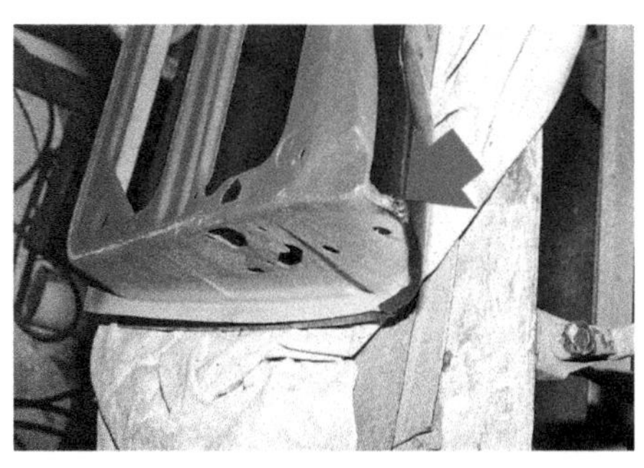

4-2-12. Once the correct height and gap have been achieved, the very top corners of the skin are tacked to the frame front and rear. You can see the skin is still separated from the frame; this is most noticeable towards the bottom of the door. Note, too, the soft sheet between the door skin and the workbench.

4-2-13. The next step is to lightly fold the four corners of the skin onto the frame using, in this case, a hammer and plenishing dolly. Note that the dolly is held tight to the outer face of the skin to prevent the skin's edge from bulging (slightly) away from the frame as the skin is turned over.

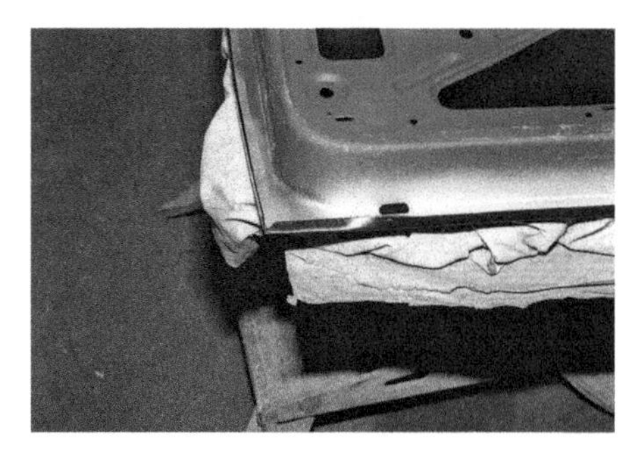

4-2-14. All that is initially bent over at each end of the bottom lip.

There is a possibly preferable alternative to repairing your doors. If you approach any of the premier TR restoration specialists, you will find that most have good quality, rust-free TR doors from a variety of models. The cost varies from model to model and, is bound to be more for an older car, but for the £50-£75 price of a replacement, it is rarely worthwhile trying to repair your own door.

Do still take care, though, as you cannot assume that any component

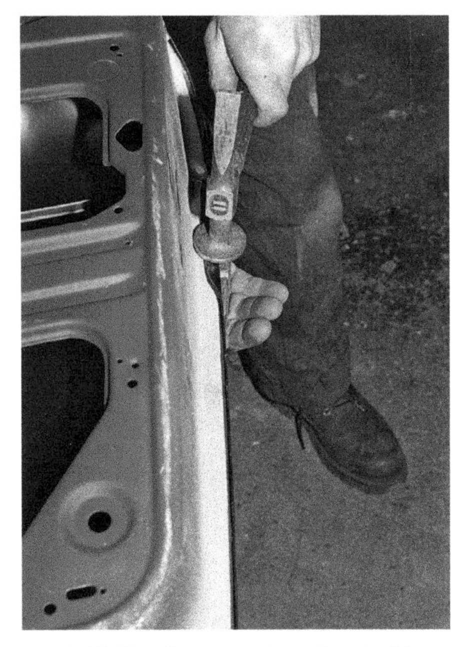

4-2-15. Starting at the top of each side, work down the depth of the door. Run down the length of the edge you are working on, turning the lip over halfway, three-quarterway, and then fully.

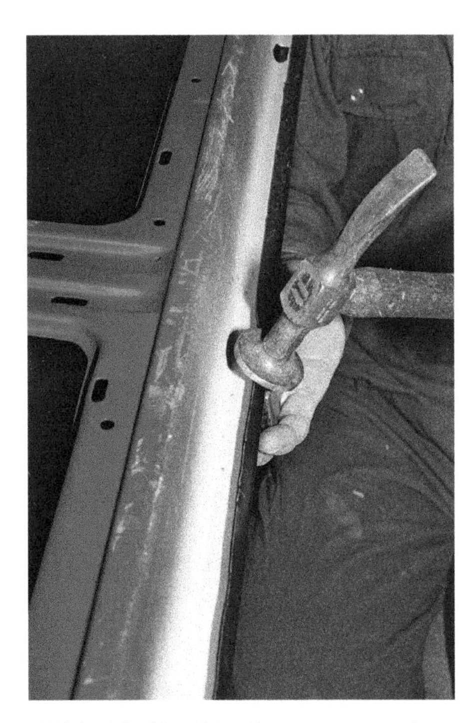

4-2-16. Finally, follow the same procedure along the bottom lip – here we are progressing the half-turned lip into a three-quarter fold.

is fine, be it a door or something else, just because the vendor tells you that it came from a Californian or other

4-2-17. When all the lips of the skin are turned over, go right round the skin ensuring that the 'clinch' of the skin is tight onto the frame, but do not weld anything until you have offered the completed door to the car. Chances are very high that you'll have to twist the door to get it to sit nicely along the original sill and wing panels. Only then should you consider applying a few tack welds (with the door still in place on the car) to hold the new skin to its frame. You should reserve final door welding until you are fully satisfied with panel/sill/door alignment on that side of the car.

warm-climate car. Firstly, the car may have spent most of its life in a more hostile environment, and only a short while in California. Secondly, the best of the ex-warm climate cars and parts have already been shipped and sold. Not everything shipped in today is of poor quality, but you do need to see and examine carefully any replacements you plan to purchase.

With your doors sorted by one or other means, it's time to look at the best sequence for repairing the body tub, initially assuming it will be by the less daunting, one-piece method, but also on the basis of a two-piece approach.

ONE-PIECE BODY REPAIRS

This is done with the bodyshell secured in place on its chassis.

1 – Refit your replacement or repaired doors to the body and line up the front door gap/alignment. Don't forget that every door will need space for a door seal to be fitted at a later date; there are those who have had their cars finish-painted, only to find that they cannot close the door with the door seals in place ... This situation is not helped by the fact that the door seals available today are fractionally larger than the originals. I suggest you use a few short lengths of the latest door seal section as spacers to ensure your doors will close before you progress too far, and certainly well before you even think about painting! As far as is possible, position the door hinges in the centre of their adjustment slots to allow for some subsequent fine-tuning.

2 – Although it sounds premature, you really need to check the reference/relationship of window glass to the windscreen. If you have followed my earlier advice you will not have removed the windscreen, but if you inherited a car with its screen in the boot, refit the windscreen frame at this point. Refit the widows to both new or rebuilt doors. A small degree of error in the way the door hangs can dramatically affect the way the drop-glass/sidescreen marries up to the screen. Clearly there is some adjustment in the windscreen's position, and this needs to be finalised, along with the door position, whilst maintaining your front door gap. Carefully mark the door hinges (perhaps with a light Junior-hacksaw line) so that each time you replace a door on its 'A' post it goes back in the same place. This is important and should be an oft-repeated reference/relationship check made as every panel in or around the door (e.g. 'A' posts and/or sections of the bulkhead) is finalised, so do not be in too much of a hurry to remove the windscreen frame. As an aside, but in the context of welding and grinding in the vicinity of a piece of glass, do ensure that the glass is masked to prevent weld splatter or grinding debris pockmarking it. This can happen with amazing speed and, in the case of grinding, from an amazing distance, and can totally ruin a good piece of glass in seconds.

3 – Ensure the body is in a position on the chassis you are happy with. For example, if the rear chassis/body spacers need increasing or decreasing, carry out the necessary adjustments.

4-3. Crossbracing the door aperture. This is probably the most common way of achieving the necessary rigidity of the body before removing it from the chassis. Several alternative bracing methods are described in the main text, and photo 3-11shows more excellent bracing.

With an IRS car the chassis may have drooped over the years, and if you have not already corrected the fault, you will need to do so before chassis and body are finally reassembled. Now is the moment to adjust the spacers to achieve the door gaps you expect, even if the body has yet to be removed from its chassis. With the door gaps to your satisfaction, securely crossbrace the body across the inside of the door frames ('A' post to 'B' post). The door brace method you choose will make a difference to the ease with which you have access to the shell, and how easily the doors can be offered up to the door aperture. Throughout the book several alternative methods are explored; a very simple one is shown in photograph 4-3.

4 – There are very few details within the restoration of a TR that are "right" or "wrong". Opinions and sequences vary. For example, many a one-piece restoration has been carried out on the basis of changing the floors after the sills have been replaced. Indeed, some of the photographs supporting this chapter will confirm this. However, assuming the floors and sills are to be replaced, many would argue that this is the time to cut out one floor and replace it and then cut out that same side's sill and replace it. Whichever sequence you decide is best for your restoration project, try hard to leave the doors in place for as long as possible, but since this is not always practical, do offer the door up to the car at frequent intervals. Clearly, the point of retaining the door in place is to ensure that the sills are tacked in place with a nice parallel gap between the bottom of the door and the sill.

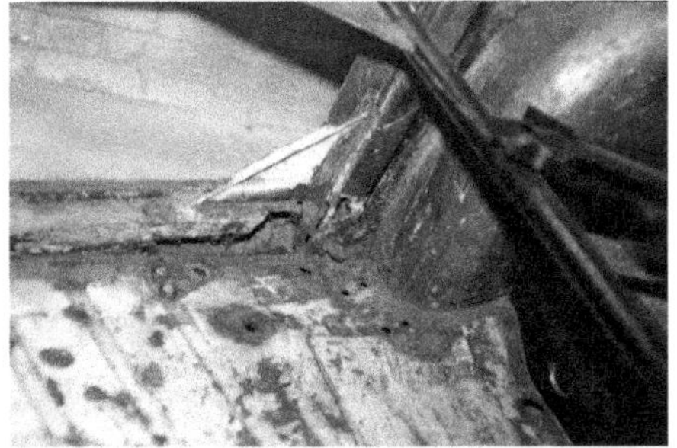

4-4-1. Time to take a close look at the right side 'B' post. Not surprisingly, it, too, looks in need of a great deal of work.

4-4-2. The situation, as is often the case, looks even worse as we strip the corroded metal away, searching for solid material from which to start a repair.

4-4-3. This is beginning to look like one of those jobs that are best not started in the first place! However, this example has not corroded too far back from the outer edge.

5 – After the floor and sill of one side comes the 'B' post. Again, close reference to the door gaps and adjacent panels is vital. There is nothing to stop you temporally tacking the 'B' post to its door, and you can even tack a couple of temporary panel off-cuts (say, about 3ins/75mm square) across the door and 'B' post gap. This will hold the various angles in place until the 'B' post's adjacent panels have in turn been affixed. An overview of a

4-4-4. One step at a time: a piece of mild steel has been folded up to provide two right-angled flanges, the first of which will bring the wheelarch back out to its correct edge position; the second will provide the securing face for the inner wing panel seen here. In order to form this radius, the second flange of the repair piece has been slit every few inches.

4-4-5. Having served as an essential reference, it is now time to take out the completely rotten sill.

4-4-6. Time to concentrate on the 'B' post repairs, after getting the sill in place in order to give something to build on.

typical 'B' post repair will be found in photographic sequence 4-4-1 to 4-4-9.
6 – Then it's the turn of the 'A' post on the same side of the car. A typical

4-4-7. Second step inside the 'B' post.

4-4-8. This shot also gives an excellent overall picture of the nearly completed repair.

4-4-9. Final pressing tacked in place. We would, of course, have checked the door gaps and fit long before now, but this is the last opportunity to ensure the door fits its aperture correctly before the final right side 'B' post welding takes place.

photographic sequence is 4-5-1 to 4-5-9.

TWO-PIECE BODY SEQUENCE

Note: this sequence is only applicable if you are to replace the floors and sills/rockers, and/or fit a replacement 'half' body to repair accident damage or a badly corroded half body.
1 – The bodyshell could have previously

4-5-1. The right side sill outer footwell panel does not look too bad in this shot, although I would guess from the weld that has been pretty liberally applied that there has been a local repair here at some earlier date ...

4-5-2. ... and a shot from the inside confirms all is far from well!

4-5-3. The first remedial step is to leave as many local reference panels in place as possible, but cut out all rusted metal where the first repair will be. Note that the floors are still in place to provide references for the sills – although they clearly will have to come right out before this side of the car can be considered restored. However, there was no choice but to remove the sills, the bottom of the 'A' post and the bottom of the inner footwell panel. In fact, the jury is still out on the top of the outer footwell panel, but it is best left in place to provide a reference for the 'A' post.

been cut into two halves, or separating front from rear could be your first

4-5-4. The next step is to fit the sills, and here we see the first reconstructive step for the 'A' post base.

4-5-5. Still following the one-step-at-a-time approach, replace the outer 'A' post base, which securely ties this side 'A' post to its sill.

4-5-6. In fact, the outer footwell panel was considered unacceptable and removed. This picture shows how right the restorer was! I imagine the owner gave a little thought to also replacing the panel that forms the base of the footwell. It can just be seen on the right of this picture. A great deal of work is necessary to remove this particular panel, with much welding thereafter, so unless the corrosion is dire, edge repairs are usually favourite.

4-5-7. The alternative scenario (picture taken from another car) is that the outer footwell panel can be satisfactorily repaired, in which case the next step would look like this.

4-5-8. By either route, the repaired front 'A' post, footwell, and sill should look pretty much like this.

step. Either way, the cut line is very important. At the front the cut comes across the floor and sills just behind the 'A' pillar, and straight across the car, as shown in photograph 4-1. After

4-5-9. The reasons and details appear in the text, but this is the recommended revised drain for the plenum chamber that passes an extended tube down through the top of the outer sill, and out through the inside of the inner sill. This would be a logical time to make these changes, and to finalise the front right side.

the obvious preliminaries, the front of the body can then be removed from its chassis. The same procedure is carried out at the back of the car. If your plan calls for chassis refurbishment during body painting, be sure to clear all body mounting bolt positions (photographs 4-6 and 4-7).

2 – Clean up both halves of the shell where the floors join each half.

3 – Bolt the new floors to the chassis and then hang front and back halves loosely on the chassis. Remembering our initial purchasing checks, check that the general shape of your body is starting to become 'fish-like'. Fit the doors (not old or scrap ones) to the 'A' pillar and carefully set the front door gaps to the front half of the car and front wings. The next step is to jack the 'A' pillar/doors forward and upward until the bottoms of the doors line up with the inner sills, and the door glass with the screen frame. At this point a few light tack welds might be used to join the front of the body to the floors before you re-check both sides for the fish shape. The rear half of the body then becomes the focus of attention

4-6.

4-6 & 4-7 (above). These pictures do not tell us what the full restoration plan is to be, but this IRS car is clearly having some major body restoration carried out. We can see that the floor and both parts of the sill have been cut out, revealing in 4-7 the left side trailing arm suspension. From these views the chassis does not look too severely corroded, although the body mounting point, visible just forward of the 'B' post, will not be adequate to even secure the floors for the short while necessary were this a 'body off after repair' restoration plan. Shortly, we will see this may not be the plan at all ...

by initially moving it towards the doors to create the correct equal and parallel door gaps. Some sideways or even a twisting movement of the rear half may be appropriate to get things right. Then, some more light tack welding and more checking will be necessary before the outer sills are moved backwards and forwards, up and down, until they are perfect, when they, too, should be tacked in place.

From this explanation I hope you will appreciate that the respective relationship of each pair of 'A' and 'B' pillars with their respective door is vital; if this is not correct, any other work is just a waste of time.

4-8. Here we see the new left-side floor in place. This will have been the first step. The rear inner wing has an excellently executed small repair to the bottom of the 'B' post, and the new inner sill shows up well. Note the plug welds in place along the top lip of the sill, and the tack welds along its join with the new floors and the original inner wing. The next step will be to seam weld the wheelarch to the floor, although it may be prudent to carry out this operation as intermittent stitch welds until the weld becomes one long seam. The sill may need spot welding. The bolt head we see in this picture goes down into the chassis mounting bracket that required repair in an earlier shot.

4-9. The bottom lip (arrowed) on this right side footwell front panel has had to be replaced to give the new floor something to weld to. It's far from obvious, but it looks as if the lip and the floor have been plug welded already, effecting a very nice repair. I mentioned earlier that this panel is very difficult to repair, and that edge repairs such as we see here are therefore the favourite solution.

4-10. The left side of what is a TR5/250, judging from the holes in the door skin, shows the sill and some excellent repairs replacing the 'B' post back and inner wing. A short while later some primer was applied and the re-skinned door re-hung. Though the door will have been re-skinned some time back, it will be offered up now to check that everything is still nicely aligned before proceeding.

4-11. The right side rear of the 'B' post, with, you will note, the three wing mounting captive nuts in place.

4 – Weld or bolt your door gap crossbraces in place (photographs 3-11 and 4-3), and weld the floors and sills to the front and rear halves of the original body.

5 – You can temporally weld a couple of panel off-cuts to keep the doors aligned with the 'B' posts – once the 'B' posts are to your satisfaction, of course. This more or less brings your body rebuild to the same point as the one-piece body restoration method described earlier.

4-12. New floor, sills and inner wing repairs completed for this side of the car. Bearing in mind that we should be doing one side of the car at a time, it looks as if it is now time to start the other side.

4-13. The raw metal was primed before seam sealer was applied – once welding was complete. Nothing unusual about that, but this picture also prompts a couple of interesting additional points. Firstly, you should prime the inside of both halves of each sill before fitting the outer sill. I wonder if you can see the primer paint through the three oval holes in the inner sill? Secondly, the bracket to the rear of the handbrake can pinch inwards, in which case it requires adjustment, a strengthening bracket, and welding before painting. This one is correct, meaning the pressing has either been re-adjusted or was never a problem on this car.

ONGOING MATTERS

There will still be hours of inner tub repairs to carry out. Photographs 4-8 to 4-23 explain how you can expect to progress with the main body tub.

A relatively inexperienced restorer should carry out panel replacement at home on a slow but sure, one panel at a time approach along the lines of photographic sequence 4-23-1 to 4-23-19. Don't forget to tackle one side of the car at a time, and remember to keep

4-14. A typical scene that greets almost all TR restorers when they remove the wings from their car. In this case it is a TR250, and many cars will be much worse than what we see here. The edge where the inner wing will eventually be refitted has never had more than a thin lick of red oxide paint, and has rusted. It will need careful preparation before painting if the restoration is going to last. Note, too, the absence of any form of underseal/protection on this inner wing. A new boot/trunk floor is in evidence from the run of weld just above the chassis member.

4-15. The comparative TR6 rear inner wing. There are numerous similarities, but some differences, too. The most noticeable is that this car has had, at some earlier time, the wing/fender faces painted, the significant benefits of which can be seen here as there seems to be minimal rust.

4-16. The good news is that this Michelotti car has not had a shunt up the rear since there are no creases in the inner wing. The vulnerable spot is arrowed. The slight 'cut-out' above the rear crossmember is part of the IRS car's design.

4-17 & 4-18 (above right). New left side Michelotti inner wing with a new left side lamp housing and new rear valance. Ensure you fit the lamp housings inside the inner wing as per these pictures. They're all too frequently welded to the outside of the inner wing. Such a mistake is surprisingly detrimental to the subsequent fit of the wing/fender. Note the circular additional hole for the reversing/back-up lights. There are two unusual holes in the inner wing which will need closing before finishing.

4-18.

4-19. The new valance will not come with the two rectangular bumper iron holes in place, so you'll have to follow this example and not only cut the holes, but cut them smaller than the size you need and stretch a strengthening lip right round each. Note that the rubber boot seal is temporarily in place to ensure that the boot/trunk lid closes properly.

offering up the adjacent opening (e.g. boot/trunk lid/doors, etc.) panels where relevant.

In this review we started by establishing the relationship of the door/glass/windscreen, and step-by-step we have carried that relationship on to each successive panel. This approach is mainly to ensure that the maximum reference points are given, and that you do not start to progressively build in a small error.

If you are unfortunate enough to have inherited a giant pile of bits, start by fixing the front bulkhead to the chassis, and, even if the 'A' posts are to be replaced, hang both doors in order to get some orientation. Tack the bulkhead in place even if you subsequently intend to replace some sections within the bulkhead, and you may even tack the doors (with the appropriate door/sill gaps) to their respective sills.

However, assuming most readers will avoid the 'pile-of-bits' scenario, do progress by lots of little steps. Prior to fitting every new panel, take off the very minimum of old body each time in order to retain the maximum of reference measurements. Your difficulties will be increased by the poor fit of some reproduction panels, which only underlines the importance of progressing your car one panel at a time.

One solution to many of the panel misfits is to weld a (usually small) extension to, for example, the rear lip of a wing. It may not be necessary to weld the same to the other wing; in fact, the other wing may need a sliver cutting from the rear: many TRs are happily driven round with slight differences from one side of the car to the other. These differences are undetectable to the eye and, with an eventual slight film of body filler, will provide the perfect base for painting.

The front end is much easier to restore as there are fewer panels

4-20. The comparative rear panel and light mounting for the TR6. The much higher rear panel is obvious, as is the simpler light fixing arrangement.

4-21. This is typical of the sort of rear inner wing/fender repair you can expect to do. A spot welder has been used here, but plug welds will be perfectly satisfactory in a home restoration.

to handle and less room for error. Basically, the front consists of a pair of inner wings with a pair of triangulated panels at the bottom. Bulkhead repairs are not too difficult either. The floors and doors will help you align bulkhead repairs, particularly when using the chassis as a jig. Most panels on the bulkhead are more or less flat and can be formed at home. The inner wings can be purchased, if need be, although, in the majority of cases, it is only the edge of the inner wing that is really

4-22. Moving to the front of the car, the control box mounted just to the right of the fusebox (in the very top right corner of the picture) reveals that this car has an early type alternator (before the rectifier was incorporated into the alternator). It's a simple wiring job to allow a later alternator to be used instead. Quite typically, the battery box area is pretty rotten and will need replacing. New battery boxes are available since they are identical to those incorporated into new TR6 bodyshells, and can be obtained through TR Bitz, which has made a duplicate tool. If you have difficulty buying one, however, it is a fairly simple fold-up job from sheet metal.

rotten, and you can make a single curvature replacement strip at home without too much difficulty.

Most home restorers cope with frontal repairs provided they frequently offer up the bonnet/hood, and make the rear edge of the bonnet and front lip of the bulkhead the datum/reference point. Use the bonnet you are going to fit as the yardstick for positioning inner and outer front wings.

If the 'A' posts require repair, then this is the moment to do it, ensuring you do not distort the existing door hinge positions. Before finally welding your tacked repair (or replacement 'A' posts), do (yet again!) offer the doors up to the car to check that the sill to door gap and the glass to screen fit remain satisfactory. At the risk of sounding monotonous, this is what I meant when I suggested you keep in place as many reference points as possible while you work. This technique can avoid your getting to the end of your rebuild only to find, say, that you cannot shut the doors with the windows up. Some home restorers even discover a major problem only after painting, often because they lost reference points quite early in the body refurbishment.

4-23-1. This picture is intended to illustrate not only how to replace a battery box, but also to emphasise the importance of tackling one panel at a time when setting out on a body tub restoration. The following sequence ignores the fact that the next panel is corroded and will have to be replaced shortly, after it has first been used as a reference point to accurately place the preceding panel. This picture gives a clue that there are several related panels to sort in addition to the battery box!

4-23-2. This view from the underside shows we are clearly into replacing or repairing the plenum chamber, too. It runs across the front scuttle behind the battery box.

Then there are the wings, boot/trunk and bonnet/hood to cope with, and I hope that photographs 4-24-1 to 31 illustrate many of the loose panel repair problems and solutions.

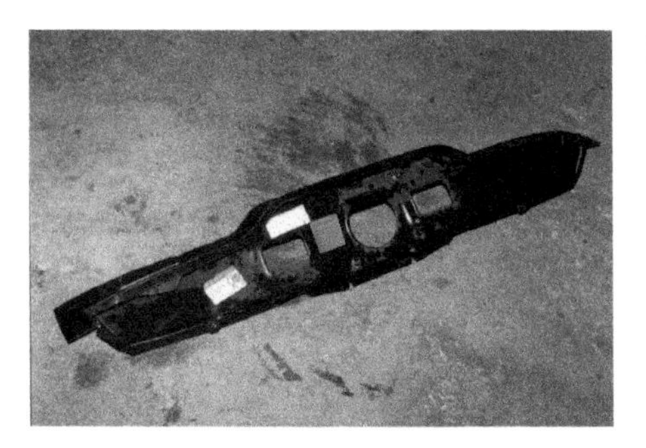

4-23-3. It's possible to buy new plenum pressings, which is definitely the way to go when the corrosion is as extensive as that shown in picture 4-23-2.

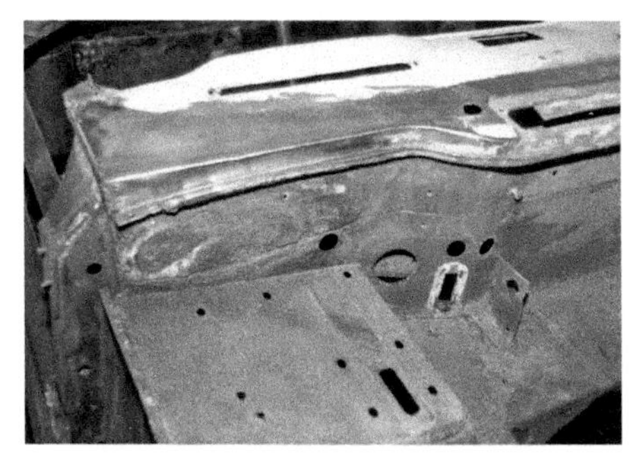

4-23-4. From photograph 4-23-1 it was clear that the right side inner wing would have to be replaced. You cannot see from this shot, but the plenum has also been carefully removed. We have cleaned up the top of the footwell, and the bulkhead/ firewall, and found them to be corroded.

4-23-5. Note that only a part of the bulkhead/firewall is out. Note, too, that panel distortion has been kept to a minimum by drilling out the spot welds, so that even when off the car the panel (it is parked on top of the scuttle for a while) it can be used for reference if required. You can now see that the plenum is missing.

4-23-6. This really is as far as panel removal should go before some reference points are re-established. Note that the carefully drilled spot welds have eased removal of the corroded footwell top.

4-23-7. Time to re-establish some reference points with the footwell top and part bulkhead/firewall in place, even though we may yet remove the next panel.

4-23-8. Still reconstructing the top of the bulkhead/firewall.

4-23-9. Time to replace the inner wing removed a few pictures ago, in order to fix some minor panel edges and remove a corroded corner of the scuttle ...

4-23-10. ... and repair the metal around the windscreen mounts.

4-23-11. We really are getting to the battery box now; spot welds have been drilled out right around the whole panel ...

4-23-12. ... and the rotten panel removed – carefully, so as not to distort the mating panels.

ADDITIONAL DETAILS

- Welding is a gradual process. Do not lay too much weld down too soon. Almost without exception, panels are best initially held in place with a few pop rivets, or self-tapping screws, whereupon panel fit should be checked again. A small amount of tack welding might then be in order. Only after several adjacent panels are definitely fitting well and the most adjacent opening panel – say, the boot/trunk lid – has been tried in place, do you get a bit more enthusiastic with the welding torch.

Remember, too, the old maxim: "If you can't clean it, don't weld it". In

4-23-13. Two steps in one picture. We have fitted and welded the battery box, and moved on to the next phase of corroded metal removal – the other part of the bulkhead/firewall. It was not absolutely essential at this point, but the left side inner wing/fender has been removed, too.

4-23-14. The replacement bulkhead/firewall looks good. The wiper rack panel looks as if it was transposed from the old panel.

4-23-15. Time for a clean-up of plug welds (to get them flush with the panel surfaces as spot welds would be), panel edges, and any final welding that wire brushing and grinding reveals.

other words panel edges that require welding must be cleaned to bright steel – no rust, grease or paint – both sides.

After initial light tacking and rechecking other panel fits, progress the welding very carefully and evenly around the area you are working on –

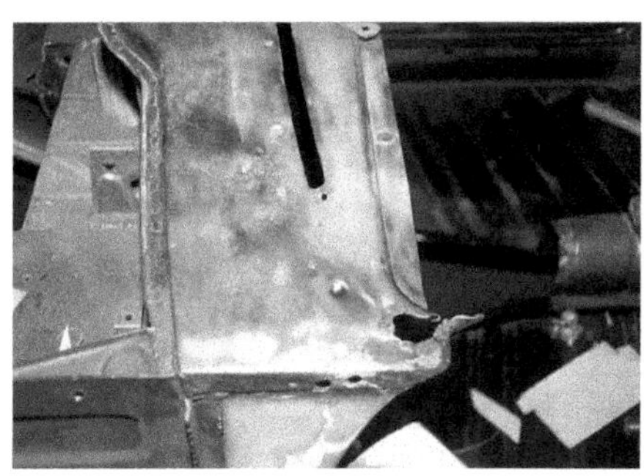

4-23-16. We can just see that the left side inner wing has been fitted but we still have the left side windscreen mounting to tidy up.

4-23-17. Which, unfortunately, reveals some very corroded scuttle that had to be cut out. However, we can see that the new plenum pressing has been welded in place after sealing along its bottom joint with Sikoflex, and painting.

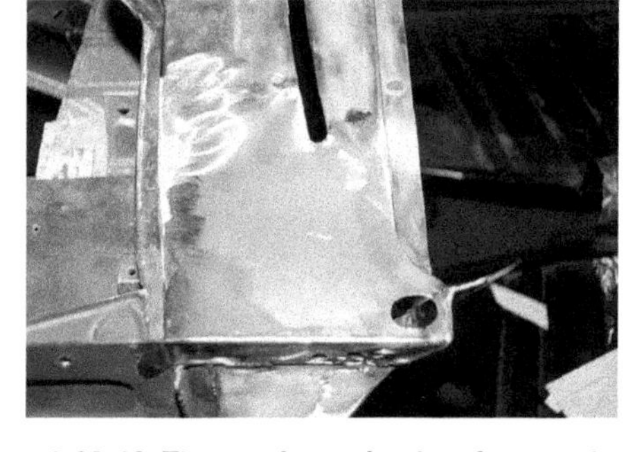

4-23-18. The repair section has front and side lips to fold and weld, as well as rear curves to form, so this was not a five minute job.

say, the back half of the car. Move from left half to right half, front (of the back) to the rear, progressively increasing the welding roughly equally all around the back of car in order to avoid excessive cooling in one area which can pull panel fit out of alignment. With the tacking completed to your satisfaction, remove the temporary rivets/self-tapers and

4-23-19. An overview of the finished inner wings, bulkhead, battery box, and plenum area.

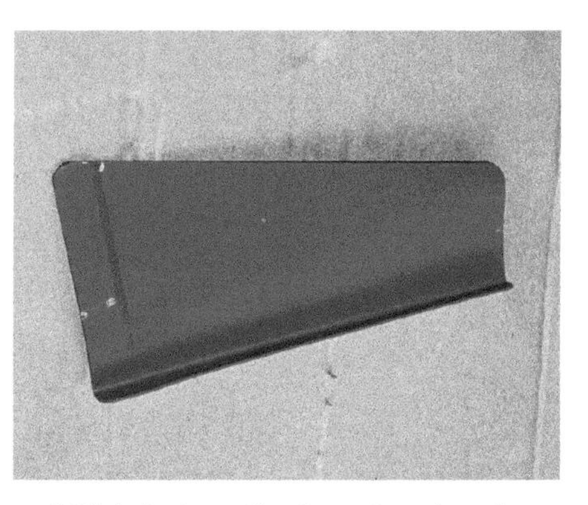

4-24-1. A strengthening edge piece for the top outside corners of Michelotti bonnets/hoods, available from Revington TR. This may well be the most frequent repair required as part of a bonnet/hood restoration, but it's far from the only problem you'll encounter ...

4-24-2. This bonnet is corroded towards the rear edge, to the extent that a repair section is required: here, the size/shape of the new piece has been carefully marked ...

complete the welding, still remembering to move from left to right to avoid laying too much weld too quickly in one area.

- Slide the floors in from the back if you are having trouble getting the new floor panels into the shell.
- When you remove the old floor, the

4-24-3. ... and cut out. Remember to allow plenty of extra material for the rear lip when cutting the replacement.

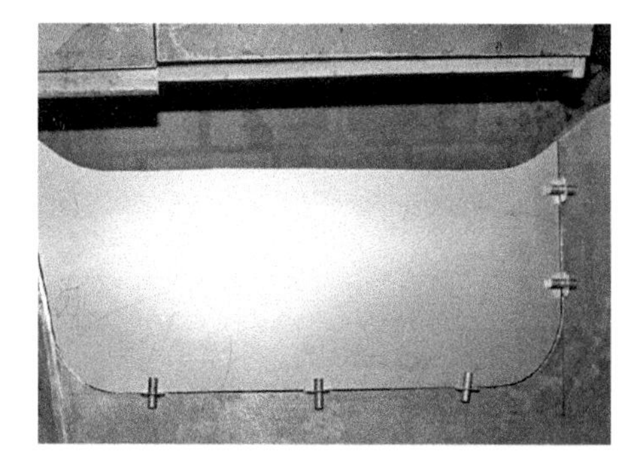

4-24-4. In some circumstances you can overlap panels that are to be joined and plug, seam, or spot weld them together. However, when both sides of the joint are easily seen, this method of joining panels becomes unacceptable. In such cases it's vital that the repair patch is butt welded, and restrained absolutely flush with the original panel. "Intergrip" clamps, available from Frost Auto Restorations, pass a thin piece of material between the two parts requiring butt welding. A small piece of rod holds both panels flush with each other this side of the joint, while the clever arrangement for exerting clamping pressure from the other side can be seen in picture 4-25-1. Obviously, the clamps should be removed once around 50 per cent of the joint has been welded.

4-24-5. The welded in piece can still be seen, but a thin skim of filler will render this repair quite invisible once sprayed.

4-25-1. The bottom lip of the Michelotti boot/trunk lid is a well known corrosion spot, which usually means that the rotten section has to be cut out. Here, we can see the lower 2in (50mm) of skin have been removed, along, of course, with the return lip. A replacement piece has been prepared (note the regular curvature and lock hole), allowing sufficient material for the return lip to be turned over in due course. The replacement bottom has been butt clamped (as explained in photograph 4-24-4). The first welding stage is to butt weld about 50 per cent of the new repair section to the original skin, and to space out welds in order not to concentrate the heat, before removing the clamps, finishing the welding, and dressing both sides of the weld.

4-25-2. Next, we need to tack in place the bottom pressing. Obviously, it is essential to align the inner and skin lock/handle holes. As mentioned in the main text, it's possible to open the Michelotti boot/trunk lids without a key. You should put a (non-original, of course) packer/spacer between the skin and the pressing at this point to improve boot security.

4-25-3. The boot frame has been replaced to give the panel as much rigidity as possible, and the lid offered to the car to ensure we are moving in the right direction, and have not created any unacceptable panel gaps. This close-up shows all must have been satisfactory for the bottom lip has now been turned over and tacked in a few places. No doubt the panel was offered up once again before the bottom part of the corroded right side lip was removed.

major component holding the 'A' post in the correct position is gone. The 'A' post will drop, and drop substantially; it may also move out slightly if unrestrained. Place a piece of wood under the bottom of the door and jack it to re-establish the correct position of the 'A' post. A trolley jack would seem ideal for this job, but don't forget that a hydraulic jack has a tendency to sink over several hours, so if you need to retain the position for a while, a screw jack may be best.

The 'A' posts do seem to have a mind of their own as, regardless of how hard you try to avoid it, they will subsequently drop. Consequently, I suggest you weld them, say, 1/8 inch higher than you think you need, and certainly never on the low side.

• Unfortunately, all sills/rockers come with a slight bow in them. Ask your preferred TR specialist about the quality of its sills, remembering that you will not find perfect examples. Even the best have a slight droop in the centre which must be corrected upon

4-25-4. The next stage is to remove and replace the top part of the right side lip, and offer the lid to the car to ensure all the shapes and gaps are okay.

4-25-5. Then it's the left side's turn. Note the old lip has been retained and used as a guide for the replacement lip.

4-25-6. Step-by-step work (with frequent checks to establish that the panel still fits) is slow. Here is the finished boot/trunk lid with the edges/lips replaced completely round three sides.

4-25-7. Take care when tightening the boot/trunk hinges to the 250/5 panel and its frame; you can distort the skin unless you insert several washers as spacers.

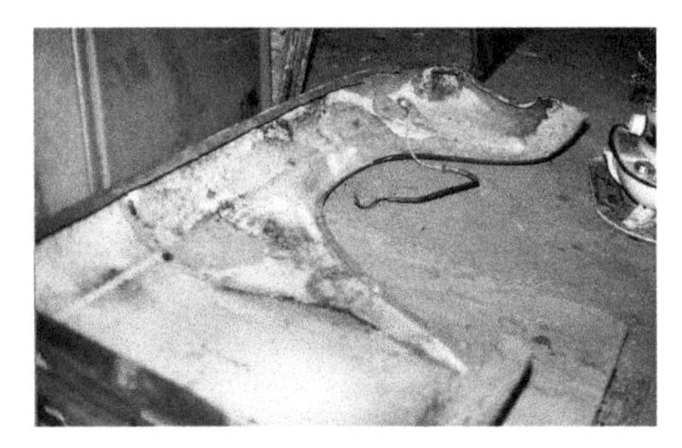
4-26. A Michelotti car wing, which does not look too bad. Note the side/parking lamp lead/wire from the front of the wing that tells us this is a TR4A, 5 or 250.

4-27. This TR6 wing is clearly being stripped of paint using a chemical stripping agent such as Nitromors (in the UK). Stripping the bodywork to bare metal is strongly recommended, as, not only does this obviate the likelihood of a reaction between differing paints when it comes to respraying, but also reveals the material (solid or otherwise) which was hidden by paint and filler. Furthermore, you will get a far superior paint finish on a bare metal foundation. A word of caution, however. Only use chemical stripper on single removable panels such as a wing/fender. Loose panels with seams, welded or otherwise, are not such a good idea as the chemical can get into the seams and never be properly removed, to the detriment of the metal and any subsequent coats of paint. Never use chemical strippers on the main body tub: it takes longer, but the safest way to strip paint is with a heat gun and putty knife.

4-28 (right) & 4-29 (below). We have seen several pictures of this car as the body restoration progresses, and in photograph 4-28 the car is in the final bodywork stage with the loose panels trial fitted. The panels should be removed and painted separately, as picture 4-29 shows (note the loose panels against the far wall). The door openings will have had bracing securely fixed in place before the car was removed from the chassis.

4-28.

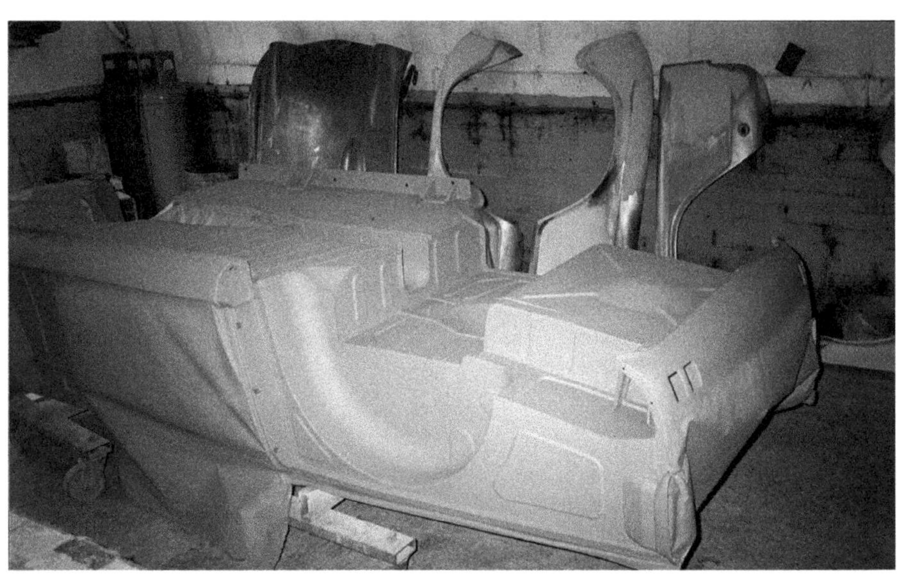

assembly to the car. Fortunately, this task is not too difficult, but necessitates a long, straight piece of stout timber. Before you actually start to fit either sill/rocker, you should rust-proof the inside of both inner and outer sill by degreasing the inside faces and lips and applying a weldable zinc-rich paint. Anything that provides a 99% zinc film will probably be great but "Metaflex Zink" will certainly protect the inside of the sills. You will obviously help long-term preservation of the sills if you apply several coats, in line with the manufacturer's instructions. Yes, the subsequent welding operations to the lip will, without doubt, burn off much of that paint, but at least 50% of the lip will be protected.

You should then apply a rust-inhibiting gloss paint to just the inner faces (omit the lips) of the sills. The longer this is left to dry the better to allow the paint to dry and the solvent to evaporate. Jack the wood/sill upwards until the bottom of the sill is in line with the door, but not to the extent that the sill's outer shape starts to dish inwards, and clamp with mole or panel grips. It's probably better to have to apply a thin layer of filler to a slightly sinking outer sill than to dish it by over-enthusiastic jacking. Weld, check, remove the timber, check and weld some more.

Your sill ends may be too big, and you should have no hesitation in grinding off any excess. Remember that you will have to get the wings over the sill ends, so any oversized or 'fat' sills will make wing fitting very difficult.

- If you have screwed-up the door glass to screen alignment, you can get your local motor glass/windscreen company to cut (to your pattern) a pair of laminated window glass(es). The original glass is flat, so a laminated replacement is feasible but expensive. However, this solution is for use only in desperate circumstances, since you will not be able to smash the glass to make an emergency exit.
- Underneath the rear body mounts there are aluminium spacers. With IRS cars, the number can be varied to suit the door gaps, and up to three shims is fine. If you need more shims, say 5 or 6, your chassis is hogged, as was described earlier, and, whilst your door gaps may be exemplary, you will have great difficulty fitting the rear bumper when the time comes!

This need not concern you if the body has to come off the chassis for painting. If, however, that step is not part of your restoration plan, sadly, the solution is difficult but worthwhile in the long run. You must support the body whilst you do the following. Remove all the spacers (but not the rubber pads), cut all but the top flange of both rear chassis legs about 50in (1250mm) in front of the rear of the car. With the chassis more easily moved, jack the chassis until it is again supporting the

4-30. By contrast the sealer tells us that this TR6 is clearly going to have the rear wing/fender bolted in place before spraying. Not only will the mating surfaces of wing and tub remain unpainted (just as Triumph originally assembled TRs!) but also, since the sealant looks like the modern and very strong Sikoflex, this wing will never come off in one piece again. Dumdum is recommended as a never-setting TR wing/body sealer for use after painting.

4-31. Now that's progress, with a new floor, primer painted, and the fuel tank replaced. This particular picture is included for two interesting reasons: firstly, to demonstrate that great care is needed when replacing the rear deck, for the new panel is not as comprehensive a replacement as it might appear. The strengthening panel beneath the rear lip (arrowed) is not included with any new panel you buy – consequently, you'll need to carefully zip-cut away the spot welds to salvage the strengthening panel. Mind you, I do not believe you would be the first to elect to leave the whole of a good rear 'U' channel in place, and to shave a little off the back of the new rear deck to compensate! The second reason is the 'rivnuts' along the underside of the front raised lip of this rear deck. Normal 'Surrey-Top' cars have plain holes in these locations, but it is common practice to fit rivnuts for cars that are converting to soft-top. If the car was going from soft-top to 'Surrey' one would drill out the rivnuts, so from this we know that this car could have left the factory with a Surrey top!

body and the door gaps are correct. Re-weld the chassis and, just when you feel sure you have things perfectly aligned, watch everything drop when you remove the support! Now you understand why all the spacers were removed, for now you need to re-jack the body until the door gaps are once again to your satisfaction, insert the appropriate spacers (probably one or two) and re-mate body and chassis. By this route you stand a chance of getting your door gaps right, and mounting your rear bumper at a later date without too much trouble.

• The tolerances are tight in the area of 4/4A/250/5 rear lights, and great care should be taken at every step of the body reconstruction. Do keep offering up the rear lights and checking fit at every step. You have been warned!

• Rust first forms along the top edge of where the outer wing bolts to the inner wing, and where the inner wing joins the wheelarch. The top deck joint to the inner wing and outer wing forms a rust trap of three layers of (unpainted) material. Etch prime all faces, cover in a two-pack primer and gloss before seam sealing the joint on final assembly.

• Obviously, cars are susceptible to accident damage at the front. All TRs carry the bonnet hinges on the inside of the front inner wings. The TR4/4A, in particular, has a tendency to open out, due to accident damage, the passage of time or when this area of the car is rebuilt. The problem is easily recognised by the excessive gap between the front wings and the bonnet. Rectification requires that the wings/fenders be pulled together, aligning both with the bonnet. Fix a suitable fastening to the inside of each inner wing and attach a strong twin loop of rope around both. Twist the rope turnbuckle-style until the wings/fenders are once more properly and evenly spaced. Get a helper to secure the twisted rope so that it does not unwind until you want it to. It is vital you ensure that each time you offer up the bonnet it is the back edge that you position (with the desired gap) abutting the front edge of the scuttle to provide your reference. Believe it or not, the inner wings will bend inwards under the turnbuckle's pressure, although you may need to persuade one wing to move a little more than its opposite number if you are to enjoy symmetrical (and very visible) front gaps.

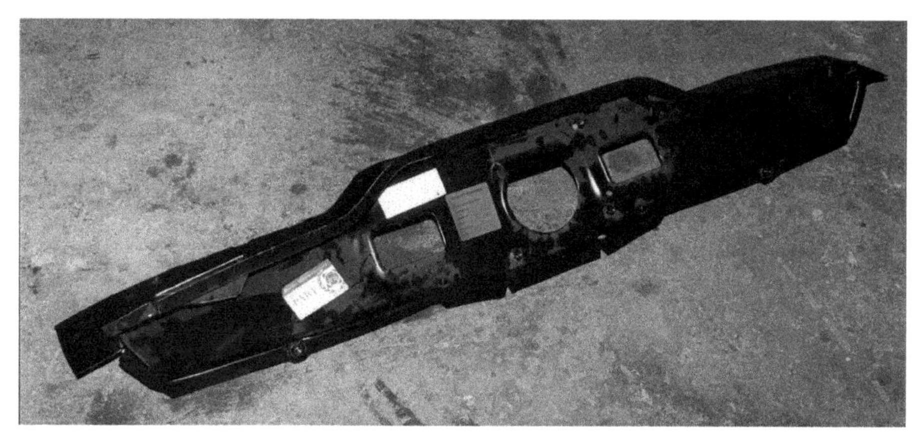

4-31. Whether you have a closable flap or open grille, it is inevitable that the plenum chamber beneath will have been under attack from the tin worm for many years. A new replacement with good corrosion pre-treatment may be your best solution.

• The plenum chamber (photograph 4-31) is the compartment beneath the front scuttle panel (just in front of the windscreen) through which air passes when the vent flap is open. Later cars are fitted with an air vent that is always open to the elements.

When the vent is open the plenum inevitably collects water, which should drain away into the inner wing cavity through two rubber tubes. It has long been recognised that standard flexible drain tubes are not the high point of Triumph design, since the original tubes stop at the top of the inner wing and allow water to run down the inside of the inner wing. Clearly, this is unlikely to increase the life of the steel in that area, while the inaccessibility of the tube makes it impossible to unblock the tube without removing either the wing(s) or splashguards. The tubes can and do get blocked, in which event water either gathers in the plenum, and/or falls onto the inside of the car floor/carpet.

The other weakness of the design is that the gap between the outer sill/rocker and the base of the wing becomes blocked. Water rots the top of the sill, enters the sill and does more damage. In any event, it's not a good idea to allow water to drain over your newly-repaired inner wing/sill. Although not original, you would be wise to fit longer drain hoses that lead the water away from the corrosion-prone areas.

Two modifications are available: one comes in kit form from Moss and feeds the water out into the wheelarch through an extra hole drilled in the bottom of each splashplate. Sounds fine, but the kit seems expensive to my mind and, in any event, when the car is in motion air can blow up the forward-facing drains and inhibit the escape of water. So option two might be favourite: it's shown in photograph 4-5-9 and, again, hinges on an extended drain tube. However, this method takes the tube straight on down through the top of the inner sill/rocker, where it turns in towards the centre of the car to exit the inner sill just below the line of the floor. It is probably a good idea to cut the end of your drain tubes at an angle that encourages water to fall down and backward.

• The foregoing idea has additional merit in that you should be thinking about rust-proofing as body repairs proceed. Granted, you should not actually apply the Waxoyl or whatever protective treatment you are planning until the shell has been painted, but the wax injection holes could be positioned now to advantage – before priming and other coats of paint are applied. The inside of the sills/rockers are going to require some holes, probably three down their full length, and if these are drilled from under the floor into the inner sill, they can be plugged with an unseen rubber bung after wax injection. To get to my main point, the front wax injection hole each side could double as the outlet for your plenum drain tube, provided you position these correctly.

NEW TR5 & 6 BODYSHELLS

As I mentioned earlier, there's a growing scarcity of good, ex-Californian bodyshells. Currently, they do still pop-up, and if a good, dry climate TR250/6 shell is available, it is definitely the route to follow. However, one must assume that they will become progressively harder to find, and make alternative plans.

The following suggestion applies to the TR5/250 restorer only, for the differences in the 4/4A shell make it completely unsuitable. The first impression is that the TR6 is a different car to the TR5/250, but with wings, boot/trunk and bonnet/hood removed, the tubs are very similar. The front inner wings, front panels (upper and lower), back half of both rear inner wings, and the rear back panel, all need changing, as I hope you will see from some of the illustrations shown, but, thereafter, you do have a solid 'new' TR5/250 body tub.

Early in the exercise you would have four surplus TR6 wings to sell, as well as a bonnet and boot lid, which should not prove too difficult and would go some way towards offsetting the cost of your new TR6 shell. It's a pity Heritage do not offer a bare 6 body tub to allow this. However, that is not current policy.

Incidentally, Heritage has approved outlets in the USA, so if the suggestion appeals to any TR250 owners, then go for it! Heritage tell me it built a prototype TR5 shell some five years ago. Since then it has been too busy to take the project further, and production of the Mini is going to occupy it for the foreseeable future. Nevertheless, Heritage is ever-hopeful of finding original tools from the TR5 era, and each one found makes current and prospective TR bodyshell production easier and improves the quality of parts.

The subject of TR6 shell quality warrants expansion, as, initially, they were of poor quality. Heritage explained to me that the TR6 was the first, non- monocoque built shell its workforce had built, and this, together with the proportion of non-original soft tooling it had been forced to use, had contributed to the overall poor build quality. Since then, several original tools have been found and numerous alternative manufacturing methods introduced, and all agree that build quality has improved. If, therefore, you do contemplate a new TR6 shell for either 5 or 6 restoration, do ensure you at least get one of the later/better quality units.

Remember, they are still not perfect, but if you only purchase one that has no seam sealer in the engine bay and has a smooth finish in the deep depressions behind the headlights (where the bonnet hinges are bolted), you will be purchasing the best available. The early and unsatisfactory TR6 shells had seam sealer around the inner wings, amongst other under-bonnet places, and had very crinkled bonnet hinge mountings. If you decide a new shell sounds attractive – and who doesn't? – buy it from a Heritage-approved dealer so that you have redress if the shell is sub-standard.

Ask what the 'build-date' was. Do not buy a shell unless it has a recent build date. Follow these suggestions and your 5/6 restoration will be off to an excellent start.

SAND BLASTING

Too many restorers shot blast the body tubs too early, making an already weak shell very fragile indeed. Leaving the sand blasting until you have completed the major repairs has two benefits. Firstly, your shell is at its strongest and, secondly, you are preparing the new (as well as the original) metal with a great 'key' for the subsequent painting operations.

So once the body tub is complete and all the predictable new panels properly welded in position, you should contemplate, for the first time, sand blasting. There may be the odd small area you are uncertain about, and want to see the result of sand blasting before deciding to what extent new or repaired material is required.

The sand bounces off thick underseal, which, consequently, must be mostly removed. Use a blow torch (the type with a disposable gas canister) very carefully, or better yet an electric paint stripper, to soften the old underseal. This must then be scraped off to leave no more than a very thin black smear of underseal – which the blaster will remove. A putty knife or paint scraper will be essential, and be prepared for a real mess! If you are not comfortable with what is left after your first scraping operation, you can use solvent/thinners/white spirit and rag once things have cooled down. If you leave the blaster too thick a coating to remove he will go through the adjacent metal before the blasting material can remove the thick underseal.

Do ensure the operator understands which areas you want blasted and those you do not. A really good operator will tell you what areas he will not blast. Experienced blasters usually avoid the flat parts of the larger panels – where they will only want to do the edges for fear of stretching the flat areas and causing rippling and distortion. Even with this point discussed and agreed, it is still a good idea to mask-off the scuttle and rear deck to protect from blasting. Do not allow outer panels to be shot blasted; wings, doors, boot, bonnet should not be blasted as they invariably end up rippled, although I have successfully had the very edge (say, 1ins or 25mm) of boot and bonnets blasted clean of rust.

Transporting the inner tub on its chassis is safest but might mean having the body blasting completed before you start to finish paint the chassis.

Most blasting companies will also have the facility to prime paint the body after blasting, which is an idea with some merit, particularly if you live any distance from the blaster, but there are also some dangers to be aware of. Certainly, priming of the blasted body must take place quickly after blasting, an hour later would be good, but you must not allow more than four hours elapsed time between blasting and prime painting as an oxide film (invisible at first) starts to form. Even if you know there are areas to repair after blasting, it's still important to get the whole body primed quickly.

However, bear in mind that your blaster can cover areas of poor blasting with a coat of primer, which may not be compatible with your subsequent paint applications. If you are so far from the blaster that you feel it best to have him prime your shell, do pre-agree the paint he will use, and do inspect the blasting work before he primes the car. Obviously, you will need to remove primer from around subsequent local repairs, but you must refinish (brush painted, if you like) the stripped area as quickly as possible. We will examine the most preferable primers and other painting detail in Chapter 6.

Chapter 5

Chassis restoration

You should remove the engine manifolds before contemplating a body-off lift as they will foul when the lower inner wings try to pass the engine. The six bumper dumb iron connections that pass through the body, together with the brake, clutch and fuel pipes, handbrake and speedometer cables, must all be at least disconnected, and preferably removed. Obviously, the seats and carpets will also need to come out (if not already done) to allow access to the body mountings.

You will appreciate from our discussions on preparing a restoration plan that there are two methods of separating the body from the chassis. Let us assume you have a reasonable solid shell, and the floor and sills require some attention but not replacement. The first step is to consider how you will proceed.

You don't have to, but I would suggest removing the bonnet (which, on some cars, is quite heavy), boot lid, doors and all four wings before you even contemplate lifting the shell off its chassis. This has two advantages: it reduces the weight of the body and gives you a choice of how you are going to brace the door apertures, which is an essential step for one-piece body tub removals.

Some do this by bolting dexion angle across the top inside of the door gaps from hood mountings on the 'B' post to the bottom windscreen mounting bolts. Frankly, I think this method leaves room for movement and does not get my vote. Photograph 4-3 shows some cross braces achieved by tack welding mild steel flats. This seems better, but using steel angle for the cross braces is an even stronger and therefore preferable option. This method is more likely to hold things rigid and also allows easier access to the inside of the tub. Another way is to use a slightly angled brace that bolts to two front wing mounting bolts on the *outside* of the 'A' pillar and two hood frame mountings on the *inside* of the 'B' pillar. I have seen professionals use 1 inch (or 2 inch) diameter tube (suitably flattened at each end, of course), but a nice sturdy piece of mild steel angle (with the angle removed at one end) has attractions.

Another professional approach can be seen in photograph 6-3. My favourite method is to bolt an RHS or angle iron cross brace each side of the door apertures from the screen mounting hole at the front to the hood mountings on the 'B' pillar. Properly done, this route allows you to offer up the doors in due course without having to remove the braces. The choice is yours, although you will not need to bother initially if the body shell is going to need new floors and sills, and you decide to cut the shell in two and take the front and rear halves off in two parts. This really reduces the weight to be lifted when removing the body from the chassis!

You will need four people to lift a whole bare body tub off its chassis in one piece, and six if doors, wings, boot and bonnet panels are left in place. In both cases it is as well to have an extra pair of hands to disconnect/cut the one cable, pipe or attachment that always get missed and only reveals itself as you try to part shell from chassis. Do not be concerned if some body mounting bolts have to drilled out or sheared, as in photograph 5-1: that will be par for the course.

The now bodyless chassis is never a pretty sight, but let's hope it's no worse than your initial pre-purchase assessment. Swallow hard, but please do not be surprised: the chassis was never primed and received no more than a quick blow-over of black paint – and that was 25 or 40 years ago, depending on the car.

Make a start by taking lots of

5-1. Few TR chassis are a pretty sight when they see the light of day for the first time in maybe 30 plus years ... As a matter of fact, this 'just separated' chassis does not look too bad, and is better than many. Nevertheless, it was still necessary for Mark Price to grind off the heads of many of the body/chassis attachment bolts in order to release the body of his TR250. He now needs to carefully drill out the centre of each of the remaining bolts. Sometimes the heat and vibration generated by drilling will release the rusted bolt, but more often than not you'll need to open out the initial hole with a second carefully selected drill size that leaves the female thread intact, and then re-tap the hole in the mounting brackets. Any threads that do not clean up should be completely replaced by welding a new nut in place of the damaged one.

5-2 & 5-3 (right). Buyer beware (one). The first picture reminds you of a typical TR accident-damaged chassis leg, creased or bulging between the rear of the front suspension brace (shown in photograph 5-14), and where the central chassis cruciform joins the main chassis leg (just in front of the body's 'A' pillar). Clearly, this photograph was taken with the body off the chassis but, as I explained earlier, this bulge is something you should have detected when you carried out your pre-purchase inspection, even with the body in place. The crease is evidence of some earlier frontal impact. Photograph 5-2 shows how the same section of chassis should be: dead straight and flat.

detailed close-up photographs before removing anything from the chassis. When you do start removing parts from the chassis, use lots of welding wire to hold sub-assemblies, brackets and related shims together. Have plenty of tie-on labels to hand, too, and religiously identify each welding wire sub-assembly as you take it off the car (i.e. LH or RH, front shims, outside LH side, etc.). Throw nothing away. Keep even the grottiest rubber bush or tatty shim, wired into its respective place in the sub-assembly. Have some strong bags to place the relevant bolts in before wiring the bag to the sub-assembly. If in doubt, use a centre punch to 'dot' brackets to chassis positions.

As removal of parts proceeds you'll get an idea of how accurate your initial assessment was. Use a small hammer to establish the extent of corrosion. When you have everything off the chassis, it would normally be time to get it sand blasted. However, IRS chassis damage such as that shown in photograph 5-2, and/or corrosion such as that in photograph 5-4 can be so extensive as to make shot-blasting a waste of time and money. Strangely, these later IRS chassis are more vulnerable to forward creeping corrosion than their earlier counterparts, and are much more likely to be in need of plating and welding. As a result they are less readily available, and are worthwhile repairing up to a point.

So, where is the cut-off point, bearing in mind that the corrosion moves forward from the rear? Certainly, corrosion at or forward of the chassis mid-point (photograph 5-4) means that the chassis is best written off. Use the cruciform or 'T-shirt' panel as your measure. Significant corrosion in front of the cruciform is very bad news (we'll look at your options if this is the case in a moment).

Let's assume you've not had any unpleasant surprises and your sand blasting operation needs to go ahead as planned. This can usually be done locally, though it's best to avoid local shot-blast companies which deal mainly with structural steel and/or 1 inch boiler plate. They could be too aggressive with the blasting operation and good parts could be destroyed. Ask the local TR restorer (who could even get work done for you), or your local TR group about where such work could be satisfactorily completed. This usually costs around £100 and is money very well spent. Don't try any DIY: wire brushing and most other "home remedies" just polish off the top surface of rust.

A last important point about the sand blasting operation is you should ensure that the chassis is painted within hours of shot-blasting. You must not allow the surface of your newly sand blasted chassis to form even a very light film of corrosion. (This point is discussed in more detail towards the end of this chapter.)

The sand blasting operation will reveal the areas of damage or corrosion weakness that require attention. The beauty of the TR's chassis construction is that most enthusiasts can very successfully tackle chassis repairs,

5-4. Buyer beware (two). A repair patch removed from the bottom face of an IRS chassis. Note the remains of the cover/ patch curled up to left of the picture, revealing the corrosion underneath. The repair would have been up to UK MoT standard, and would not have been unsafe, providing the welding was good. It would, however, have led to acceleration of the rust under the patch. Consequently, it would have been better in the long run to either cut out the whole section of chassis (which is difficult with the body in place), or remove the lower flat part of the chassis section only. This car could have been sold with a full MoT to an unsuspecting buyer, yet clearly further repairs would have been required in the not-too-distant future, demonstrating the need to look at the chassis before purchase.

5-5 and 5-6 (right). Buyer beware (three). The first picture was from our pre-purchase examination, and shows the bottom 'Tee-shirt' plate of an IRS chassis swollen by corrosion. An outrigger/rear suspension mounting arm exits the photograph bottom left, while one main chassis leg exits the top of the picture. All the fore, aft and side loads brought about by acceleration, braking, and cornering pass through this absolutely safety critical focal point on the TR4A, 250, 5 and 6. With the body off, the corrosion in both pressed 't-shirt' plates will be even worse than you expected, and the picture bears witness to the vulnerability of this crucial part of IRS cars. In fact, this bottom gusset has rotted away almost completely, as will have all the internal interconnected chassis members above this plate – thus reducing to nil the structural integrity of this area of the car. All of this is bad news, but what is even worse is that someone has previously tried to make the car appear roadworthy by welding a top gusset over the top of the rotten material, as evidenced by the part corroded strip along the rear suspension mounting arm. Who is he kidding? Even a thicker than average gusset, when welded to nothing, provides no structural integrity whatsoever, and the owner risks a rear end chassis failure with possibly horrific consequences. This, therefore, is a crucial area to examine very closely indeed when carrying out a pre-purchase inspection. If you miss this point you will certainly be in line for an expensive body-off repair at a later date, and could be putting life and limb in danger in the meantime.

particularly since there is a good selection of repair sections available. Even if you have a superb chassis – ex-Californian, maybe – it's prudent to go over it with a fine tooth comb checking for cracks. The chassis normally corrodes and cracks through fatigue in fairly predictable places (listed below). Vee grinding and MIG welding will normally correct most cracks. Usually additional plating to the main chassis members is not required, although some strengthening and additional gussets will be; we will evaluate these a little later. Even those readers considering a "concours" restoration would be well advised to think about strengthening the front suspension and steering attachment points that I detail. If no cracks or corrosion are found, this is still a once-in-a-lifetime opportunity to upgrade the chassis in several known weak spots to prevent future problems. So you can be sure, one way or another, that there will be welding work to carry out.

The chassis is a load-bearing structure, so, from a safety point of view, all welds need to fully penetrate the materials and provide a long-term solution. Access to the problem areas could not be better with a bare chassis, so this is your chance to get the welding done well to guarantee a permanent solution. If your welding is not up to this standard, then it will be cheaper in the long run to get the repairs and modifications done by a TR chassis specialist, or obtain a replacement chassis. If you complete half a repair now and the problem re-occurs a few years down the road (pun intended), it is likely to be much more difficult and expensive to effect a proper repair with the body in place. It will certainly spoil the paintwork in the area that requires further repair, if nothing else.

MAIN AREAS OF CORROSION

When we were examining the car prior to purchase (Chapter 2), the importance and vulnerability of the cruciform or 'T-shirt' just forward of the differential was discussed. This is one of the IRS car's major problem areas, and I hope photographs 5-5 to 5-7 will help emphasise the importance of this part of an IRS chassis.

It is difficult establishing the true extent of the corrosion within the enclosed 'capsule' where the box section (which carries the trailing arms) is attached to the chassis main rails. It is essential you grind off both top and bottom stiffening plates, thus opening up the intersection of the various box sections. You can now properly inspect the condition of the sections previously hidden by the diamond stiffening plates. You may be very surprised by what you see, and will almost certainly find that the box section metal under the stiffening plates is missing. This means that there was, in fact, no structural integrity at this crucial intersection of load-carrying chassis members, in spite of your having closely inspected this area prior to purchase of the car, and numerous MoT inspectors having looked closely and declared the car roadworthy!

An early sign, incidentally, that all is not well in this area is when doors pop open as the car/chassis twists when travelling a bumpy road. Older, rusty cars have been known to lose a trailing arm box section completely

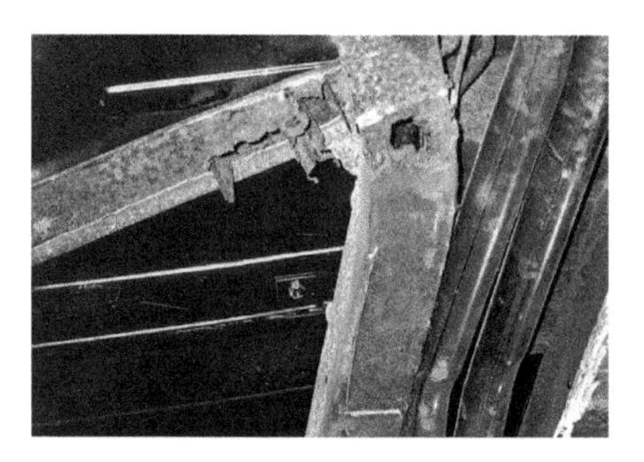

5-7. This shows the underside of an IRS car with the lower centre cruciform, or 'tee-shirt' pressing removed. The right arm of this chassis has had the lower face repaired already, and is, in fact, the left side trailing arm mounting member. The repair looks very good, but would have been even better had the corner of the plate we see in the top right of the picture been extended backwards (towards the top of the picture) by about 2 inches (50mm). This would have ensured it did not end too close to where the tee-shirt pressing will eventually finish. The two centre tubes are, of course, the exhaust pipes, that should be removed to allow proper re-welding of the replacement tee-shirt pressing to the chassis – unless, that is, this restorer intends to plug weld through the tee-shirt panel to the chassis.

when the corroded and fatigued metal separates from the cruciform. Without putting too fine a point on it, this area is vitally important to the car, so do the job properly and get the stiffening/T-shirt' plates off and have a real good look!

Assuming you need to replace one or both of the box sections that carry the rear suspension trailing arms, it is important to position the new box sections at a true 45 degree angle to the longitudinal axis of the chassis when viewed from above. It is, of course, quite possible to either measure the outboard point of the box section or use a plater's protractor to measure the 45 degree angle. Probably the simplest way to get the new angle correct is to make up a (large) steel template using the old box section angle as your guide (before the pressed stiffening plate or box sections are removed, of course!). Check that the template 'fits' both sides, and mark the 'top' very clearly.

A further suggestion that might be considered helpful is to take your light gauge template down to your local engineering works, and have them guillotine you an identical template from 3mm mild steel plate. You can then tack weld your template to your new box sections to retain the true position, while you first tack and subsequently weld the new trailing arm mountings in place. Just a thought. I know the rear suspension is shimmed, so there is some leeway, but I would like to get this part of my chassis rebuild spot-on.

The trailing arm chassis members have some subtleties you should be aware of. The inside of the box section is more complex than it looks from the outside, because it has an internal stiffener/spacer to prevent the front and rear faces of the box section closing together when you 'pull-up' the bolts that carry the trailing arms. This spacer has an important secondary use as it also spreads the stresses from the mounting bolts.

There are two types of trailing arm box section, and both have the matrix in place, but option 1 is designed for those who have elected not to remove the diamond-shaped pressed stiffeners. These box sections come with the rear closing channel loose, as shown in photograph 5-8. Consequently, it's possible to position the front main channel of the box section inside the cruciform 'cavity' created by removing the old/rusted legs, and weld it to the rest of the chassis members from the inside of the new main channel. Then, position the rear (smaller) channel 'half' of the box section and weld that in place, too.

The second option for the proper spacered trailing arm box section comes assembled and welded, but is only practical to properly weld in place with both diamond stiffening plates removed. In a bare chassis restoration I do strongly recommend you choose option 2 and fall back to option 1 only when repairing a complete body-on chassis.

Now we come to replacement of our two diamond-shaped pressed stiffeners, or 'T-shirt' panels, as they are known. We discussed a 'hogged' chassis in Chapter 2, and you are now at the point where, unless care is taken, you can spoil hours of work by hogging your chassis.

The first prerequisite is that all is set up as per the chassis drawing in the official workshop manual, and both plates are solidly tacked in place. Hogging is actually caused by shrinkage brought about by hot metal cooling after welding. So you need to counter this unavoidable shrinkage by welding both diamond stiffening plates in short runs only (say 1 inch long) and in a strict diametrically-opposed sequence.

5-8. The all-important rear suspension mounting legs showing the equally important, but oft overlooked, internal stiffeners/spacers. When purchased the smaller top channel section is only tack welded in place to allow it to be removed so that the internal edges of the main channel section can be fully welded to the structure of the chassis. When that is done, replace the smaller channel section and seam weld the edges to the main channel. An alternative fully-welded option is available.

To explain this last sentence. Say you start with a 1 inch run on the bottom of the chassis on the left side, this must be followed by a 1 inch run on the top right side. Then a 1 inch run on the bottom right side, countered by a top left side (diametrically opposite) run. This will involve a lot of chassis turning, and the process takes time. Nevertheless, don't be tempted to short cut this technique or you will twist and/or hog your chassis! Perhaps it would be prudent to every so often get out the tape measure to ensure that the main dimensions are being maintained. I would certainly confirm these dimensions when you have completed one cycle of welds (i.e. one 1 inch weld in each corner of the plate). Accuracy to about + or – 1/4 inch is acceptable, but

5-9. The main diff bridge showing two welding repairs. Firstly, note that the right side pinhead has been welded, but without the extra square strengthening plate advised in the main text. The second repair shows where pressure from the rear coil spring has cracked (arrowed) the front edge of the main diff bridge. As a result of the crack the end of the bridge actually bent slightly upwards. The welded repair is clearly visible after the outer end of the bridge had been knocked back flat, but, if this was my car, I think you would also see an extra 6in (152mm) long strip of L-shaped 16swg mild steel welded along the face and lip. If this section fails, the coil spring comes up through the body, and you fly off the road moments later.

5-11. Rear diff bridge mounting pin, this time viewed from the top. You can see that the pin clearly cracked right round the top. The repair for the rear diff bridge pins is quite different to that used for the front pins, in that you should make or buy two plates about 1.5 inch (40mm) square that span the front to back width of the bridge. They are the smaller rectangular plates/washers shown in photographs 5-14 and 5-15. with a central circular hole that will be used to plug weld through to the top of its pin. Obviously you must first locate and completely seam weld each plate to the rear bridge so that the hole is indeed central to the pin before the final plug weld is completed. This repair or preventative step spreads the stresses from the diff pin over a much larger area than originally thought necessary by Triumph.

any greater discrepancy and you need to have a beer and think about what you are doing.

The last 4-5 feet (1.2-1.5 metres) of chassis may look particularly sad after shot-blasting, and is vulnerable on all the IRS cars but particularly so in the case of the TR6. This TR had a rear valance design that seemed to hold road spray and exhaust gases in the proximity of this rear chassis section particularly well, which accelerated corrosion, of course. Not to worry, the tail end of the chassis is available (along with the boot support cross tube) in various lengths, so find out the lengths in stock, cut off the most appropriate length to remove all the corroded section and weld the repair section in place. Take account of the fact that the chassis tends to slope (slightly) downhill from this rear section, meaning that water runs forward from the rear towards the differential and rots the chassis at the outer edges.

If an IRS car has been frequently used over rough roads, the differential bridge can split in-board of the coil spring, as depicted by photograph 5-9. The repair is simple (without the spring in place, of course). The bridge should be knocked down to form a flat channel again and the splits vee'd out and welded.

5-10. IRS rear diff bridge (or rear cross member). This is the outside end where one coil spring sits up into the 'cup', which has been badly repaired. We can see the edge has been welded, with no apparent effort to replace the corroded lip on which the whole of this corner of the car rests. The strength of the spring mounting depends upon the integrity of this lip, for without it the end of the bridge can and will fold upwards allowing the coil spring to escape the bridge and poke up through the body! New diff bridges are available, and should be fitted if the rest of the chassis is sound.

5-12. Taken with the chassis upside down, this picture shows how badly the main bridge diff mounting brackets can crack. Note the separation of what is, in fact, the right side bracket, and how important it is to reinforce it by 'boxing' it in if you are to prevent a reoccurrence.

Photograph 5-10 illustrates another problem with the extreme ends of the main diff-bridge, where the coil springs 'sit'. As the photograph shows, the spring-cups

5-13. A front diff pin. There are two pins hanging down from the front diff bridge. While both are vulnerable to cracking – particularly in the higher-powered PI UK cars – the right side pin is particularly prone to coming loose, due to propshaft torque becoming focused at this point. It is usual to fit non-original strengthening plates to prevent the problem, or, as in this case, repair it. The picture is taken from beneath the car with the diff removed, and shows the repair well advanced with both new side plates welded in place, and the far side welded to the original cracked mounting plates. Clearly, the next task is to weld the nearest new (inboard) side plate to the original bracket, thus closing the very visible gap and 'boxing' in the whole pin. The final step is to weld together the cracked original bracket.

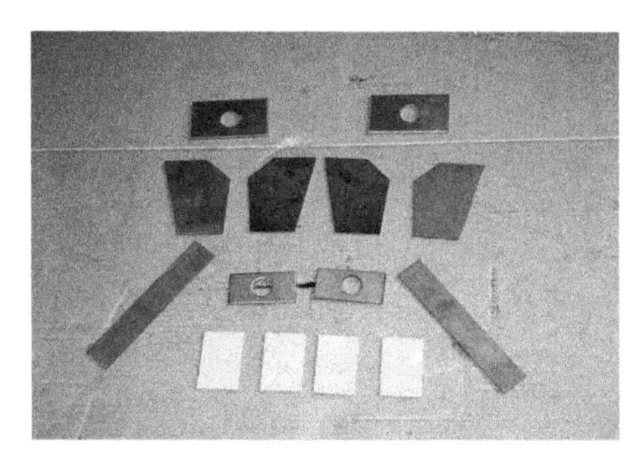

5-14 and 5-15 (right). Two differential pin strengthening kits, both, in this case, from Revington TR. The former, with its large strengthening plates for the top of the main diff bridge, is for body-off repairs, and is referenced RTR7012/1. The kit shown in photograph 15 – Revington number RTR7012/2 – has smaller main diff bridge plates, and is for body-on repairs. If you are considering a home/DIY job, remember that the repair work involves cutting the two top plates in half, followed by much overhead welding.

5-16. This 'wider' photograph shows both front and rear diff pins, one rear driveshaft and the part-complete repair with the cracked original bracket and the new inboard unwelded gusset clearly visible. Periodically, check that the pins are still solidly affixed to their brackets, although you will be in no doubt when a pin pulls out of its diff bridge as the differential will thrash and bang about as torque is applied to the rear of the car! Prevention is far better than cure, so if your car is running, do take a close look with lots of light from time to time. If your car is undergoing restoration, do away with the need for periodic inspection by using the strengthening kits shown above.

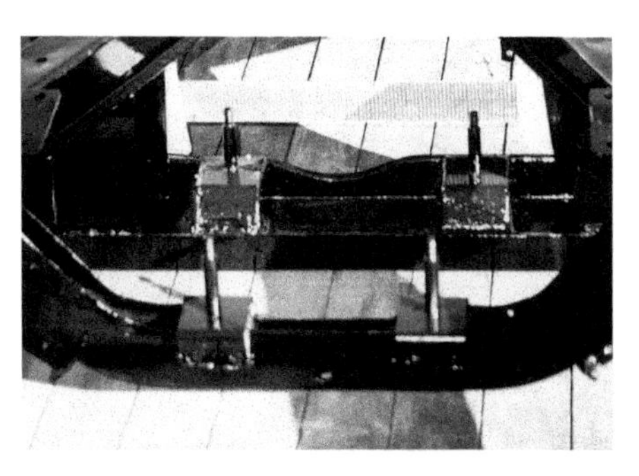

5-17. The four IRS diff mounting pins after strengthening by additional boxing. In case you are wondering, I took this picture when the TR6 chassis was upside down.

5-18. And here is the finished job.

are susceptible to corrosion and it is important to repair this thoroughly otherwise the spring will come out through the body while the relevant rear corner of the car will sink to the road. A new diff-bridge is your best solution if yours looks even remotely like this example.

The four differential mounting-pins are known weak points, and tend to crack away from their respective chassis mounting points. One solution is shown in photograph 5-9 although, as you will see very shortly, even better solutions are now available. Photographs 5-11 to 5-13 show examples of the most frequent problems.

The most vulnerable pin is the front right side pin, since it takes most of the torque load as a reaction to that transmitted through the differential. The higher the engine power/torque, the more chance there is of the front right side pin cracking away from its base. Today, most cars have had some sort of repair to a differential mounting pin – if not all four of them, and, more often than not, the repairs have been of the "get the car mobile and sell it" variety.

The pins are available new and, whether or not the pins have cracked, this is an opportunity to carry out a chassis improvement modification which will ensure that the problem never re-occurs. Photographs 5-14 and 5-15 show typical repair kits that are available from all the premier TR specialists. The plates are best pre-welded to the head of the pins, thus effectively providing four double strength mounting points for the differential. Simultaneously, you should create four complete 'turrets' by boxing in the lower brackets, as shown in photographs 5-16 to 5-18.

The smaller bridge that goes over the rear of the differential deserves attention, too. It spans outwards to bridge between both rear chassis members and carries the rear lever arm shock absorbers. Whether or not this

5-19.

5-19 and 5-20 (above). The first picture is a general view of the front left hand turret, whilst the second gives a closer look at the top of the turret, in particular a badly repaired top suspension mounting plate. The top plate and dome-shaped shock absorber mounting are prone to cracking, and need to be very closely inspected for cracks, which must be properly prepared and welded. This is but one area where sand-blasting your chassis is particularly helpful, for many of the smaller hairline cracks will only become apparent after blasting. It is important to seam weld right round the periphery of the top plate, as has been attempted here, except that this is a safety critical part of your chassis and the quality of the welding is vital. It is also important to seam weld right round the 270 or so degrees of dome to plate – which has not been more than toyed-with here. Also visible in this pair of photographs is damage to the angled rearward brace that is attached to the chassis some 2 feet (600mm) to the rear of the turret. Such damage is typically the consequence of an accident, and it is well worth examining this cross-brace for creasing and repair.

bridge is cracked, it is usual to take this unique opportunity to reinforce this bridge by 'boxing-in' the channel section. Obviously, if yours is corroded or cracked you must first effect a solid repair before further strengthening is undertaken. Even if you are planning rear telescopic shock absorbers, this is nevertheless a valuable strengthening exercise. The telescopic shocks fit to the bottom of this bridge and, consequently, the loads still have to be safely spread throughout the chassis.

5-21. The lower/rearward ends of the front turret brace, showing clear evidence of significant corrosion and possibly accident damage: definitely an unsatisfactory repair.

Moving forward to the front suspension, not only do we have the sort of corrosion shown in photographs 5-19 to 5-21 to combat, but you will recall that when examining the chassis prior to purchase, we were very interested in the single channel part of the chassis just below the body's A-posts. We were looking for signs of accident damage, since any chassis wavering might not be just a simple matter of correcting the distortion, but could signal a more serious problem with the front suspension turrets and geometry. The situation is exacerbated by the fact that it is not really possible to establish the extent of the problem, or to correct it, without a proper jig. So, if accident damage is suspected, checking and correction could be a matter best left to an experienced professional chassis specialist. In fact, the amateur could create more problems at point B than he solves by trying to jack-correct a distortion at point A. So, creases in the single-leg section of the chassis may colour your whole approach to chassis renovation.

Front suspension corrosion is usually obvious once the body is off the chassis, but start by looking at the slanted back and downwards square brace just behind the turrets. These braces are susceptible to corrosion at the lower 2 inches or so, just above where they are welded to the chassis. Moisture clearly tends to sit in the bottom of the rolled hollow (square) tube and corrodes outwards. You therefore need to bang them both quite vigorously to ensure they are absolutely solid. Repair is quite easy in a variety of ways, but do ensure both turret braces are solid enough to carry out their intended function for many years to come.

Perhaps not so much a corrosion issue, around the top of each turret there is a shallow, bell-shaped pressing that accepts the top of the front coil springs. You will note 3 fairly short lengths of weld, positioned roughly at 90 degree intervals around the pressing; these very often crack, as are shown in photograph 5-20. It's advisable to weld right around the periphery of the bell/turret junction – about 270 degrees.

Next, remove the bolted cross member that joins the two front turrets just in front of where the engine was located. They can be damaged by a side impact, in which case wrinkling will be evident. If yours is damaged, find one from an undamaged chassis and use it as a jig to correct your chassis/turrets. Note that some have 2 bolts each side, whilst the latest cars have 3 bolts each end.

Finally, in addition to accident damage and corrosion issues, several other items are now accepted as weak spots in the front suspension of IRS cars, and you would be well advised to disregard originality and consider the following relatively small, but very important, improvements. I'll also mention some other changes you could think about when the chassis is under the spotlight and you are planning your strategy.

MODIFICATIONS & IMPROVEMENTS

Before you put the welding set away you should give some thought to the rear gearbox mountings. These vary from gearbox to gearbox, and if you plan to re-install the original gearbox then, clearly, no changes are called for.

However, you may wish to add an overdrive to your gearbox, you may elect to fit a later 'J' type gearbox and overdrive in preference to the earlier 'A' gearbox (discussed in more detail in Chapter 9), or you may be thinking of a proprietary 5 speed gearbox conversion. The latter conversion will, unfortunately, have to wait until a later book, but you can get information from TR Bitz now.

Some of these conversions require

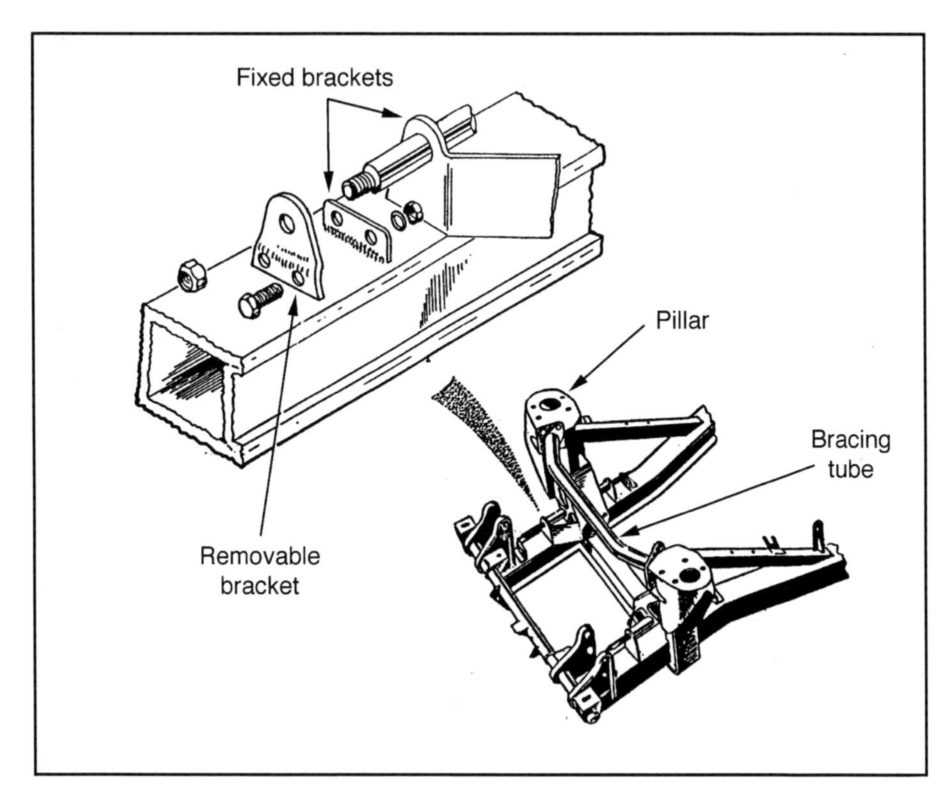

D5-1. General and detail (right) view of early TR front suspension lower mounting arrangement. (Courtesy *TRaction* – the magazine of the UK TR Register.)

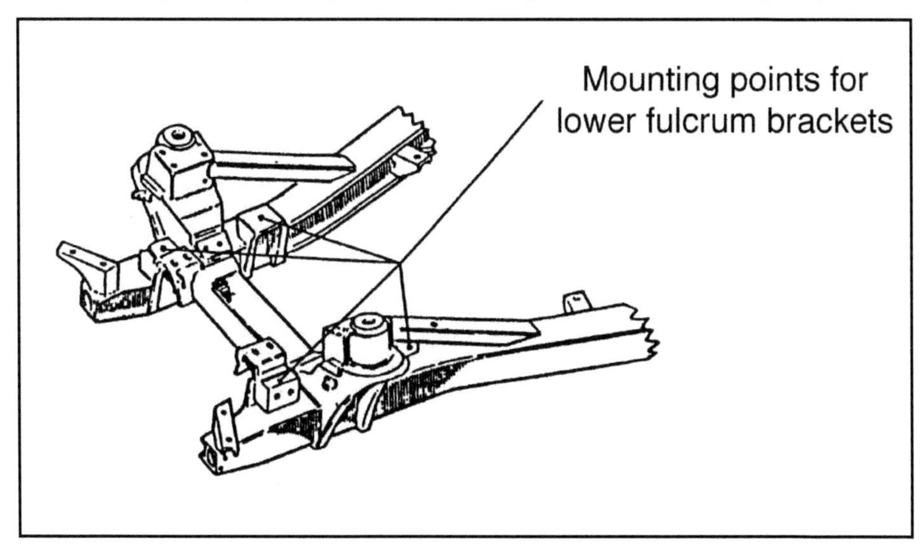

D5-2. The four lower mounting points for the front suspension in the IRS cars. (Courtesy *TRaction* – the magazine of the UK TR Register.)

alteration to the rear gearbox mounting, so this is obviously the point at which to tackle the changes. Note that, although not popular, it is easy and fairly quick to change from 'J' type mounts to those that will accept an 'A' gearbox. The reverse of this may be far more popular, but the time and complexity involved in making the chassis alterations should not be underestimated. Again, you will be motivated by your own particular aspirations; those with originality or concours in mind may well elect to switch to the earlier gearbox, or overdrive unit, although the additional reliability and availability of the 'J' unit makes it the more usual choice. Variations of the 'J' type gearbox and overdrive unit were used on Triumph Saloons and Dolomites Sprints, so the 'J' unit may also be the cheaper option as well as the most reliable.

LOWER WISHBONE MOUNTINGS

Up until the TR4A the lower suspension fulcrums were, in effect, welded directly to the chassis frame, making any adjustments, particularly after an accident, very difficult. Indeed, an oxy-acetylene torch was necessary to heat and bend the erring component back to what looked like the correct position before you could try again.

The front suspension pillars were manufactured with a pair of fore-aft holes near the base of the pillar, through which passed a pretty substantial steel pin. A pair of plain diameters flanked the central part of the pin and carried the front and rear lower wishbones, whilst a pair of two-piece brackets secured the pin either side of the central pillar (shown in drawing D5-1). Accident damage can result in the wishbone mounting becoming twisted in relation to the chassis side member, and partially or even completely broken away from the chassis. The fulcrum pins are thought to be no longer available, but TR Bitz has them especially made and they are usually in stock.

With the introduction of the '4A' two pairs of brackets were welded each side of the front suspension towers. These can be seen in outline in drawing D5-2 and in more detail in D5-3. It will be noted that the brackets now allow for shims to be used to achieve the required geometry. However, initially (in the TR4A) these fulcrum brackets were attached to the frame bracket with just one $^{3}/_{8}$ inch stud, while the stresses fed into the rear brackets were distributed throughout the chassis via only one gusset welded to the top of the frame. Both these arrangements have subsequently been proved inadequate and an Achilles heel of the IRS TR chassis – particularly the rear bracket. We touched on these in Chapter 1 where you will recall looking down on each rear bracket through the inner wings.

The brackets at the rear of the turrets are susceptible to the tensile stresses imposed by reversing swiftly and then braking hard. In fact, the

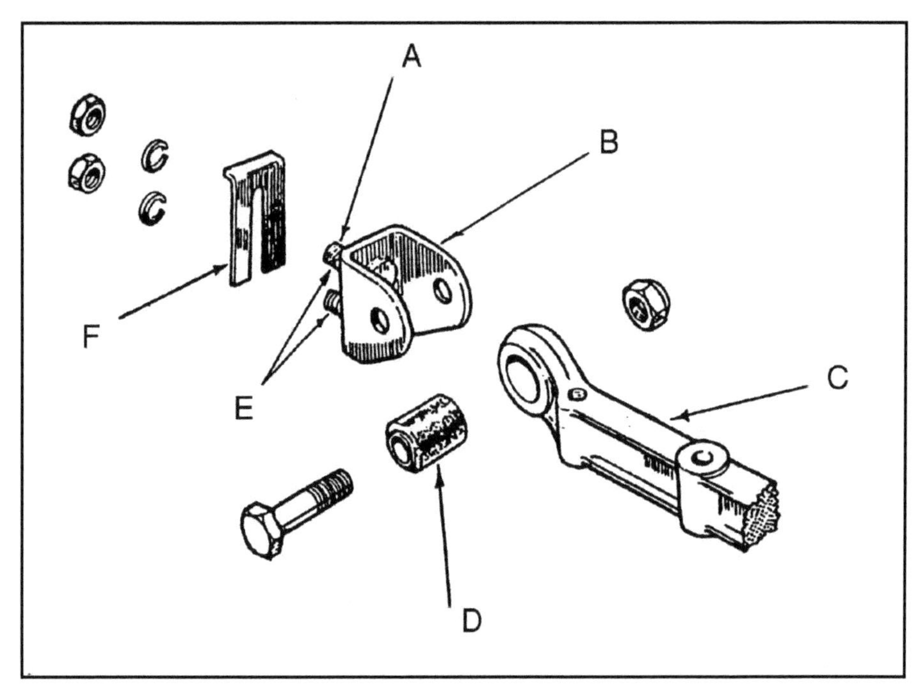

D5-3. The lower mounting bracket details for the TR4A (1 stud) and the TR5/6 (2 studs). A. The 1 stud of the TR4A. B. Lower fulcrum bracket. C. Wishbone arm. D. Bush. E. The 2 studs of the TR5/6. F. Shims. (Courtesy *TRaction* – the magazine of the UK TR Register.)

5-22.

5-23.

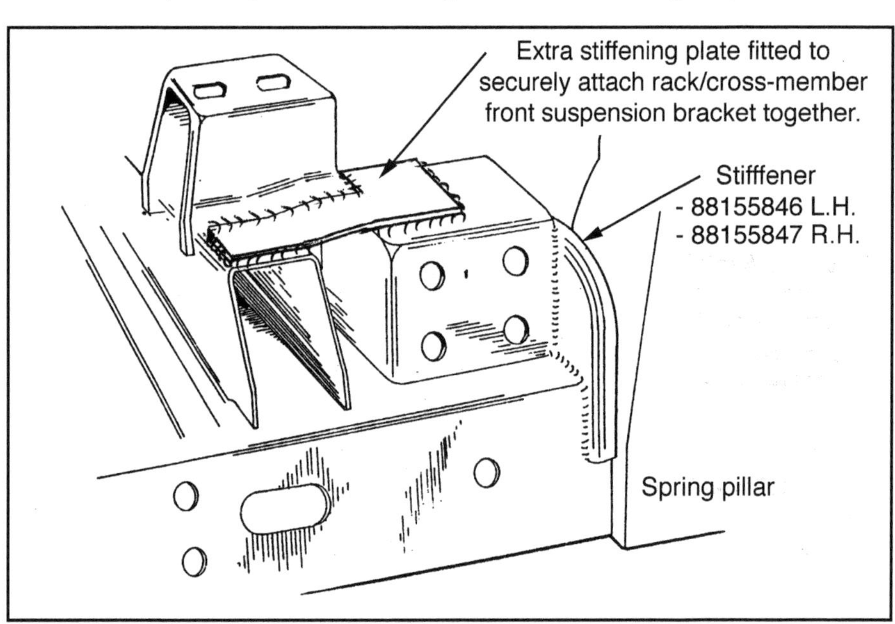

D5-4. Strengthening the front lower left side pivot bracket. (Courtesy *TRaction* – the magazine of the UK TR Register.)

5-22, 5-23 and 5-24 (above). A sequence of 3 photographs showing initial progress in repairing and strengthening the left side of a TRs front chassis area just behind the bottom rear wishbone mounting bracket. Photographs 5-26 to 5-28 show the complete picture.

whole front suspension mounting arrangement proved marginal on the TR4A and with the added length, weight and stresses of a 6-cylinder engine, the rear lower brackets were very suspect on the TR5, TR250 and early TR6s. If your car is a pre 'CR' or 'CF' model, the modifications shown in photographs 5-22 to 5-28 – which many will regard as non-original – will be of interest. In the interest of long-term safety, you would be well advised to incorporate the relevant strengthening kit(s) into your rebuild. TR6s built after November 1972 already have the improvements incorporated in the original build, although it almost goes without saying that every aspect of the front suspension deserves very close attention.

The front lower wishbone attachment brackets on IRS cars also need to be checked for cracks or accident damage. The brackets themselves generally give less cause for concern than do the rear brackets, but they could have sustained earlier accident damage. However, the front

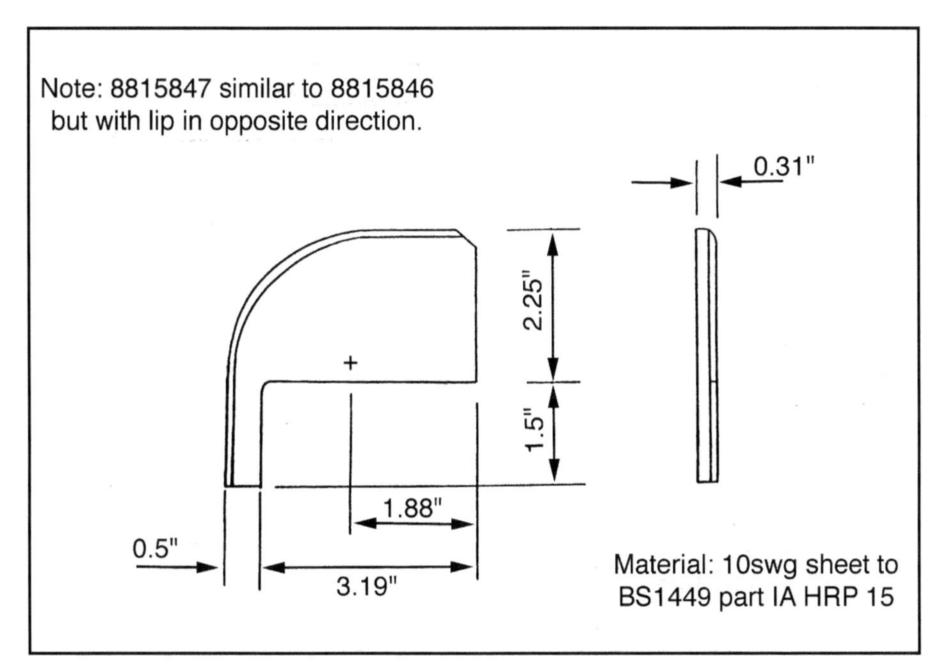

D5-5. Dimensional details of the front suspension stiffeners. (Courtesy *TRaction* – the magazine of the UK TR Register.)

5-25. All premier TR restoration specialists will supply you with kits of strengthening gussets. This is one quarter of Revington TR's front suspension kit RTR7017, plus an example of its substantial lower wishbone attachment bracket. Note the pressed radii on the edges of the gussets for considerably added strength. This particular kit can be welded to the car with the body in situ.

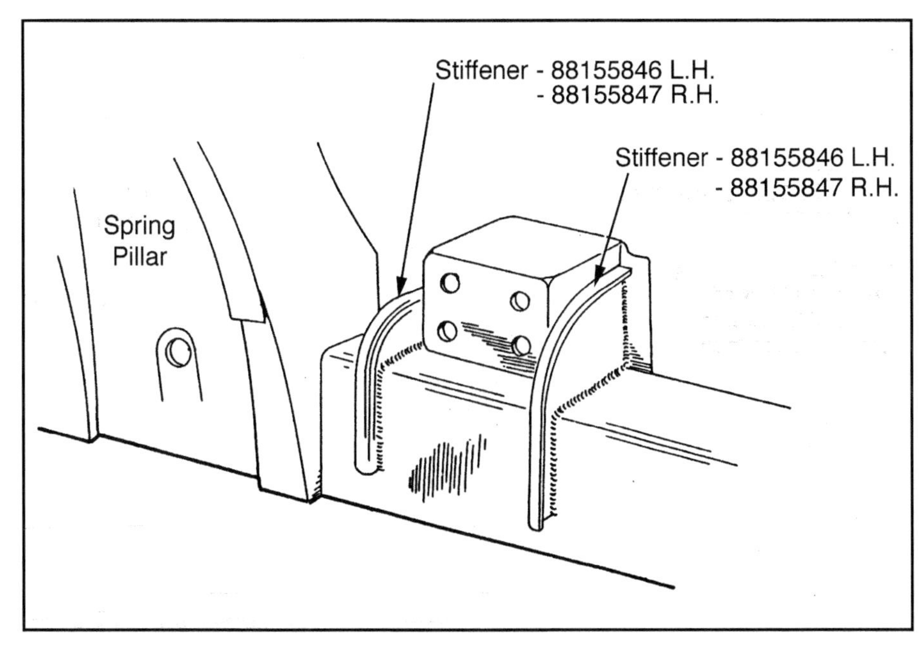

D5-6. Strengthening the rear lower left side pivot bracket. (Courtesy *TRaction* – the magazine of the UK TR Register.)

wishbone brackets are stiffened to the steering rack cross member by only a few millimetres of weld, and it's 90% certain that, after shot-blasting, you will see that this small weld has cracked. You can, of course, re-weld the joint, but it's better to add the new plate or gusset shown in drawing 5-4 and weld it so that it securely joins and mutually stiffens the rack cross member and the front lower wishbone attachment bracket. Photographs 5-29 and 5-30 demonstrate why the extra plate is required.

STRENGTHENING THE LOWER MOUNTINGS

Strengthening kits with pressed gussets are available from premier TR restoration specialists, although you will get an idea of what's needed from drawing D5-5.

The first step requires that the strongest and most satisfactory mounting brackets are welded to the chassis. Dealing with the TR4A first, two changes are recommended; firstly, remove the 4A lower fulcrum brackets and carefully check the frame and original welds. After correcting any problems, prepare to drill an additional hole below each original 4A hole in preparation for fitting post-1972 TR6 suspension brackets. The late 6 brackets can be identified by four mounting holes (drawings 5-4 and 5-6) and the square backing washers they use. The latter can be bolted through the original 4A holes and then used as templates to drill the additional hole in each frame bracket.

Assemble with the backing washers *inside* the frame brackets to increase frame strength in this area. Unfortunately, the TR5 and TR250 require rather more drastic measures and I recommend all four frame brackets be cut off and replaced with the post-1972 TR6 brackets and square support washers. This is a very difficult job with the body still on its chassis, so if you are carrying out a body-off

5-26.

5-26 and 5-27 (above). The first picture shows the bottom rear wishbone mounting bracket on the left side of the car. Two strengthening plates have been added; one in front and one behind. The second picture is a back view of the bracket showing the additional rear stiffening plate that runs right down the rear of the bracket and turns under the chassis frame. It is fully seam welded right round the periphery. Note the rectangular access hole to the two wishbone fixing bolts.

5-28. The bottom front wishbone mounting bracket on the left side of the car. One strengthening plate has been added to the rear of the bracket between the turret and the steering mounting crossmember. A rectangular plate has yet to be added to tie the steering mounting to the top of the wishbone mounting.

5-29 and 5-30 (right). Picture 5-29 shows the latest TR steering rack (note the 'D' washers welded to the rack and rubber cushion mounting) used on TR4A, TR5 and TR6s with its traditional, but most undesirable, cracked mounting (arrowed) where it should be attached to the lower wishbone mounting bracket. Photograph 5-30 shows, from a slightly different angle, the front lower wishbone mounting bracket in the centre, together with the steering rack mounting. This is the same problem without the steering rack in the way, and exposes the full area where the extra rectangular square stiffening plate shown in drawing D5-4 (page 62) should be welded in.

5-31. And here is the rectangular plate in place in what is a completely new chassis for a TR5.

restoration do take this opportunity to attend to this. If your TR6 already has the twin-stud pivot brackets you can probably relax, although you may well need to add the extra gussets and/or the square support washers.

Certainly, from November 1972 all CR and CF models reflected the latest thinking, and, although the changes I am suggesting may not be original to your car's particular year of manufacture, they were an original Triumph design and were incorporated in later cars for safety reasons. Need I say more?

The second step requires we add strengthening gussets to the brackets themselves. The weakness in the chassis mounted bracket was realised by Triumph, which issued designs,

5-32-1. The sequence in which you proceed with an IRS chassis repair isn't always crucial, provided you remember the need to retain "reference points". However, it is important to get the angle of the rear suspension mounting leg correct and some explanation of this is included in the main text. It is equally vital to ensure that the integrity of the welding between each chassis leg and its respective chassis member is first class before you give any thought to tacking the t-shirt pressings in place. Here we see that both tee-shirt pressings have been cut longitudinally in half, and the left side of the chassis is being used as a mirror reference while the right side of the chassis is being repaired.

part numbers and fitting instructions for six 1/8 inch (3mm) stiffeners shown in the drawings detailed earlier. To carry out the modification, you need 2 off 88155846, 2 off 88155847, 1 off 88155531 and 1 off 88155532 – to quote the official Triumph part numbers. All can be purchased from your

5-32-2. With the inside of the right side rear suspension mounting leg securely welded to the inside chassis member, it is time to repair the outer end ...

5-32-3. ... and not a bad idea to do the same with the left side, too!

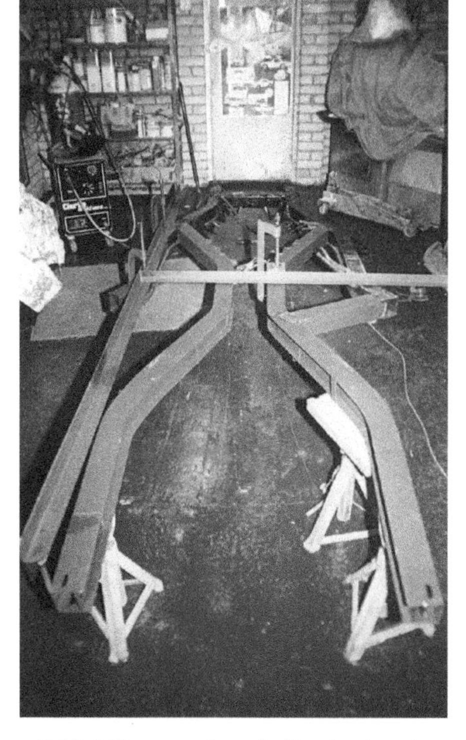

5-32-4. If you are tempted to do chassis repairs and have to 'let-in' a complete section, do use straight edges to keep your work on the straight and narrow! The clamped right-angled 'jig' is obvious, but did you notice the small vertical spacer in the foreground, tack welded to the bottom rear of this IRS chassis? The left side trailing arm mounting is already in place, but the right side clearly is the next step – followed by new 't-shirt' pressings top and bottom.

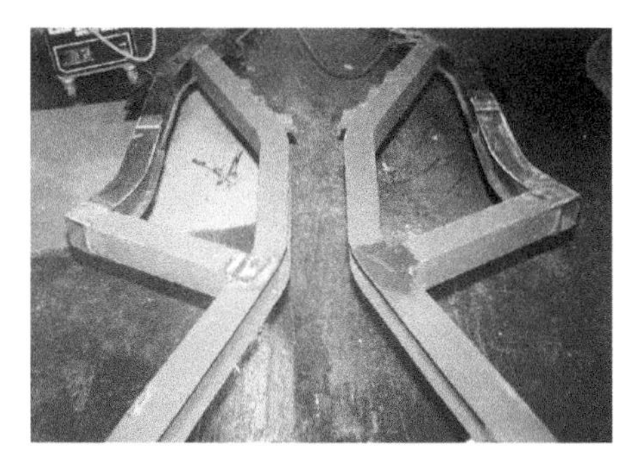

5-32-5. Some closing pieces can be added to the open chassis sections almost whenever it suits you. However, it is clearly vital to have the workshop manual chassis dimensions to hand, and to align the two halves very carefully indeed ...

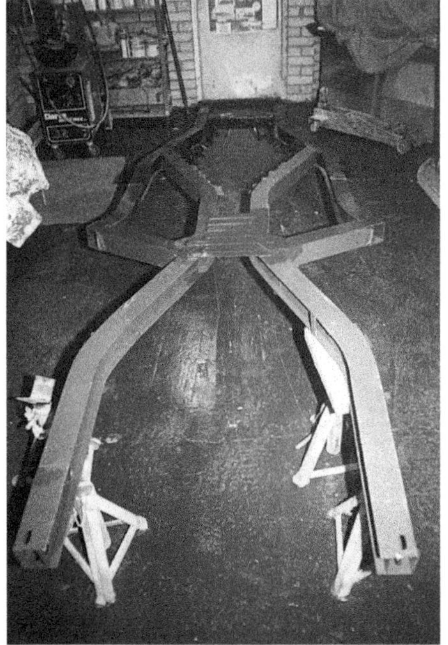

5-32-6. ... and securely tack the top t-shirt pressing in place.

5-32-7. Turn the chassis over gently and tack the bottom t-shirt in place. Double- and treble-check the dimensions.

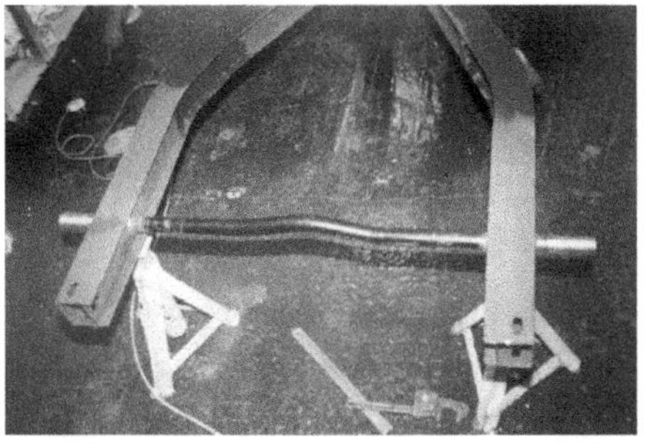

5-32-8. Give the rear of the chassis some stability by welding the rear cross tube in place.

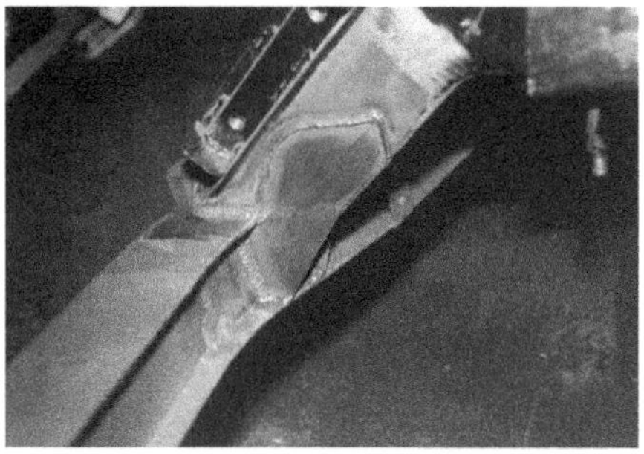

5-32-9. Securely weld the front and rear diff-bridges in place, and recheck the chassis dimensions. Note the superb welding quality that is (ideally) required shown in this picture of the left end rear diff bridge.

preferred TR specialist in a ready-to-fit kit, and all need to be affixed *after* the existing stiffeners have been removed.

With the welding complete, you should ensure that every tapped hole is clear; some drilling and re-tapping is inevitable.

A REPLACEMENT CHASSIS

A replacement chassis may only be necessary in a very small percentage of cases, but if your chassis has to go off to a specialist repairer for (any) one reason, it seems sensible to consider all your options.

The alternatives and respective costs are approximately -

(a) Brand new chassis – approx £2000

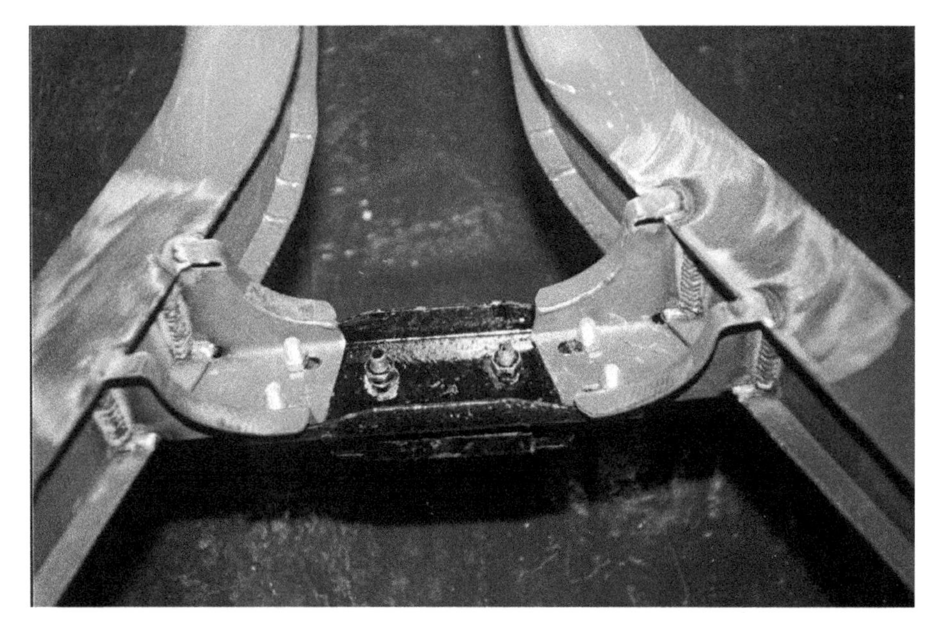

5-32-10. Fit and check that the gearbox mounts are correct using (here) the 'A' box mounting bracket.

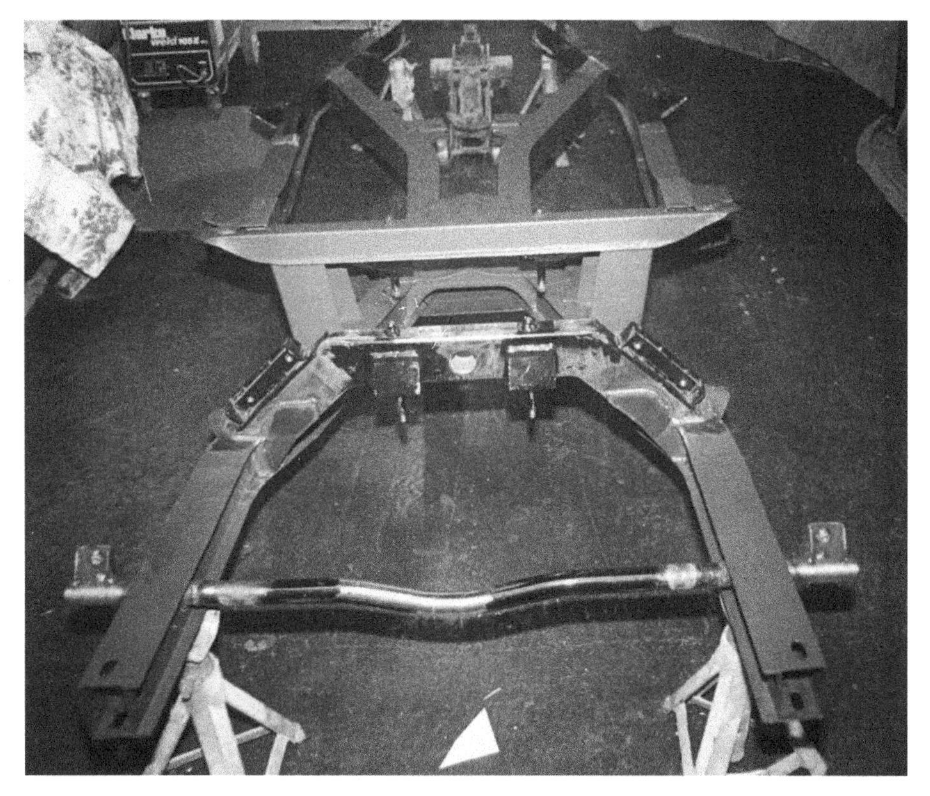

5-32-11. Weld the tee-shirt pressings very carefully. Be sure to follow the "diametrically opposite" sequence recommended in the main text and only weld in shortish runs, not much longer than 1in (25mm) or so, at a time, to minimise distortion from weld cooling. Check your dimensions. Finish welding the chassis.

(b) Ex-Californian/dry State – approx £1500
(c) A professional chassis repair or exchange – approx £1000

A brand new chassis sounds great, and has the advantage of being stiffer than the original, so is the way to go if you can afford the cost.

5-32-12. Make and fit the extra gussets within the tee-shirt pressings for your TR.

You should approach your preferred premier TR specialist or the TR chassis manufacturer listed in Appendix 1. However, potential concours contestants should note that there are (minor) differences to the original chassis which, if noticed by your concours judge, will be marked down on the basis of being non-original.

Californian chassis are in such wonderful condition because, of course, the climate is dryer and humidity low. Furthermore, the 'dry' states largely have no need to salt their roads – to the further benefit of the cars generally and their chassis in particular. So, if you have concours aspirations, go for an ex-Californian chassis. All the spot welds are 'correct' and the chassis cannot be faulted from an authenticity point of view. If tempted to buy privately do take care, for there are cars that have spent the majority of their lives in less agreeable (climate-wise) states and been shipped to be broken or sold as 'Californian'.

If your chassis is as clearly corroded as that shown in photographs 5-6 and 5-7 then you have no doubts as to the fact that everything in that area will need replacing. Your only quandary is whether you will do the work yourself, whether you will ask one of our specialists to do it, or whether to buy one of the replacement chassis we have just been discussing. You may find the photographic sequence 5-32-1 to 5-32-12 helps you decide which route you want to follow. I would reiterate that the integrity of the rear suspension mounting sections and the associated chassis members are absolutely crucial to the safety of the car and if you are in any doubt whatsoever as to your abilities, I do urge you to let experienced hands carry out this work.

Chapter 6

Painting, corrosion protection & metal finishing

THE CHASSIS

The repaired chassis is best hand-painted once welding is finalised. Lay a large, heavy-gauge plastic sheet (the kind used as a damp proofing membrane below concrete floors) on the floor, then position the chassis at about waist-height on a pair of trestles and hand-paint the top and one vertical face. Once the paint is dry, turn the chassis over and paint the remaining surfaces.

You might be surprised to know that galvanising is not recommended. I've heard it said that two men can carry a chassis to the galvanising tank, but four will be required to carry it away! The zinc builds up inside and outside each section, but sadly this is not a guarantee of longevity. The acids involved in the pre-galvanising preparation don't seem able to completely escape the chassis box sections, and are probably trapped in the tiny gaps between the folded/welded flanges. Whatever the reason, many galvanised chassis have rotted away from within, leaving the well-intentioned owner with a zinc 'shell' that provides no strength to his TR. Furthermore, the process is carried out by immersing the chassis in molten metal (mostly zinc), and the heat can distort the chassis before it's married to its body. The final nail in the galvanising coffin, from my point of view, was the discovery that zinc, when welded, gives off very toxic fumes. You're going to fix all the chassis problems before galvanising, of course, but what of the future? Accident damage may need to be repaired, for example, and modifications or upgrades could involve chassis welding.

If you don't wish to carry out the finishing process at home, or you feel an even better chassis finish is essential, then your only alternative is a polyester or epoxy powder-coated finish. Powder-coating is, as its name suggests, a process by which the finishing material is applied in very fine powder form. It is discharged from a special 'gun' and electrostatically charged in the process so that it 'seeks' your earthed chassis. The whole chassis is then put in an oven that fluidises the powder, which then flows to form an excellent corrosion- resisting and smooth-surfaced finish. Another piece of good news is that the electrostatically charged powder is, unlike paint, attracted to the edges of your chassis, which is a characteristic that makes it a particularly suited to protecting chassis-like products. The expense is one disadvantage, though, and you had better first get estimates for the work beforehand. However, there is a perhaps more serious disadvantage to consider. As I just mentioned, the electrostatic charge makes the powder seek the nearest earthed metal, but this makes it very difficult to get the powder to penetrate right into the bottom of tight corners before it is 'pulled' to a nearer piece of chassis. As a result, the actual powder covering at the base of any right-angled areas of the chassis can be minimal. Nevertheless, the polyester coatings in particular are much more resistant to chips from stones, and, unlike paint, will usually withstand steam cleaning. The latter is a consideration if you live in a rural area and need to steam/pressure clean mud from the underside of your car. Paint will quickly be washed off, polyester will stay the course, and probably offers the best long-term chassis protection. Polyester coating will also smooth over a slightly pocked chassis, but is impossible to repair – although you can paint over any spots that have been damaged by welding, etc.

Painting is the most cost effective and, therefore, most popular method of post-repair chassis protection. If you're to have the chassis sand blasted, remember that it's imperative you have the resultant bright surface protected within a few hours (four maximum) of blasting. If the blasting takes place on a wet or humid day the chassis is best protected immediately, so that the blaster blows off the chassis and paints it within the hour.

A sand blast contractor will usually spray-paint the chassis, but, if you're doing the work at home I'd recommend brush-painting it; spray-painting a TR chassis at home is deceptively difficult. It's difficult to handle the chassis in the average garage in a manner that gets a good covering of paint on all surfaces. Furthermore, the volume of paint used to spray a chassis is far greater than you would ever estimate, and it will probably require a high-build primer (which will have to be rubbed down between coats) to 'fill' the inevitable imperfections. A very expensive and time consuming exercise, not to be recommended.

Although you have several options open to you when it comes to paint, do not use a high-build paint. The ideal is a thin surface finish that will prevent corrosion in the short term, that adheres very securely to the virgin shot blast or cleaned/degreased surface, and should be easy to wire-brush off if welding is subsequently required. The use of paints designed for stoving means that when you weld, say, an inch or two away from the boundary of the painted surface, the heat effectively 'stoves' the paint.

Alternatively, contact a marine chandler, and procure a brushing metal primer and a gloss black brushing top cost. Ships chandlers are very experienced in dealing with the harshest of environments, and any products they recommend should be ideal for protecting your TR chassis, particularly in a winter/salt laden environment. Ask International Coatings Yacht Division (UK 01962 711177 or US 1 908 686-1300) to send you its free booklet that explains, in some detail, how to prepare and paint steel boats. International's marine grade paints are of the highest quality, are safe to use, and can be brushed on to give an excellent surface finish.

Do not use just any old black paint, but rather find one advertised for use on steel/ferrous products. Most DIY stores will stock ICI's Hammerite, or the probably preferable smooth surface version called Smoothrite. Chassis degreasing and/or the inevitable brush-cleaning needs to use ICI Hammerite thinners. There must be no oil, grease or loose areas of rust on the chassis prior to painting, but, with the appropriate paint, some minor rust spots/depressions should be 'killed' by the paint.

PAINTING THE BODYSHELL AND PANELS

Home application problems

This may seem an inappropriate moment to mention a couple of details from towards the end of your restoration, but it's worth just giving the final body rust protection and the hood frame a moment's thought. Before moving on to the highly visible body protection, it's wise to consider where and how you intend to inject rust preventative wax into the cavities. You should drill any access holes that will be required now! Furthermore, as is explained in more detail in chapter 17, the hood frame is best checked and repaired as necessary on the car. Heat and/or penetrating oil will possibly be needed, so the hood frame is best sorted before the body is painted, and certainly before trimming takes place, which is why I choose this moment to jog your memory to fit and correct your hood frame!

This book is intended to help readers complete much of their TR restoration work themselves. However, modern paints, expensive specialist equipment, and legislation designed to protect the environment and paint shop employees from some very unpleasant chemicals, have put most bodyshell re-painting work beyond the amateur. Consequently, right from the start of the chapter, I suggest that bodyshell re-painting is an unavoidable, and expensive, subcontract job. The new techniques are outlined below to help you understand why it's advisable for you to leave much of the re-finishing of the bodyshell to a professional paint shop, although there are still some things you can do at home to minimise the costs.

European emissions legislation tightened significantly from 1st January 2008, and it's now illegal in Europe to supply body shops with non-compliant automotive paint, and all auto re-finishing paints need to meet the Directive on Volatile Organic Compounds (VOC) standards. The VOC Directive specifies strict limits for the percentage of organic compounds that are released into the air during the coating process, making VOC compatible two-pack primers containing ever-higher solid contents, thick water-based colour-coats, and ultra-high solid two-pack lacquers mandatory requirements. In 2010, the legislation in Europe became even tighter. Thus, the solvent-based and thinned cellulose and thermo setting/plastic finishes that most of us are familiar with will disappear as vehicle re-finishing progressively relies more and more on two-pack primers and finishes containing acrylic and isocyanate components, and water-based colours.

Water-based paints have been in use within the UK auto-manufacturing industries since about 1990 (in the USA for a little less time) and for automotive repairs for 10 or so years on both sides of the Atlantic. So, although the legislation is new and the widespread use of water-based finishes for classic repairs/restorations relatively recent, the paints are not new or untried. Appearance-wise the overall finish is as good as the solvent-based paints they replace, some say better, and this is particularly so if you're spraying the whole external shell.

If you're repairing/spraying a single panel or two, a front wing and door, for example, the colour-matching task for any spray-shop using any type of paint becomes more difficult. For a start, two consecutively built TRs that left the production line in the same colour, one being exported to California and one to a rather less sunny corner of the UK, will be several shades different by the time restoration is required. Thus, the original Triumph colour code will be little better than a rough guide. Furthermore, the paint shops might get a very good colour-match in natural light, but under artificial lights, street lighting for example, there may appear to be a colour difference.

Some commentators have suggested that this is a new problem brought about by water-based colour-coats. I disagree; it's an old problem, and one that modern technology in a properly equipped paint-mixing facility makes easier to resolve, given

6-1. A pair of spray-gun nozzles for comparison. On the right is the conventional nozzle anyone used to spraying solvent-based paints will be familiar with. On the left is the very different and much bigger 'ring' (my term) required to spray the ever-thicker water and two-pack paints necessitated by modern legislation. All the components need to be manufactured in stainless steel to avoid corrosion.

the variety of pigments available for water-based paints. However, the paint supplier will need to measure the colour spectrum (light) given off by an original panel, and then make up a very small initial sample of paint. This is put in a small lightbox and viewed under four different wavelengths of light to reveal any colour matching anomalies under all lighting conditions. Any necessary corrections are incorporated into the full paint mix, which should make for perfect colour matching under all lighting.

Water-based paints spray more easily than solvent-based paints, and they don't 'sag.' Solvent-based paints contain 80-90 per cent solvent, whereas water-based paints are virtually solvent free. Furthermore, the manufacturers say that the special pigment composition of water-based colour-coats ensures that the finish remains fade-free, even under intense UV radiation. They dry to a uniform matt or satin finish (depending upon the manufacturer) that relies on the final (lacquer) coat for the gloss.

6-2. The first steps in painting the underside are shown here. This shell is clearly inverted, but at home, provided the suspension is in situ, there is much to be said for using a roll-over jig to turn the shell 90 degrees on its side. The etch primer has been applied, followed by careful seam sealing around all seams. Did you spot that the outer sill on the left side of the picture (actually on the right side of the car) has been masked and left unprimed? It's best left until the final finishing process takes place. Unless you're aiming for concours competitions, you could hand prime, paint, and seal the underside at home.

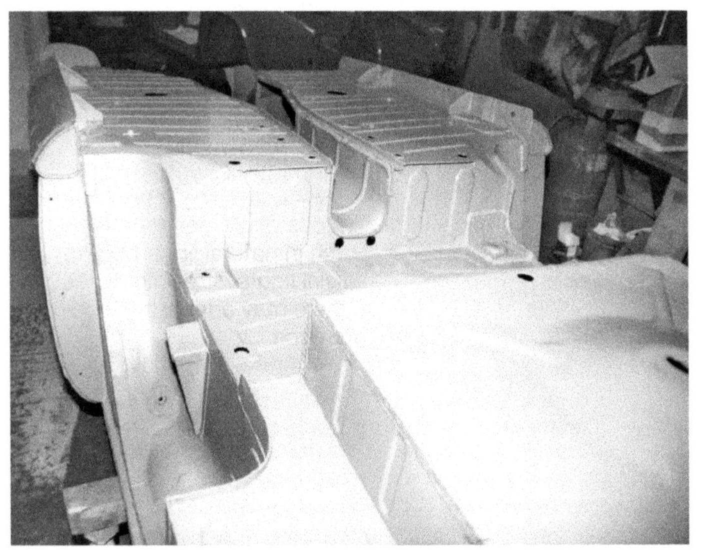

Two-pack (chemically hardened) finishes require an external air-fed breathing system for the sprayer if the paint is to be applied safely. Furthermore, the toxins can get into your system through the skin, hence the need for full gloves/gauntlets and protection even during mixing, and also through the eyes, necessitating a sealed FULL-face mask/helmet when spraying. Failure to utilise the necessary breathing equipment can result in the two-pack coatings setting-off in your lungs, as well as the inhalation of carcinogenic isocyanates.

As a result, application booths need to have the appropriate air changes, circulation, and filters to prevent overspray getting out into the atmosphere. The professionals, even those working in paint shops categorised as 'low-usage,' have to have annual checks of both booths and employees, while the local health inspectors may also pop in from time to time to measure the emissions from the booths. Furthermore, today's water-based finishes are very thick, and so may require a more powerful air compressor than you might expect, but will certainly need comprehensive water vapour heater/dryers that are much more sophisticated and effective than the water traps of old. The quite different (and expensive) spray gun seen in photograph 6-1 is also necessary for the actual paint spraying. No home workshop will be equipped with any of this equipment, and, while it may be possible to hire the compressor, possibly even the air-supply, the booth and filters are always going to be a problem.

There are other consequences to consider, in that, perhaps influenced by the necessary equipment and increased regulations, the number of garages that undertake re-spraying seems to be declining. I would imagine this will contribute to increasing the cost of this part of a restoration. As already suggested, fewer home bodyshell re-paints seem likely once stocks of solvent-based paints are used up, and, consequently, the cost of a restoration will escalate as the shell will have to go to a professional paint shop. The overall result may be that, ultimately, fewer restorations take place. However, let's explore the revised painting process so you can make up your own mind.

The painting process

Etch primers are the essential first step, particularly where painting bare metal is concerned. The acid primer has a passivating and corrosion-inhibiting effect on the surface of the base metal. The professionals will use a two-pack version, but amateur restorers can still get cellulose/solvent etch primers. Their adhesion and properties are significantly inferior to the two-pack primers, and, since, you're going to

6-3. A typical etch-primed, filled and shaped wing/fender, in this case on a TR7. You're best preparing and priming the removable panels off the car, as has been the case here, and, for non-metallic finishes, masking off the interior of the car at the panel joints (i.e. back from the shut lines) as we see here. Metallic paints require that the car be finished with the removable panels fixed in place in order to uniformly align the metallic particles. This difference affects the position of the masking which, with metallics, moves outwards to become level with the shut lines.

come to the two-pack application block sooner or later, you need to ask yourself whether you can accept the poorer home-applied foundation for your restoration. However, there will be occasions when a solvent-based etch primer will be adequate; for example, when painting a small panel or an internal compartment, such as the engine-bay. In such cases, Upol Acid-8 will do the job, and has the added advantage of being available in 1-litre cans and in aerosols, and I believe will be available for some time to come.

After etch priming, you will need to seal the shell's seams, as demonstrated by picture 6-2. This is particularly important on the underside of the car if you plan to paint the shell after remounting it to the chassis.

The next step will be filling and shaping the visible panels. Almost all home-restorers will be familiar with the application and subsequent sanding/shaping of body fillers – and this labour-intensive process has changed very little and can still be accomplished at home. Picture 6-3 illustrates what I mean by filling and shaping. However, what has changed is the need to guard against applying a water-based paint straight onto body-filler or onto a primer coat that allows the water-based paint to pass through to the filler. This is because the filler is porous and will absorb the paint's water content, trapping it in contact with the underlying substrate with disastrous consequences. So, you need a proven sealer/primer, which probably means a two-pack one, in which case the painting process definitely becomes a job for a professional. Currently, JM Autocolour makes and can supply a solvent-based VOC-compliant primer, which you may care to enquire about, but check that it provides the essential seal between the water-based colour and the filler.

For those mulling over the concept of re-finishing your bodyshell up to the filling/shaping stage at home, consider that you might not find a local paint shop prepared to paint your externally prepared (etch-primed/filled/shaped) shell for you. The preparation accounts for about 90 per cent of the finished paint quality, and so most high quality paint shops concerned about their ongoing reputations will say, not unreasonably, that without their preparatory work they are not prepared to paint the car. However, where applicable, there should be much less resistance to your preparing the inside of the shell (i.e. the engine bay, interior and boot/trunk) and have the paint shop prime and finish-paint these internal areas. All this needs to be planned in detail with the paint shop concerned, of course, and your mask-lines, etch priming, and sealing mutually agreed.

You then have a choice of having the paint shop prime and finish the whole car, or help you reduce the possibility of external paint damage by painting the car in two parts – if you are prepared to make two trips to the paint shop. This is clearly not ideal. However, if you take the internally finish-painted shell home, you can then assemble most of the engine-bay, electrics, and interior of the car before returning the car to the paint shop for external shaping, preparation, masking, and finishing at a later date.

The selection and application of two-pack primers is an art in itself. The so-called 'high solid content' is no longer legal, and from 1st January 2008 had to be 'ultra-high' in order to bring the primer's VOCs within current legislative limits. There are plenty of primers to choose from, and all flat-down reasonably easily, although the ultra high built primers are slightly more difficult and time consuming to apply than the relatively soft solvent-based primers that most of us are used to. Once hard, you can wet or dry flat them. There is, however, a complication when it comes to selecting your primer, because you can avoid a lot of extra colour-coat expense by selecting a suitable coloured primer. Modern colour-coats may be water-based, but they are expensive, so there are obvious attractions to selecting the right colour primer. However, they are available in a limited range of colours, and, needless to say, it's essential to have the experience to know which colour primer to apply! As an example, a dark/midnight blue car would probably require less finish-coat paint if it were applied over a black primer, but, if available, a blue primer would be less likely to reveal any subsequent stone chips.

Bear in mind that, should you elect to carry out some post-primer rubbing down, and go through the primer coat, it becomes essential to reseal/insulate

6-4. US (ultra high) and HS (high) clear lacquers compared. The can on the right, although purchased only a few weeks before this shot, became unusable after January 2008 because the proportion of solids did not meet the latest VOC Directive requirement.

the underlying filler by applying an additional coat of primer.

Then there's the question of paint adhesion in some locations. Most classic cars are finished in cellulose or thermo-setting/plastic paints, including metallic finishes. Such finishes are a combination of colour and gloss components that, with today's technology, is actually applied in two coats: colour followed by a lacquer to provide the gloss and protection. While much of the original external paint will likely have been removed when panels were replaced, repaired, rubbed-down, or sand-blasted during the body restoration process, there will likely be some old paint remaining in the more protected places, like boot/trunk, engine-bay, and cockpit. A modern two-pack primer is usually acrylic based, and that, when sprayed directly over cellulose or thermo-setting/plastic finishes, can cause swelling, wrinkling, and adversely affect both the original paint finish and its adhesion. So, an epoxy resin primer is, therefore, an essential foundation/sealer over earlier paint finishes before application of the colour primer essential for water-based colour-coats.

When it comes to applying the colour-coats, their thickness makes them spray well, and they are easier to apply than a solvent-style paint, but you need the correct gun. There are some additional difficulties to bear in mind. Water-based colour-coats are usually/ideally applied wet-on-wet. As an amateur sprayer, this concerns me because I lose my opportunity to rectify my (inevitable) errors between coats. However, water-based colour-coats can be flatted after the first coat is hard, and before applying the second colour-coat. You do the flatting, only if necessary, dry, using a 3000-grade special pad. However, one useful rectification trick is that you can use 1500 grade water-based panel-wipe on imperfections about 20 minutes after second-coat painting (to allow the coat to flash-off); that will resolve many problems before the application of the essential lacquer coat.

Another problem relates to the varying 'depth' of coverage with colour. Some water-based paints cover well, and the best may require only one full coat followed by a light, second, 'half' coat. More usual is two full coats of a good-covering colour, with a third 'dusting' coat in order to get an even coverage. The 'half' final coat is particularly helpful for metallic finishes, as might be required by the later, 'wedge' shaped TRs. Dark blue is an example of a colour with good natural depth to it, which could be ordered in an appropriate quantity for two-and-a-half coats of the area involved. However, other colours, some reds, for example, are very difficult to get any 'depth' to, even over the best available colour primer, and may require perhaps eight to ten coats of finish colour! Thus, the volume of paint required, the application time, and significant extra material costs can vary because of your choice of colour.

The lacquer finish is currently a very thick two-pack clear-coat. It is, perhaps, frustrating for the home restorer to learn that solvent-based clear lacquers are available for industrial use, but they are not allowed to be used for automotive repair. A typical lacquer/hardener mix ratio is 2:1, and, by selecting the correct hardener from an extensive range, the same transparent lacquer can be used for complete coverage of the whole car, or for spot/partial panel repairs. The thickness of the lacquer has been increased progressively in recent years, passing through the high (HS) to current UH (ultra-high) solid contents. Both types are pictured at 6-4. Thinner/solvents are (ideally) not used, but if the UH contents are just too thick, a small percentage of thinners would be permissible. Unthinned, one and a half coats are all that is required, wet-on-wet.

In cases where individual panels have been re-sprayed, there may also be an issue with differing levels of gloss/shine between old and new. Modern lacquers certainly shine very effectively, and could initially 'show-up' the original level of gloss. However, with special polishing techniques such imbalances can be equalised, but this, of course, takes time and experience, and so adds to the cost.

In summary, the application of modern paint finishes, in spite of technological advances, has become largely a job for the professionals with the necessary equipment and experience.

Transportation

The safest method of transporting the restored bodyshell to the painter is to first re-attach it to its chassis. However, that precludes the painter colour-painting the underside of the shell without a lot of extra work. While an unpainted underside would not concern me personally, potential concours competitors will likely be unhappy about it. In such cases the shell has to go to the painter without the added rigidity and support of the chassis. Plenty of good bodies have been damaged being transported to or from the paint shop, and it's worth giving this potential catastrophe particular attention. Damage can also occur when moving the shell perhaps to or from shot blasting, of course, and you'll have noted several references to moving the body on its chassis to prevent transit damage. Before we explore this in detail, it might help to understand the most frequent causes of shell damage.

The two main causes of damage both relate to allowing the rear of the shell to travel unsupported. The centre of the shell is usually securely strapped down, but, without its chassis, the rear 'overhang' of the shell bounces up and down during transportation, and stretches (fractures in extreme cases) just in front of the 'B' posts. So, it's a good idea to support the shell under the boot floor with a couple of old car tyres of a suitable width that prevent all unsupported movement. The second opportunity for equally devastating damage can occur when a well meaning driver goes to the other extreme and runs a commercial vehicle wagon strap (the type that ratchets down to secure a normal load) over the back of the car. He then, in effect, breaks the back of the car by pulling the rear down. This is particularly easy to do if the floor is unsupported.

The reverse journey also needs to be planned meticulously. I would ensure the doors, wings, boot, and bonnet panels are never re-fitted by the painter, but are instead transported home (very carefully wrapped and protected) off the bodyshell. This eases the body to chassis marriage, and allows you to do much of the mechanical re-assembly with the loose panels stored well out of harm's way! Obviously, even with the shell mounted on its chassis, do not contemplate removing the door braces until your rolling shell is home and off the transporter.

The simplest and best damage-prevention solution, then, is to remount the restored body on the restored/painted chassis in your garage. Even with the added strength of the chassis supporting the shell, I would leave the door braces in situ at least until the assembly reaches the paint shop.

6-5. These four wings/fenders have been finish-painted, including their mating faces, and now have the weatherproofing sealer applied to the mating faces in preparation for bolting on the body tub. As discussed in the main text, Dumdum is a black, never-setting sealer that most restorers would use, although this looks like something other than Dumdum since it's considerably lighter in colour.

Protecting the underside

Once you've got the painted bodyshell back, there are several other issues still to resolve, starting with the question as to how/when you'll protect the hidden underside of the repaired bodyshell? I would personally degrease and hand-paint it in a ferrous metal corrosion-resistant finish such as was suggested for the chassis. Such paints come in a variety of colours, in addition to the black used for the chassis, and the colour that most resembles the body-colour would be my choice. Subsequently, but not yet, I would spray the underside in one of the rust preservatives discussed later in the chapter, so, in truth, the underside's colour is probably less important than ensuring its complete protection. There's also the option of using 'Stone-Chip' spray, and this is discussed later in the chapter.

Then we need to consider moving the chassis/shell assembly, because it will be heavier than the bare shell and more difficult to move. So, do you attach the front and rear suspension temporarily to the chassis prior to remounting the body and before the assembly goes to the paint shop? I favour the mobility that this method provides, although it means that you'll have some additional suspension removal/reassembly work, and will need to carry out the full suspension refurbishment later in the car's restoration than might otherwise be the case.

Then there's the question of protecting the restored and newly painted chassis from overspray while the shell is painted. This can be achieved by masking the outward faces of the chassis, but there's an alternative – a product called 'Slime,' that can be sprayed on and, being water soluble, washed off once shell spraying is complete.

If you're adamant that you want the underside of the shell finished in body-colour, clearly the shell will need to go to the painter without its chassis, making not only the retention but the quality of the door braces extremely important. However, do not attempt to bring the finish-painted body home on its own. A finish-painted shell must have its chassis in place before you contemplate return transportation, so gather your helpers and marry the painted bodyshell to its (preferably rolling) chassis at the paint shop. If you try transporting a painted shell without its chassis, you will crack the paint. It helps if you pre-glue the body mounting pads to the chassis before taking it to paint shop.

Sooner or later you'll need to address the question of how to affix the body panels. The 'how' is not so much a question relating to the fastenings, but rather the sealant that one must use between tub and panels if the job is going to last. As already highlighted, Triumph created a corrosion problem when fitting the various panels prior to painting, so the panel interfaces remained unprotected. We have corrected one problem by ensuring our panels and the body-tub are at least painted prior to attachment. However, there are still plenty of opportunities for water to get between wings/fenders and body, so you'd be wise to use a sealant between the panel joints. But what sealant?

The choice is more difficult than it might seem, for the joints between the rear wings and rear deck will continue to move, and any 'setting' sealer will crack (as will any paint you put on this area after fitting the wings). So, you'll need a non-setting sealer, and accept that TRs are best not painted after fitting the wings/fenders. Black rainwater gutter sealer remains flexible and has been used on TRs, but it's essential you keep it low-down in the seams. Modern body shops use Sikoflex, and, while you can use it on your TR, and are on view in picture 6-5, I should warn you that the panels in question will be very hard to remove at a later date! Dumdum is probably the preferred alternative. It can be over-painted if necessary, although, all things considered, you'd be better advised not to paint at least the rear wings/deck after fitting. Dumdum is a black non-setting sealer and so it's best to squeeze out and remove the worst of the excess, and then use a plastic scraper to remove what's left. You can finally wipe over the joint with some petrol/gas on a rag to ensure that no more than a thin black seam remains visible on the joint.

Stone-chip protection

It's very much a matter of preference, but many excellent, if non-concours, restorations have had a coat of 'Stone-Chip' to the underside of the shell, sills, front and rear valances, and even on the inside of the floors. Stone-Chip is usually white, and can be sprayed from a gun but is available in aerosols. It leaves a slightly mottled finish that can never be flatted, but does provide an excellent chip/corrosion-resistant surface for long-lasting bodyshell protection. It can be painted over if required, and is usually applied after seam sealing, over an etch primer. When used on visible surfaces, the sills/rockers for example, you should paint over the Stone-Chip with body-colour so it's hidden from all but the closest inspection. Don't confuse 'Stone-Chip' with under-body sealers with not dissimilar names. These are usually black, cannot be painted over, and are not really recommended for classic car restorations. You'll need to agree the areas where the coating is to be applied if your painting contractor is to apply it for you.

Metal finishing

Chrome plating

Chrome plating is porous; yes, it lets in water. When our TRs were originally manufactured, the top layer of decorative chrome was preceded by two substrates: a layer of copper plate, followed by a layer of nickel plating. This additional plating was designed to provide a non-permeable coating between such water as passed through the chrome and the underlying steel, thereby preventing rust taking hold. Plating with this copper base is unquestionably best, but the cost of copper has made it progressively more difficult to find re-platers who will triple-plate on a copper base. However, if you're investing in the restoration of an immaculate TR you expect to keep for many years, then the additional cost of a copper-based triple-plated decorative chrome finish could well be worthwhile. The next best thing is to specify a double coating of nickel, which is more readily available, but you must avoid the cheap chrome-on-steel approach – it may look reasonable on the journey home, but will rust as soon as the car gets wet. Whatever the quality of your chrome finish, it's worthwhile taking some precautions in order to extend the life of your plated parts. First, keep them clean, and remove the worst of any moisture as soon as the car is put away (particularly important if there's any salt in the moisture). Second, use chrome cleaner to keep it looking bright; and third, wax-polish the chrome. The wax acts as a barrier to water and makes the chrome sparkle!

While on the subject of chrome-work, note that the quality of reproduction bumpers is frankly disappointing, regardless of which TR you consider. Consequently, I strongly recommend you stay with an original bumper wherever possible. If your original chromed parts are rusted into holes they can usually be repaired by welding-in a similar section from a scrap bumper, provided the perforation is localised or at least not too extensive. The repaired part is then polished flat and re-chromed, so you'd never know. If your bumper has been dented, even badly, it's amazing the skills the restorers can bring to repairing it. Obviously, if your (probably rear) bumper is badly corroded to the point of extensive perforations, then it may be beyond saving. However, don't make that decision yourself; take it to your favourite TR specialist or chrome restorer and seek expert advice. I've used Central Engineering Services (see Appendix) for several restorations. If your existing parts are declared 'past it,' you can at least start looking for an original replacement that is in salvageable condition. Your ideal, needless to say, is an ex-Californian part with minimal rust (a few minor dents are fine).

The method I've recommended isn't the cheapest option, and, if your restoration is on a very tight budget the 'repros' are probably the route to go. Sooner or later, though, I think you'll be disappointed.

Component preparation

A superb restoration and paint job can be made to look quite ordinary if the tiny details aren't attended to. It surprises me how many classic restorations fall into the trap of not replating door catches, bonnet fastenings, large bolts, various brackets, wiper motors, servo cases, etc. Mind you, some go to the other extreme of fitting new replacements, which certainly complement the restoration (but at what cost?). Replating many of the originally plated parts (such as those listed above) can be carried out, but it must be stressed that the concept is not worthwhile for small fastenings, such as nuts, washers, and small bolts/screws. Plating is a three-part process, starting with degreasing, and followed by bead-blasting to remove any rust. You could contemplate doing these preparatory jobs yourself, since the task of bead-blasting each component is fairly labour-intensive, and you can buy a small bead blasting cabinet for around £150 from Machine Mart. The results will be much better than trying to clean up dozens of rusty parts with a wire brush, electric drill, etc., which do little more than polish the rust! I have no doubt that the cabinet would, in fact, pay for itself if you could spread the cost with another local enthusiast. Furthermore, new large bolts/screws can be expensive, and some of the original imperial bolts are hard to find. However, most of the original steel bolts, particularly the larger ones, can be replated, to a considerable financial advantage, if they have gone through the two preparatory steps.

The actual plating process is called 'BZP' (Bright Zinc Plating) followed by a passivating process. This technique doesn't involve any significant heat, so steel embrittlement is avoided. Nevertheless, it's prudent to avoid replating any highly stressed studs or bolts (such as cylinder head studs). However, the process is definitely worthwhile elsewhere, and adds a new level of professionalism to any restoration project. Assemblies such as door locks and bonnet catches can be rejuvenated in the same way as the fastenings mentioned above, although a little organisation on your part is required if you're to maximise the cost effectiveness of the process. Plating is carried out by the load (about 10kg), and the cost per load is about £25. You'll be charged for a full load whether you use it or not, so gather all the parts for your car into one load, pair-up with another restoration project friend, or persuade your local TR restoration specialist to put your bits in with his. In the case of the latter option, I'd advise you prepare a list of the parts you're sending.

A bead-blaster is also very helpful in preparing parts for painting, and will save your paint/powder coater much time, and therefore cost to you! Whether you're doing the bead blasting yourself, or contracting it to the painter, taking your parts along piecemeal will result in a very large bill. My recommendation is that you try to get all the blasting and prime painting and/or powder coating done in one load; two at the most. This will save you hours of time and, if done 'in bulk,' the cost per part will be quite small. Do check all parts before getting them finish-painted or powder coated, and correct any thread or more major faults first (e.g. re-bushing, welding, re-setting, etc.). As you're dealing with a large number of parts, a photograph of what went to the powder coater will be helpful when instigating a search should something go missing.

RUST PREVENTION

That you must ensure future corrosion is kept to minimum is beyond doubt. How to do this (and when) is open to discussion. Waxoyl, or one of its competitors, is the preferred method. I find Dinitrol's 1-litre containers, shown in picture 6-6, convenient and more practical than aerosols. Dinitrol also has various material options, from its ML Cavity Wax to its Corromax 3125, all fitting straight onto the compressed air applicator/gun also seen in the picture. I think it's worth buying/hiring a small/medium sized compressor to both speed the task and ensure that the protection is maximised.

Apply the wax to box sections and in the inaccessible corners/areas of the chassis/shell only once you're confident all painting is complete. You can actually apply a wax coat under components (e.g. the suspension mounting brackets) that will later be covered, and thus get protection where there is unlikely to ever be any by the other route! Being a great believer in wax protection I wax the inside of the bulkhead/firewall during re-assembly of the car ... an area that you are unlikely to ever attempt with a finished car. I also apply wax to each component as it is assembled, and finish the whole car off with a final spray!

For the small cost involved, it's worth buying the compressed air injection gun, long reach tube, and flexible hose we saw earlier. Machine Mart and Frost Auto Restorations are possible sources, with Frost having the added advantage of also selling the rust preventative treatments. You must close all visible injection holes with a plastic bung, and be prepared to go through the underside/box section protection exercise every couple of years.

These days there's a clear stone repellent film that can be applied to the outside paintwork to protect cars that are likely to be subject to really tough conditions (rallying, for example). This protection can be stuck to vulnerable areas, and then pealed off as and when appropriate. It is available from TR Enterprises, as well as, no doubt, the other TR Specialists listed in the Appendix.

6-6. Compressed air rust proofing equipment and 1-litre Dinitrol Corromax 3125 canisters. The flexible extension hose for cavity injection comes with the gun kit, but is, to my mind, overly thick, although its size helps to quickly dissipate the Dinitrol.

Chapter 7

Restoring engine & clutch

INTRODUCING THE 'SIX-POT'

Derived from the Standard 8 four-cylinder engine, the six-pot (six-cylinder) was conceived as a 1998cc unit for Triumph's 2000 Saloon/Sedan. The engine had a 74.7mm bore and was also used in the GT6 and Vitesse before being 'stroked' by Triumph's development engineers, who felt that more urge and torque was needed for the TR250 and 5 sportscar.

The stroke was extended from 76mm to 95mm, generating the now-familiar 2498cc used in the TR250 and 5, later in the TR6 and also, in slightly de-tuned form, in the 2500TC, 2500S and 2500PI Saloons. You may already have heard that, even with this enlarged capacity, Triumph engineers felt the prototype TR5 was under-powered. In spite of numerous attempts to improve power and torque, however, they were unable to reach their objective until Lucas came along with the 'PI' or Petrol Injection system.

Fuel injection had been available for very up-market engines for many years. The famous Mercedes Benz Silver Arrow Racing cars used fuel injection very successfully in the 1930s, whilst some German aircraft also used fuel injection very successfully during the Second World War. The Lucas system offered to Triumph was a mechanical/hydraulic one which worked by directing fuel via a mechanical metering unit down a feed pipe to each cylinder. When fuel pressure rose to an appropriate level, the injection nozzle opened. Subsequently, the metering and injection of fuel has become universally known as Electronic Fuel Injection, but the Lucas system in question was a step earlier than the computer technology which enabled electronic control in motor vehicles.

The Lucas system was used exclusively by Triumph, and did the business by raising power and torque by some 25 per cent, giving Triumph TR5s and TR6s a significant performance boost. The rest, as they say, is history. (The system is explored in more detail in chapter 10.)

The six-cylinder engine was not only used with Petrol Injection induction, but those TR250s and TR6s exported to the USA, of which there were many, used a de-tuned version of the engine with carburettor induction to aid compliance with growing pollution legislation. There were other variations in engine specification and components, such as pistons, camshafts and cylinder heads for the Triumph Saloons – so we had better start by listing the facts related to the various 2498cc engines in order that you appreciate exactly what engine and tune you have.

Engine number prefix	Origin of engine	Maximum BHP /torque
CP	Original TR5 or TR6 PI up to late 1972	150/164
CC	Original TR250 or US TR6 to late 1972	104/143
CR	Original TR6 PI from 1973	125/143
CF	Original US TR6 from 1973	106/133
MG	2.5PI Saloon	132/153
MN	2.5PI Saloon	124/153
MM	2.5TC Saloon	100/135
MP	2500S Saloon	105/139

Six-cylinder engine numbers and tune.

If originality is important to you, you clearly need to use an engine block/number that is compatible to your specific TR; hopefully, the accompanying table will be helpful in that respect.

Additionally, there were three 2000cc engines, each with their own identification prefixes. They appear very similar indeed to the 2500cc engines (crankshaft excluded), but the availability of complete 2500cc engines is such that you hardly need to bother about the smaller capacity engines for many years. Incidentally, the ex-Saloon engines have a different front plate but will go straight into a TR250/5/6 once the front engine plate has been changed for a genuine TR one or an angle grinder has been used to take off the sides of the Saloon's front engine plate. The grinding route is a bit of a bodge if you don't remove the plate from the engine first – but has been done. This is because the Saloon's engine is mounted using the front engine plate, whilst a TR uses a mounting bracket bolted roughly one-third back along the side of the block. If in doubt, some of the engine installation photographs at the end of this chapter will clarify the point. The 4 tapped holes required each side for the TR mounting brackets are available in all six-cylinder blocks – thus allowing their adoption for TR use if required.

Before you get stuck into the task of rebuilding your engine, do consider a couple of important details. If you do-it-yourself you have absolutely no warranty fall-back. Even if you use just one machinist he will claim that any problems which may arise are the consequence of your assembly. Check out just how much you might save doing the assembly yourself; you may find the saving does not justify the risk. For maximum piece of mind and warranty, use one specialist to remanufacture your engine, and, as I have mentioned before, do not leave a rebuilt engine laying around for months before installing and running it. For those who feel they must rebuild the engine themselves, the following are some (hopefully) helpful pointers.

REBUILDING THE BOTTOM END

Your engine generally, and the crankshaft in particular, could already have been changed from its original configuration, so be aware that there are two types of crankshaft, and you must marry the correct flywheel and rear spigot bush to your particular crankshaft.

7-1. The 'long-back' six-cylinder crankshaft used in all early engines can be identified by the long spigot at the rear of the crank. Later ones were only half this length.

The early crankshafts, used up to about 1970, were the 'long-back' type shown in photograph 7-1, which must have a long tubular bronze spigot bush that slides into the rear of the crank. The later, 'short-back' crank originally fitted from car CP 50000 had a different, heavier flywheel and a much shorter spigot bush that sits inside the flywheel. If you change any of these related components, you must ensure the right mating parts are procured. If you have the later crank/flywheel, you may wish to fit a lighter (earlier) flywheel to your engine to improve performance. Unfortunately, you will have to do this by machining about 6lbs (3kg) off of your flywheel since the light/early flywheel will not fit the later crankshaft.

Speaking of flywheels, note that many home restorers forget which way round the ring gear came off the flywheel. The ring gear does wear, usually in three spots, but can be easily changed; new rings are readily available but must be fitted correctly. So, make a sketch of yours before you take it off the flywheel, noting particularly where it sits on the flywheel and which way round the teeth face. For the record, the teeth should always face the back of the car and there should not be a gap between the flywheel shoulder and the ring gear. Note that it's possible to overheat the ring to the extent that it cannot shrink back onto the flywheel sufficiently to really grip the flywheel.

As an aside, it is a very good idea to check the ring gear if and when you need to attend to your clutch. Of the three wear points, it is usual for one to be particularly worn, and if you spot this when you are inside the bellhousing it is wise to take that extra precautionary step and fix the problem before it necessitates a further gearbox-out job.

There can be few enthusiasts interested in the six-cylinder engine TRs who are unaware of the dreaded crankshaft end float problems the engine is prone to. We touched on the problem and its causes when we discussed examining your car before purchase. As we will discuss in some detail shortly, the two half thrust washers are easily accessed and, provided your thrust washers are still in place, you could well bring your crankshaft end float close to tolerance by fitting new thrust washers of the appropriate thickness.

However, sadly, many engines have gone beyond worn thrust washers in that they have been allowed to wear to such an extent that they have turned inside the bearing housing, and then dropped out. If you find one or two half thrust washers in your sump, this means that the crankshaft is quite likely to have moved so far forward that it and the block have been damaged. Sometimes the damage is minimal and the situation can be remedied, but more usually both block and crank are damaged beyond repair and replacements are called for. Hence, your need for alternatives.

The saloon 2500cc crank can be used and is available at reasonable cost. The saloon blocks are also available but do not, of course, carry the TR's range of engine numbers. The crank you have could already have been changed from its non-original configuration, so be aware of the two types of crankshaft detailed above, and note that you must either obtain a crank that marries to your flywheel and rear spigot bush, or a matched set of crankshaft, spigot bush and flywheel.

When assembling a six-pot, do not cut corners. Do everything properly: the following are examples of where it really is worth doing the job well.

7-2. The 'long-back' crankshafts had a screwed plug in the big end to facilitate thorough cleaning of the oilways before fitting. Do take advantage of this facility, which was dispensed with when the later 'short-back' crankshafts were introduced, along with a different flywheel and clutch spigot bearing.

- Naturally, whatever the source of your crankshaft, you will need to get it reground. I would advocate that you specify AE main and big end shells. The extra cost is relatively small and, by my judgement, well worthwhile. It should be a matter of routine to ask your crank grinder to thoroughly wash the crankshaft through to prevent old oil mixed with grinding swarf being swilled straight into your new bearings the first time you start the engine! However, the early 'long-back' cranks have an added operation that must be carried out. As photograph 7-2 shows, the big end journals each have a screwed plug; it is important to de-plug the crank, wash it through and then re-secure the plugs.
- Unless the teeth on the cam gear and crankshaft drive sprocket are in as-new condition, replace them. They are not expensive and the alternative is slop in the link between crank and cam, which makes consistent timing impossible. Renew the timing chain tensioner for the same reason.
- Check the con-rod big ends for ovality and, if any is present, get them honed with the caps *in situ*. The way to check is to look closely at each big end cap and its mating con-rod as you remove the old shell bearings. You are specifically looking for two things. The shell manufacturer always stamps his name and part number on the rear of each shell, which should transpose pretty clearly onto the surface of the big end cap and con-rod. If it has, the big ends are unlikely to be oval. However, if the printing is not clear and sharp, look for shiny barrel-shaped polishing marks on the inner surfaces of the con-rod and cap where the back face of the shell bearing sits. These polishing marks are sure signs of hammering the big end shells in oval big ends. The ovality may be only 0.0005ins, but if you ignore the problem your new shells will start to wear prematurely and generate 'bottom-end' noises much earlier than you would expect from a recently rebuilt engine.
- For 'belt and braces' safety it's best to throw away your existing main bearing cap bolts and fit new bolts (part number BH607241), noting that these do need new spring lock washers (WL600071).
- The camshaft normally runs in the block. If there is any scoring in the cam bearings, have the block line bored and fit Triumph Spitfire cam bearings.
- Too few owners ensure that the cam locationing plate (that fits into the front of the camshaft shown in picture 7-23-22) is a really close, 'no-play' fit. A new plate sounds a good idea, but may not be the only solution since you may also have to replace the camshaft due to wear in the mating groove. The alternative is for the camshaft to be jumping about quite unacceptably.
- There are two aluminium components within the engine that need more care than the inexperienced engine builder may appreciate. Illustrated by photograph 7-3, the parts in question are at the very front and rear of the engine, respectively called the front sealing block and the rear oil seal housing. Both have tapped threads which, by the time it comes to rebuilding an engine, will be near, partly or completely pulled out; it should therefore be a matter of course to have them 'heli-coiled'. Aluminium ideally requires a coarse thread to provide maximum tensile strength but, for some reason, Triumph used UNF (fine) thread, which may be why these tapped holes have acquired a reputation for being fragile. You will increase engine reliability and longevity if the fine threads are replaced by coarse UNF heli-coils. A silly detail, but get the heli-coiler to supply the correct replacement bolts too.

Furthermore, these blocks distort or 'bell', particularly around each hole. Each face of each block requires machining (as little as possible) to flatten it again, although this will result in the block being too low for the sump to fit flush right across the front and back. So you will certainly need to fit extra gaskets. However, take a close look at your blocks before you put too much time, effort and cost into each, for you may feel replacements are prudent and more cost-effective.

7-3. The rear oil seal housing, like its opposite number at the front of the engine, is made from aluminium. Both merit careful inspection, as described in the main text.

- The rear oil seal housing has one further important but oft-omitted detail you will be glad you remembered. The top bolt only must have a copper washer under its head to prevent oil leaking from the rear oil gallery. It also requires a special bolt (part number SH605091) in just this one location, and this is shown reasonably clearly in most parts books. The all-important copper washer is not shown in any parts book but must not be omitted.

Crank thrust washer refitting

If you have just purchased a TR and intend it to be a long-term partnership, for the little time it takes you would be well advised to immediately drain the oil and drop the sump. Experts do

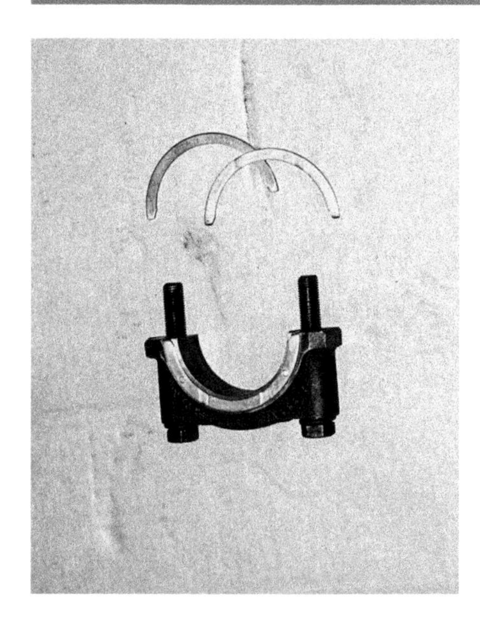

7-4. The dreaded thrust washers, although here is a really neat solution to the six-cylinder engine thrust washer wear problem, which not only allows you to double the area of the most important thrust washer, but to do so without any block machining whatsoever. This is Revington TR's solution, which will do the machining of your rear main bearing cap to allow you to pin an extra half thrust washer in place to the rear face of the cap (which takes 90 per cent of the wear). I am sure Revington TR will even do the pinning for you, too, provided you specify what thickness washer you need. One of the two brass pins is arrowed; the two extra washers shown in the picture are the 'normal' pair that will go in each side of the main bearing cap in the usual way, and are discussed in some detail in the main text.

7-5. It's easier with a DTI (dial test indicator) shown in the next picture, but this is the space available, and the gap between the timing cover and rear of the crank pulley (arrowed) is where you can measure the float in the crankshaft when checking for and monitoring thrust bearing wear.

7-6. The six-cylinder engine thrust washers pose a maintenance problem, and you are advised to regularly measure the end float of the crankshaft at the front pulley. The best method is to get a DTI (dial test indicator) to the pulley and move the pulley backwards and forwards.

it in minutes, but, even if it takes you an hour, you will get an opportunity to wash out the sump and change the crank thrust washers shown in photograph 7-4.

For the latter task, simply undo the main bearing cap about 2/3 from the front of the engine. You will find two half washers or shims above the cap, either side of the block's main bearing housing. Feed the two shims/washers round and they will drop out. It is then prudent to fit standard thickness shims/washers each side of the main bearing housing and to push the crank backward via the front pulley. Measure the distance between the front pulley and the timing cover (arrowed in photograph 7-5) with a vernier or, using a DTI on the front pulley (see photograph 7-6), zero the dial. Press the clutch to move the crank forward. Measure the gap again or, if using a DTI, read off the end float. Buy and fit the appropriate thickness shims to correct any excessive end float.

The thick washer/shim is usually fitted in place of the rear standard shim/washer, whereupon you put the cap and sump back together. White metal thrust shims are still available, but, for maximum reliability, you should only use lead indium thrust shims made by AE/Glacier. Always fit the soft (grooved) bearing faces of the thrust washers to the crankshaft and the plain steel backs of the shims (maker's name and thickness usually written on) back up to the block. Modifications can be made to extend the life and reliability of the thrust washers/shims, and we will look at several alternatives in a moment.

A sump-off routine carried out, say, every 15/20000 miles (4 to 5 years for many) will ensure that the shims rarely give trouble. Furthermore, this routine will also help to keep the clutch in the best operating order by ensuring minimal crankshaft float. We will also talk more about clutches in a short while, but I would add that it would be prudent to carry out this sump-off check maybe a shade more frequently if your car is used constantly in heavy traffic, which necessitates above-average use of the clutch.

Let us explore some thrust washer modification options:

- Some individuals drill both half shims and the top half of the rear bearing recess in the block to allow a pair of brass pins to be fitted. Obviously, this can only be achieved with the crankshaft removed from the block, and equally obviously it is important to ensure that the top face of the pin is beneath the wearing face of the thrust washer. This procedure has the advantage of ensuring that the shims can never rotate and fall out, which does sound a great idea since the consequences are likely to be expensive, as already discussed. However, as you contemplate the idea, also bear in mind that it prohibits the swift shim check and change I have just described.
- There is an option which involves machining a slot in the bearing cap to accept special thrust washers that are available with a tang (which stops them turning).
- The occasional owner considers it worthwhile to have the bottom cap recessed to take a second pair of washers/shims. These are usually pinned by countersinking the thrust washers to just below the bearing surface to take 2BA brass countersunk screws. Drill and tap the bearing cap 2BA and fix the thrust washers so that they can neither rotate nor fall out. If you use brass screws and position the heads just below the bearing material strata, you will not damage the crank if (or should I say when) the bearing face wears.
- Revington TR has, to my mind, the ideal compromise solution by using the usual pair of washers/shims but supplementing them with a third half thrust washer pinned to the rear

7-7. The later higher capacity oil pump can be identified by its alloy housing. It's always worthwhile fitting a new oil pump.

(machined) face of the rear bearing cap. A pair of small brass pins shows the pinning in photograph 7-3. The beauty of this solution is that it provides for a full circular thrust washer (double the standard area) against the action of the clutch. Furthermore, this modification can be effected with the crank *in situ* and still allows the washers to be checked and changed every few years, or as the end float dictates.

Some of these steps do seem slightly extreme just to avoid a morning's work every four or five years, but no doubt you will ask around and make up your own mind. It's a good idea to follow the best practice within the trade, which is to check the end float at every service, or as part of your annual service. If you keep a record of the end float readings on your service notes you could then watch for any progression, and almost be able to predict the wear pattern and plan accordingly. Remember whenever you hear horror stories about the six-cylinder's thrust shims that, by good servicing, you will, in fact, not have trouble, and that the shims in question are very accessible.

Whichever route you select to ensure the end float of your crank remains within tolerance, remember that one thing stays constant. The face of the washers with two vertical grooves is the bearing material, and this faces the crankshaft (the washer to the rear of the main bearing faces towards the rear of the car, while the washer in front of the bearing faces forward). They can easily be installed backwards, in which event they wear out even quicker than usual, sometimes in as little as a few thousand miles.

7-8. A welcome sight to most restorers. Note the original side-mounted oil filter is in place, no doubt to ensure there's no chance of dirt entering the engine.

7-9.

7-9 and 7-10 (above). A couple more shots of the engine showing where the ancillary parts should be fitted. Engine accessibility is outstanding with the body off. The dual vacuum diaphragms on the distributor, and the mechanical fuel pump tell us this is a US photograph, from Mark Price, in fact. Note that the bellhousing bolts all point the correct way, and the clutch slave cylinder and clutch actuating arm assembly detail.

Although not practical if you are restoring a concours car, be advised it's better to use an electric fan in place of the crankshaft-driven original fan. As daft as it sounds you can increase longevity of your new crank thrust washers by fitting an electric fan (say, a Kenlow). You can do very little about the axial thrust imposed on the crank by the clutch, but the fan also imposes a lower but much more persistent propeller effect on the crankshaft, pulling it forward against the rear thrust washer. Furthermore, the original fan absorbs power just at the time when you least need its cooling capability and least wish to lose power!

Lubrication matters

All six-cylinder engines up to about the end of 1971 used an oil pump with a cast iron body, which was superseded by the alloy-bodied unit with greater capacity shown in picture 7-7. Consequently, it makes sense to fit the later, high-capacity pump (part number 217488) to any rebuilt six-pot regardless of the originality issue.

The engine has a reputation for self-priming, so it's not essential to fill the oil pump with Vaseline. However, this does no harm and will ensure the pump primes first time. The petroleum jelly melts into the oil under heat and will be removed after 500 miles when

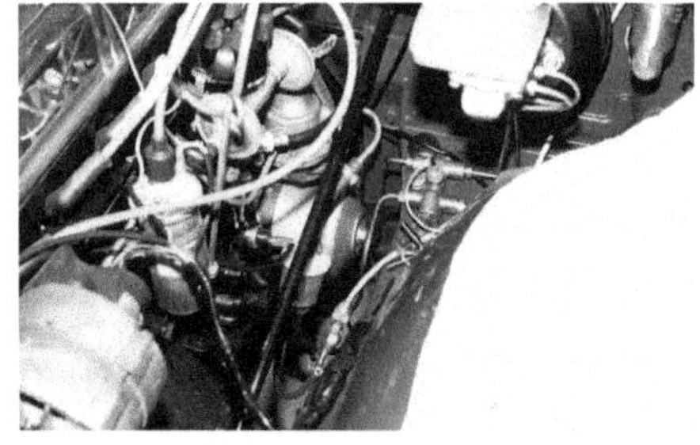

7-11. Things are definitely more difficult to get at when the body goes on. Mark Price tells me he fitted the body with the exhaust manifolds/headers in place, but one would normally remove them, or delay fitting them until the body is in place. He also tells me that, right about now, he wished he had fitted the spin-on oil filter we see in photograph 7-13!

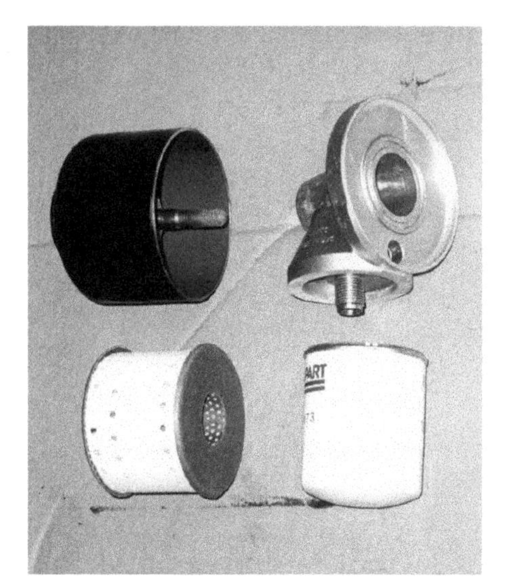

7-12. Here are the two six-cylinder oil filtration systems; the original horizontal bowl and paper element is on the left. The adapter on the right uses the same mounting bolt and O-ring, but its vertical cartridge filter is much easier, cleaner and quicker to change, and comes very highly recommended since it also gets oil to the main engine bearings far quicker on start-up. So it's no-contest, really, unless you are a concours/originality buff. I did find I needed to move my clutch slave cylinder backward slightly to enable me to get at the oil filter cartridge.

you really must replace the running-in/mineral oil with a sump full of fresh mineral oil. We will talk about oil in more detail in chapter 8, but don't forget to change the filter every time you change the engine oil.

The original six-cylinder oil filter mounts horizontally onto the side of the engine block, as shown in photographs 7-8, 9 and 10. Whilst access is superb with the body off, the original filter becomes very difficult to get at with the body in place – as photograph 7-11 shows. More than one garage, when faced with an oil and filter change, has replaced the oil but done little more than wipe a rag over the filter housing. You can often recognise an original oil filter that has not been changed as the top face of the element will be slightly pushed down into the body of the filter, sufficiently to allow the oil to by-pass the blocked filter!

Although non-standard, and therefore not, strictly speaking, a topic for this book, you really would be well advised to fit the vertical canister oil

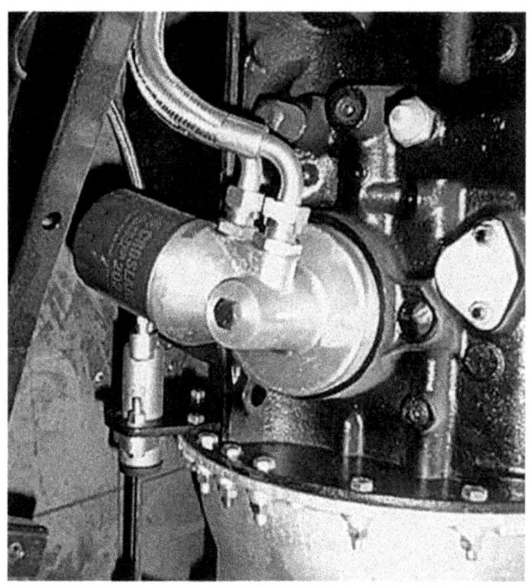

7-13. The improved 'spin-on' canister oil filter and its Mocal base casting. Note that the same fixing bolt is used as on the original side-mounted filter. You've probably guessed that the twin pipes to/from the filter body are for an oil cooler. The second purpose of this excellent TR Bitz picture is to show the proximity of the clutch slave cylinder to the filter. Note how our friends in Warrington have angled the oil filter forward to allow more space between the filter and the clutch slave cylinder to aid filter changes and bleeding of the clutch. Finally, note that the bellhousing bolts all face backwards, and the relative location of clutch slave cylinder and mounting bracket.

filter that mounts on a Mocal cast housing, as shown in photographs 7-12 and 7-13. Not only is this much easier to get to and change, it also stops the oil from draining out of the filter once the engine stops, reducing the time it takes for oil to reach the working parts of the engine when the car is started. If you fit a Mocal vertical filter mounting, an oil cooler becomes a further option. An oil cooler is not absolutely essential for the average road-going car, unless you expect to drive it very hard, very often. However, the 'spin-on' filter mounting castings, for cars with and without an oil cooler are fundamentally the same, apart from the cooler take-offs.

If you do decide upon a cooler, then you really must fit an oil thermostat also, like that shown in photograph 7-14. There is nothing more damaging to the engine than oil that is rarely allowed to reach proper operating temperature. The Mocal filter housing does have an immediate advantage

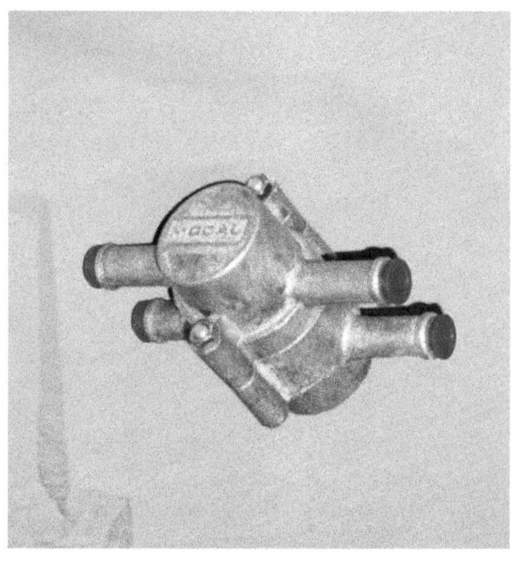

7-14. The oil thermostat recommended if an oil cooler is fitted.

in that it is now practical to pre-fill the vertically mounted filter when starting your newly rebuilt engine for the very first time, thus reducing the initial lubrication time-delay.

To further assist the engine upon initial start-up, I suggest that you liberally coat all engine components with a long-term, clinging lubricant such as STP, Wynns, Cam-Lubricant, or similar. Engine oil is, of course, better than nothing, but the first time you start a newly rebuilt engine there will be a fair (and very uncomfortable) delay before oil pressure is established, and the more viscous the lubricant used during assembly the better. Do not leave an assembled engine hanging around for week/months/years. The inside of the engine rusts while it stands and will invalidate any warranty regardless of how much clinging lubricant you use during assembly.

Prior to start-up, remove plugs, lubricate bores and crank the engine until oil is evident in the rocker shaft. Watch the oil pressure gauge and do not be tempted to start the car until you are sure the oil system has primed. If in doubt, remove the rocker shaft and crank the engine (remembering that plugs should not be fitted), and this time watch for oil from the feed to the rocker shaft at the rear of the engine.

TR engines are not usually difficult and self-prime without any problems whatsoever. If you used Vaseline in the oil pump you will certainly have little trouble, and oil pressure should rapidly build to its working pressure as soon as you replace the rocker cover and plugs.

I run an engine for only a couple of minutes the first time, letting it cool and checking oil and coolant for both level and leaks. I put the battery on charge and then, next day, give the engine 15 minutes at 2000rpm – so long as it does not seem too tight and/or overheats. Running in and oils are dealt with in detail later in the book.

Additional bottom end details

- The little ends are re-bushable and you should have the bushes replaced and reamed. For a standard engine overhaul, new pistons/rings are, of course, a must. Do not forget to warm the pistons in hot water prior to inserting the gudgeon pins.
- Apply silicone sealer, Locktite Multi-gasket or Blue Hylomar sealer, sparingly to both sides of all gaskets.
- When locating the timing chain cover, ensure you do not finally tighten the bolts until the crank pulley is in place.
- Lighten the flywheel at least to TR5 specification unless you are really into originality; the car's responsiveness will repay you. I believe balancing the 6-cylinder engine (and clutch and flywheel) to be unnecessary unless you are uprating it, or anticipating frequent use above 4500rpm.

While talking flywheels, let's consider the six-cylinder attachment bolts. The 4-cylinder TRs all call for a securing/locking tab to be fitted under the flywheel securing bolts, and, to the best of my knowledge, this has always proved entirely satisfactory, even with engines tuned for enhanced power and performance. For reasons that I have been unable to discover, but probably with cost-saving in mind, Triumph decided to delete these locking tabs when the (more powerful) 6-cylinder engines were introduced.

Now, to be fair to Triumph engineers, I have not heard of a spate of loosening flywheel securing bolts on the standard TR5 and 6s, but it is a fact that tuned 6-cylinder engines can slacken-off these bolts unless precautions are taken. Since there seems a potential weakness in this area – and in view of the work involved in getting to and rectifying a problem – I would recommend that all 6-cylinder car owners follow what has become standard practice amongst the racing fraternity. The belt and braces precautions recommended involve not only cross-drilling the bolt heads of all flywheel-securing bolts for stainless steel locking wire, but also the use of a 'Locktite' product when assembling the flywheel bolts to the crankshaft.

7-15. Apply lots of cam lubricant to the cam followers when they are first assembled.

- The cam followers (photograph 7-15) and pushrods need close inspection. The consequence of using poor quality or second-hand cam followers is that they will ruin your camshaft, so buy new, OE quality cam followers and fit them with lots of cam lubricant. The pushrods are most likely to be re-usable, but inspect for straightness and ensure that both ends have a spherical (domed) brightly polished finish.
- Use a straight edge to check that the timing sprockets line up. Shim the sprocket until they are aligned.
- Replace the timing chain and tensioner without too much thought, although the current new ones have a reputation for rapid wear. You would be wise to always fit a new duplex chain and duplex gears if your strip-down reveals and ex-2000cc single link simplex chain. The engine will lose performance if the crankshaft and camshaft are not kept in perfect synchronisation. Loss of position occurs when the timing chain slackens, particularly if the tensioner fails to fully take up the chain wear. A worn chain or tensioner can be particularly noticeable when you change from a trailing throttle or overrun to drive.
- You would not be the first TR owner to have to weld a hole worn in the timing chain cover, although if a welded cover is unacceptable to you, ex-Saloon covers are available and quite satisfactory.
- You should, of course, fit a new crank oil seal to the front (timing) cover, then loosely bolt the cover to the block. Thoroughly oil the bottom pulley collar (particularly the chamfered faces) and push it onto the crank and up to its stop. There is more on this in photograph 7-23-25.
- The choice of camshaft is not difficult for TR six-pots. Assuming we are not constrained by strict originality, for everyday road use it is difficult to beat the 150bhp camshaft used in the TR5 and early TR6s. If your car does not come into that era (about 1968 to 1972), but you wish to retain your original engine specification, the later replacement '125bhp' cams are readily available.
- Take care when buying new engine mountings. Do not use the Stag engine mountings that have a hollowed-out rear offered by some TR retailers. These allow a TR engine to sit too low, and you risk metal-to-metal contact/vibration with associated, quite unacceptable, noise. Insist upon and use aftermarket/OE ones with a solid cross-section. These cost about £6 each and do a good job. Notice, too, that TR engine mounting brackets (part number 145385) are slotted. Changing fan belts is infinitely easier if the engine and gearbox mountings are moved as far back as possible – which means all four rubber mountings need to be positioned at the front of their respective slots.

THE TOP END

Removing the head

Triumph cylinder heads often seem reluctant to part from the engine block. In spite of the temptation, do not try levering the head from its block by putting anything between block and head, as you will damage either or both mating surfaces. Do not try and 'shock' the head free, either. Lots of patience and large quantities of penetrating oil should be your first approach, so don't leave starting the head removal job until a day or so before it is scheduled for gas flowing, skimming, etc.

So exactly how do you get the darn thing off? Wind all the nuts clear of the head and build a small circular 'dam' around each stud using plasticine, blue tack or similar soft, mouldable material. Resign yourself to a week of pouring penetrating oil around each stud at least twice per day, more often if you have the time and if the studs continue to soak up the oil. After about a week, wind a second nut down onto the original one we thoughtfully left *in situ* and lock the two nuts together. You need two good spanners (no open-ended wimps here, please) to achieve a solid lock, which can only really be achieved by unscrewing the bottom nut hard up to the top nut that you have held static. Remove the top spanner and try to unscrew the bottom nut and the stud. Most should come out with some hard work, but, if not, it will be another week of patience and penetrating oil before you try again.

There is a more aggressive method which I think I would only resort to if I was in trouble with the first approach. With everything (e.g. manifolds, water pump, etc.) off the engine, the spark plugs out, the gearbox in neutral and a box-spanner/tommy bar on the front crank pulley, turn the crankshaft over until one of the two central cylinders is at the bottom of its stroke. Poke as much thin (nylon?) rope through the selected plug hole as you can, and continue turning the crankshaft until the rope compresses and unsticks the head.

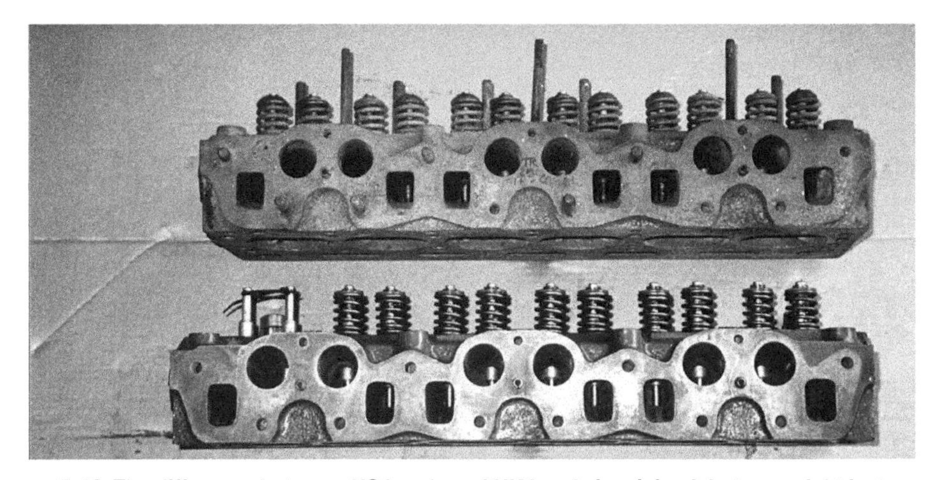

7-16. The difference between US heads and UK heads is minimal, but you might just spot that the (top) US head's circular inlet ports are a shade closer together than the UK head (below). Can you spot the roll pin inserted into the UK head by Triumph to identify its PI association? It is aligned with the bottom of the UK inlet ports, dead centre and, of course, not fitted to the US heads since PI cars did not go to America.

USA cylinder heads

Ex-USA heads have the very considerable advantage of better cooling properties, compared to a UK head. However, US heads were originally machined to give a much lower compression ratio and, if you live in the UK, do need to have their compression ratio increased. This necessitates machining, but you need to first check on the porting of the US head.

Many of the US cars (all the TR250s and the early TR6s) had slightly different cylinder head porting to standard UK heads. Since virtually all TR tuning equipment is made for the 'English' head, it's best to swop cylinder heads before starting to spend money on the special US one.

To decide which cylinder heads to reject or select, we need to establish and reject the 'narrow' inlet-ported heads. Picture 7-16 shows the alternatives, but the differences between the three pairs of circular inlet ports cast into the side of each head are difficult to see. There is a gap between each pair of ports which, if you measure from the inside of one port to the inside of its other half, will be either 9/16in (0.563in, 14mm) or 7/8in (0.875in, 22mm).

Select the wider ported head and by all means machine it to increase compression ratio. Prior to machining a low compression wide port head, it's important that an expert first checks the head to ensure it has not already been planed. This check is done by closing the spark plug hole with a bolt, and measuring with a pipette the cubic capacity of the combustion chamber. It's then possible to calculate how much metal to remove to give the compression ratio required. Most aim for about 9.5:1, and the majority of US heads need about 80 thousands of an inch machining from their mating face.

If you are not sure how to identify a low compression head, you can do so with ease even with the head fixed to the block. There is a lug sticking out on the left/distributor side of the engine. This is about 1/8in high on a high compression head but increases to perhaps 1/4in on the thicker, low compression units. It is even easier when the head is off the car as the external lug thickness can be checked, and also whether the combustion chamber has a lip to it. A high compression head will flow straight from the mating face to the concave of the chamber, whereas a low compression head will have a lip or boundary 'wall' about 1/8in high. In the UK, a late US head that has passed the above checks will also need a hole plugging where the ERG valve will have been removed.

Regardless of where the head came from, the head studs can be – indeed, often are – assembled into the block upside down. Head studs have one long and one short threaded end, and it is the shorter threaded end that must go in the block. If you reverse the studs you are, at best, unlikely to be able to properly torque the head, particularly a high compression head, and, at worst, run the risk of cracking the block when you try to torque the head. As extra identification, the bottom of the stud has a plain flat face which should be screwed into the block, whereas the top invariably has a small pip that needs to point upwards upon assembly. Note, too, that some cars – particularly later US ones which had air-conditioning and/or vacuum pumps attached to the head studs – have some longer head studs. These can also lead to head torqueing difficulties, particularly when a cylinder head that has been reduced in thickness in order to raise the compression ratio is fitted. If in any doubt, replace the studs with the short English versions where appropriate.

Be alert to the potential consequences of fitting overly long pushrods and then turning over the engine; you risk snapping a rocker arm or two. This can even occur when you refit the same pushrods that came from the engine, if you have made alterations to the cylinder head. The pushrods are different for each major difference in compression ratio. There were three lengths of pushrod,

7-17. Plain block – used on CP and CC six-cylinder engines up to about autumn/fall 1972. The top of the block is absolutely flat, as expected.

7-18. Recessed block – used on CR and CF engines from late 1972, has a machined recess around the top of each bore, as shown here. The depth of the recess was only about 0.015ins (roughly 0.4mm) but, nevertheless, it is essential to use the correct 'recessed bore' gasket with a small rear identification tag.

but those of primary interest are the longest (US) rods and the shortest (UK) ones.

If you increase the compression ratio by using, say, a low compression ex-USA head reduced in thickness, you should replace the pushrods with the UK type. You will get a clear indication of whether your pushrods are over length when you fit the rocker gear and try to adjust the tappets. If you seem to virtually run out of thread on the rocker adjusting screw, you probably need the shorter/earlier screw and/or shorter pushrods. You can buy spacers that fit beneath the rocker shaft pillars, raising the whole rocker assembly to compensate, but I am sure you agree that it makes more sense to get the job right and fit the correct pushrods for the compression ratio of the head.

Incidentally, if you're unsure whether or not the pushrod or adjusting screw length is correct for your car as you assemble the top end, stop. Do not try turning over the engine or running it, for you could, as I said earlier, break a rocker arm or strip a half-in adjusting screw. Check with an expert first and fit the correct parts.

Don't increase compression ratio beyond about 9.5:1. Although a petrol injected engine is very unlikely to 'run-on', running-on (or 'dieseling') problems with unleaded fuel (95 RON) await if you are running carburettors and increase the CR too high. It's possible to relieve some running-on problems by using an anti-run-on dump-valve, and we will explore this in chapter 8. Running-on can be due to a hot spot in the head, and this could be one instance where you might be glad it's the engine-builder's responsibility to rectify it.

When refurbishing the top end of a TR engine, take the opportunity to have the head converted to run on lead-free fuel. There's more detail on lead-free fuel and the work involved in a lead-free cylinder head conversion elsewhere, so here I'll confine myself to saying that an engine rebuild or head overhaul definitely signals an opportunity to have the conversion done. It is a specialised job, as the original exhaust valve seats have to be machined out to make way for hardened inserts. Until fairly recently, these used to be fitted frozen in order to shrink them to a size that can be pressed into the head and stay put even at the elevated temperatures within a cylinder head. However, technology has improved and the task is now much simpler, thanks to modern sintered materials.

New valve guides must also be fitted during the head reconditioning work in order to allow for accurate valve seat machining. Do not try hand-cutting valve seats that have been made from modern materials; you will hardly scratch the surface! You can get phosphor bronze inner sleeves that fit inside the original guides but, although more expensive, I would recommend fully replacement phosphor bronze guides. The phosphor bronze material is needed to more quickly dissipate heat from the valves and lessen valve stem friction. Virtually all of the operations

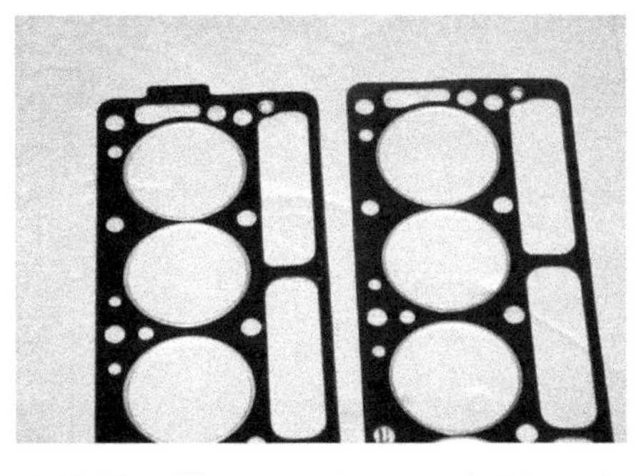

7-19. The different gaskets used with each of the aforementioned cylinder blocks. Easily seen is the extra tag at the rear of the gasket for the recessed blocks of post-1973 CR and CF engines.

associated with reconditioning and modernising the cylinder require specialised equipment not found at home, so are sub-contract tasks.

Just prior to assembling the cylinder head to the block, take precautions to prevent a couple of irritating, rather than serious, six-pot weaknesses. One peculiarity of this engine is its tendency to weep oil mist from between the cylinder head and the outside of the block above the distributor. At about 0.2/0.3in, the block land looks thick enough to properly seal the cylinder head/gasket/cylinder block joint. However, this seal seems reluctant to hold and a mist can escape. This turns to oil which, in turn, runs down the outside of the block behind and adjacent to the distributor. It's worse if you fail to extract the crankcase but can, in any event, be avoided if you run a thin bead of silicon sealer along the land on the outside of the pushrod chamber before fitting the cylinder head gasket.

As with any car, it's vitally important you fit the correct cylinder head gasket for your cylinder block; that is to say, the cylinder block you actually have in your car or on the garage floor, as the case may be. It is immaterial what the book says your car should have by way of a cylinder block. The early six-cylinder engines had what most of us think of as a normal flat face for the head to sit upon, as shown in photograph 7-17. Whilst it might work for some, you certainly should not buy cylinder head gaskets on the basis of the year or model of your car, but on what you see with your own eyes, because later 6-pots have recesses cut into the top of the block around each cylinder! These can be seen in photograph 7-18. So, if you have a recessed block you must fit the appropriate 'recessed block' gasket, which has a small extra tag sticking out of the rear of the gasket. The comparison can be seen in photograph 7-19 but the extra tag will even poke out of the back of the engine when all is assembled. If you fit the wrong gasket to a recessed block you will be able to start the engine, although it is unlikely to run for more than 20 minutes before it starts to blow water everywhere. If, on the other hand, you have the earlier 'flat block' and mistakenly fit a gasket designed for the recessed block, you

7-20. It will probably be rather more cost-effective to buy these ready-assembled rocker shafts than buying the parts individually, and assembling the thing yourself. Rocker arms are cast steel, however.

can look forward to a month or two's service before the engine also starts blowing out water.

Always replace the rocker shaft with a new one and check the rockers on the new shaft. Any play in a rocker on the new shaft means a replacement rocker is also required. If six or more rockers are worn, I would replace the whole set without wasting any more time. You can actually buy new rocker shaft assemblies as shown in photograph 7-20.

A frequently advertised, non-original modification is shown in photograph 7-23-35. This feeds additional oil from the oil galley adjacent to the distributor, up via an external pipe to a pre-tapped but normally plugged hole in the cylinder head. Like many things to do with motor cars, it is very much a matter of opinion whether or not this modification should be incorporated within your engine rebuild. The oil consumption of many engines can be increased, and the rocker shaft is relatively easily, quickly and cheaply changed. However, the crankshaft and main bearings are a shade more difficult and costly to access. Personally, I would have all available oil lubricating and cooling the crank. In any event, use lots of engine oil when re-assembling the rocker set; and engine oil mixed with STP is even better.

Another peculiarity of the engine is a tendency to shed the small, slightly tapered plug (part number 137811) that goes into the end of the rocker shaft. It rarely does any immediate damage, since it usually drops down an oil drain hole and sits in the sump. However, in time, the oil pressure in the rocker shaft drops, lubrication to the rockers (particularly the front ones) is reduced and rocker wear increases. Eventually you will get rust and squeaking noises in the front rocker arms and accelerated wear in all of them. Always fit a new plug to the end of your new rocker shaft, fit it carefully and ensure it is still in place every time you service your car.

Other top end detail

- Plugs – recommend NGK BP6ES gapped to 0.030in.
- Timing – this is sometimes incorrectly stated, but the firing order is definitely 1-5-3-6-2-4
- Manifolds – an extractor manifold and exhaust system is usually required to complement any head tuning in order to maximise the benefits. I recommend a stainless system, even in a car that is to be used relatively infrequently. Mild steel systems do cost less but rust from the inside, even when the car is idle.
- The next topic really only applies to repatriated TR250/TR6 US models. You could find your Stromberg carburettors are mounted on the very 'square' inlet manifold shown in photograph 7-21. If so, ideally you need to change the manifold to a later, more swept, one as shown in chapter 11. However, the later swept manifold will only fit the wider/later ported heads. The different porting was shown in photograph 7-16, and if your head is a TR250 or an early (about pre-1969) TR6, the porting will originally have been the slightly narrower type – which means you'll have to stick with the 'squarer' inlet manifold.

Stromberg carburettors are not as inefficient as their reputation suggests, but they are significantly more difficult to adjust and are best replaced by a pair of 1.75ins SUs. The original air filter needs changing at the same time.

We will look at carburettors in

7-21. The earlier and less efficient 'square' inlet manifold. If your cylinder head is a wide port UK version, you're better off fitting the later, more 'swept' inlet manifold. If you have the narrower inlet ports described in the main text, you will have to retain this shape of inlet manifold. It's worthwhile checking that the width of the inlet ports on the manifold are the same as your cylinder head, since parts have been so interchanged over the years it's not safe to take anything for granted.

some detail in chapter 11. However, since we have left very little of your US induction system, we had better find a complete system that will (largely) just bolt-on, but nevertheless make your TR250/6 a bit more interesting than when in US trim.

The induction system from a Triumph 2500TC uses a pair of 1 3/4in SU carburettors already fitted to the more effective 'swept' manifold, and it bolts straight on to the TR six-cylinder (UK head ports). This change will liven up performance slightly and improve it still further if you also fit a pair of SU 'BAE' size needles and K&N air filters. This arrangement is described and pictured in chapter 11. You can make this change for a cost of around £100. You can go to triple Webers to further enhance performance, and this will be a topic explored in some detail in a future book.

THE CLUTCH

Not to put too fine a point on it, the clutch on larger TRs has proved a nightmare for many a TR250, 5 or TR6 owner. Early TR2s to TR4s use what is a now an old-fashioned 'spring' type clutch, which, nevertheless, gives very little trouble. It has a good and consistent life and pedal pressure remains constant throughout.

The six-cylinder car's clutch is not such a happy story, and is the Achilles heel of the six-cylinder cars. The whole system seems to have been designed with minimal tolerances, and even with the best of cover plates, once some wear occurs, you will experience difficulties.

The clutch seems most susceptible to pedal stiffness, roughness, and/or clutch drag, making gear (particularly first) engagement difficult. The make of clutch generally (and the cover plate in particular) is relevant and also important is the security of the actuating fork on its shaft. Clutch drag is caused by the pressure plate not separating enough from the driven plate when the clutch pedal is pressed, and there are several possible external reasons for this which we should eliminate before jumping to the (probably correct) conclusion that it caused by a malfunction within the bellhousing. Before running through our checklist, it's helpful to know a little TR5/6 history, if only to understand why a change of master cylinder bore size could be the solution.

The primary reason for this unenviable clutch reputation is that, originally, the cars were fitted with a 0.75in bore master cylinder, which worked fine. However, some customers complained that the clutch had a very heavy pedal, which restricted the use of the car. Triumph consequently fitted a 0.7in bore master cylinder from mid-1970 which certainly made the pedal lighter, but also completely removed the safety margin from the slave cylinder's travel. So later lighter clutch pedals had to have the absolute maximum stroke to achieve adequate slave cylinder travel to fully disengage the clutch. Any wear in the clevis pins and/or in other mechanical parts and/or any carpet/felt under the clutch pedal now restricted slave cylinder travel sufficiently to cause a problem. The 0.75in bore master cylinder is slightly (about 7 per cent) heavier and, from the point of view of drivability, the smaller cylinder is better, but necessitates all other clutch parts being kept in pristine condition.

If you have been driving a car with a worn/heavy clutch, check every link in the clutch chain, noting that it needs only one worn fork and/or clevis pin, particularly with a 0.7in master cylinder, for your slave cylinder to have insufficient travel to allow comfortable selection of first or reverse gears, even after you have fitted a new clutch assembly! As part of your check of every detail in the clutch chain, you may even find it necessary to drop the pedal box to replace the nylon pedal bushes with later phosphor bronze bushes. If you are carrying out a full restoration, this is a very worthwhile upgrade.

How can you tell whether you have a 0.75in or 0.7in master cylinder? The size is cast into the side of the reservoir, along with the name 'Girling', which reminds me to recommend you use only the genuine Girling product, however tempting, cost-wise, the alternatives might be.

Here is the checklist you should go through before seriously contemplating a gearbox-out/clutch job:

- Check for excessive crankshaft end float, which will dilute forward movement of your clutch mechanism.
- The slave cylinder pushrod needs to move at least 0.625in (16mm) and be connected to the middle hole on the TR's external lever.
- If your pushrod definitely and consistently travels that distance, you can almost certainly eliminate the master cylinder as a potential problem. However, I would check that there is no significant slop in the slave cylinder clevis pin hole, wear in the clevis pin and/or cross shaft bearings, as any/all

dilute the transfer of linear movement into the shaft's angular movement.

- Check that the slave cylinder is installed the correct side of the mounting plate, and also the right way up. Photographs 7-10 and 7-13 show the correct relationship between engine bracket and slave cylinder, but fundamentally, the flange on the cylinder should be to the rear (i.e. the differential side) of the bracket. The bleed screw should be at the top. Incidentally, the possible consequence of mounting the cylinder the front side of the bracket is that, once the clutch mechanism wears a little, the pushrod becomes so extended that the piston is forced out of the end of the slave cylinder. The result will be absolutely no clutch whatsoever at, no doubt, the worst possible moment.
- Still working externally, on the basis that you have got full slave cylinder movement and none of the problems outlined above, you would do well to ensure that the gearbox mounting holes are not worn and that the two (3/8in diameter) locating bolts have not been forgotten when last securing the bellhousing to engine. They should have been the first bolts inserted the last time the gearbox was married to your engine. A problem with either of these (not uncommon) fixing details can cause engine/gearbox (and therefore clutch) misalignment. It is worth mentioning at this point that Stag owners have also experienced alignment difficulties with a virtually identical gearbox (with consequential clutch problems). The Stag Owners Club (Headcorn, Kent, England) has designed a floating clutch thrust bearing and special carrier. It removes the need to precision align the bellhousing with the engine and, in view of the similarities between the two gearboxes, could be one route you should explore if you are having problems with your six-cylinder TR. I'm not aware of a TR using a Stag clutch thrust carrier and bearing, but it has to be worth considering.
- On the other hand, if you're not getting the full 0.625in slave pushrod travel we need to look upstream for the cause. Excessive wear in either/both master cylinder fork or its clevis pin must be your first check, and if that proves free of play/wear then ensure the master cylinder is operating satisfactorily and, for that matter, that the slave cylinder is okay. If in any doubt, fit new seals, or replace one or both, as that elusive 0.625in of slave pushrod is an absolute essential if you are ever going to give your TR's clutch a chance to perform.

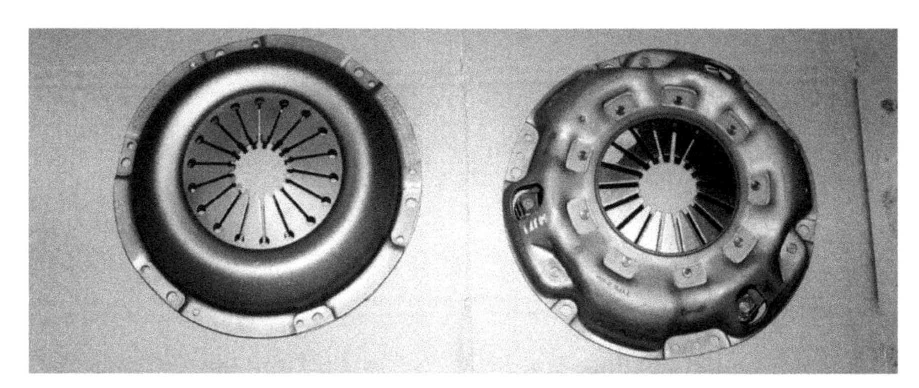

7-22. Two of the preferred TR six-cylinder clutches. The left one is a rebuilt Laycock clutch cover (note the plain cover), only available from Revington TR, TR Bitz and TRGB, who rebuild the original Laycock units and sell them strictly on an exchange basis. The cover on the right is a Borg and Beck unit with its slightly different but quite serviceable diaphragm fingers.

- I helped my situation by fitting a screwed (and therefore adjustable) pushrod to the clutch master cylinder shown in chapter 18. This helps locate the pedal where you can achieve full pedal travel, but will in no way improve the stroke of the slave cylinder. If you have a 0.7in bore master cylinder but elect to fit the earlier, larger bored master cylinder to maximise slave cylinder travel, you will need a small adapter (readily available from all TR specialists) to marry the early master cylinder to your later hydraulic pipes. This is also photographed in chapter 18.

If all of the foregoing are satisfactory, and the magic 0.625in slave movement is regularly achieved but the clutch is still heavy or dragging, and/or you are having to force the car into gear, it will be necessary to remove the gearbox in order to check and correct an internal problem.

I don't intend to go through a general/comprehensive list of what one might term 'usual' clutch problems, but if you elect to fit a new clutch to a 6-cylinder TR, you need to do so with greater than usual care. There are some especial TR details we should explore, if only to ensure you avoid the most common mistakes when reassembling your clutch – and hopefully, shortly thereafter – the gearbox to engine.

You will, of course, be fitting a new clutch assembly as an automatic step when the time comes for reassembly. No? Then a few words of explanation are called for. All of the clutches offered by various manufactures are the diaphragm type, but the OE clutch was made by Laycock and is still the only one that seems to do its job consistently well and last for a while. It is, unfortunately, no longer available new, but Revington TR, TR Bitz and TRGB do a service exchange Laycock cover plate strictly on an exchange basis. Not all aftermarket clutches work well, even when new, and you should avoid all but Borg and Beck or AP/Lockheed aftermarket clutch assemblies.

The next preferred alternative to Laycock is a blue spot Borg and Beck cover, and you will find the two cover plates shown in photograph 7-22. You may find that, if your car has already been fitted with a Borg and Beck unit at an earlier date, the clutch you remove has a yellow identification spot. This would have been precisely the correct TR clutch for its day. Since then, B+B has carried out some product rationalisation and the six-cylinder TR clutch has been made common with a Ford Transit with, you've guessed it, a blue identification spot. The 'blue' clutch is a fraction heavier than the original unit, but of no material difference. However, there was a period when 'green' B+B clutches were erroneously thought suitable for all cars from the TR4A onwards. You could, therefore, remove a green spotted unit from your car, but will be pleased to hear that you should be better off with the 'blue spot' cover and its lighter diaphragm pressure.

AP units can, in fact, work well when new, but it would seem that their tolerances are so close to the TR's needs that, after some use, the cover plates drift outside the TR's usable separation band and start to give trouble. Another consideration is that, perhaps because of their multi-purpose use, all aftermarket alternatives are noticeably stiffer than the OE Laycock unit. Provided you have a good left leg, that does not sound too significant does it? But consider what happens when you press on a very firm clutch. You are, of course, pushing equally hard on the six-cylinder's notorious crankshaft thrust washer, which we all know wears quickly enough already and generates excessive crankshaft end float.

Furthermore an overly strong clutch plate is probably also directly related to another, not infrequent, problem in that the pin which holds the clutch fork onto its gearbox cross shaft seems more susceptible to breakage when subjected to the extra pressure a very strong clutch pressure plate puts upon it. I have heard that pin breakages account for as many as 60 per cent of six-pot clutch problems. The pin is hardened and tapered and, for some reason, very rarely breaks off cleanly. More often than not it breaks at 45 degrees. The problem with an angled break is that it allows the folk to move on its shaft, but only by about 15 degrees. As strange as this may sound, this movement is rarely enough to allow an absolutely positive diagnosis without stripping the car and testing the pin. Indeed, more than one TR has been stripped, a new clutch fitted and the car put back together with the prime reason for the car's problems uncorrected! The slight movement of the fork on its shaft allows the clutch to partially operate, but it will probably feel very 'heavy'. On strip-down, the fork will appear secure on its shaft, which is why numerous TR gearboxes have been re-married to the engine with a new clutch in place but no worthwhile improvement in operation.

In fact, the pin is so difficult to get at and the proper orientation of the fork/shaft so important that it warrants at the very least the 'dual-pipe test' whenever the gearbox is out. The cost of the pin is such that it's worth replacing as a matter of routine every time you take a peak inside your car's bellhousing. You'll be wondering what on earth the dual pipe test is. Take two pieces of pipe or tube, each about 12in (300mm) long. Fit one over one leg of the fork and one over the shaft's external actuation lever, and test one pipe against the other to see if the fork really is securely pinned to its shaft. Even 2 or 3mm of movement is too much and necessitates a new pin and a re-test.

We will need to address how to remove a broken pin shortly, but the good news is that, apparently, the B+B and AP aftermarket clutch covers are improving. Nevertheless, if you can get hold of one, it is worthwhile paying the extra cost (about double) for an exchange Laycock unit.

Overall, the six-cylinder TR clutches are bad news, but there's worse to come ... Even the best of these clutches do not enjoy a long life by any standard; TR specialists are not surprised when a replacement clutch is required after 5-10,000 miles. They are surprised, however, to hear of an owner who has enjoyed 50,000 miles from his clutch, for this is pretty rare these days. It's absolutely essential you assemble the clutch to give the best possible service: the following highlights the specific pitfalls that await the inexperienced TR six-cylinder clutch assembler -

- Always fit a new clutch cover every time you separate gearbox from engine. Try to use a Laycock unit, but use a B+B set and an AP assembly as second and third choices.
- Aftermarket covers are made from thinner material than are OE clutches, so the original bolts that fix the cover to the flywheel can be too long when an aftermarket cover is fitted. Make absolutely sure your securing bolts do not prematurely bottom-out inside the flywheel before the cover is totally secure. To properly affix an aftermarket cover, the bolts must be no more than 3/4ins (19mm) long, measured under the head.
- I made reference a little earlier to checking that the cross shaft bearings in the bellhousing are not worn. If there is any axial play, now is the time to replace the bushes and, after my earlier diatribe, forgive me for reminding you to securely fix the fork on the shaft with a new pin and carry out a 'dual pipe test'. My pin came with tiny 'wire' holes in the head to allow stainless security locking wire to be fitted. You can find details in chapter 9 as to how to increase the security of the fork on its cross-shaft.
- The thrust bearing that opens the clutch sits on a carrier, which, in turn, slides up and down a front cover/oil seal. The carrier must be a nice smooth sliding fit on the front cover. There must be no tight spots, nor is excessive play acceptable, and the two sliding faces must have a very high smooth surface finish completely free of all marks and scoring. The carrier groove must not be worn, damaged or otherwise misshaped.
- Replace the crankshaft rear spigot bush, but check the new one on the front of the gearbox input shaft first. You can get the old one out by filling the space behind the old bearing with grease, and, using a piece of rod as a piston through the bore of the bush, hydraulically eject the bush.
- It is prudent to try the new driven (internally splined) plate on the gearbox first motion (splined) shaft. Position it the correct way round (with the longer splined boss facing the gearbox) and ensure the plate moves up and down the first motion shaft smoothly. Assemble it under the cover plate on the flywheel the correct way round and centralise using an old shaft or a clutch alignment tool (Moss part number MM387-220).
- Remember that, on one hand, grease is the kiss-of-death to a clutch, whilst, on the other, careful application of a little copper-slip or light molybdenum grease to the moving mechanical parts is necessary before the gearbox re-marries the engine.
- Right at the beginning of this section I mentioned the importance of the two 3/8ins bolts being the first to align the bellhousing to engine. That still holds good at this point, particularly after all your hard work and expense. Do not forget that the slave cylinder pushrod goes to the central hole in the cross shaft drop crank. Before finally bolting the engine to the gearbox it is probably worth the few minutes it takes to re-check that the vital movement of the pushrod is still at least 5/8ins, and that the angle of the external lever is (ever-so-slightly) forward of vertical before the pedal is pressed. By the way, the bellhousing bolts should face

backwards – nuts to the rear of the car.

- Having just completed a clutch change you are unlikely to be thinking about the next time you have to complete the task. However, you will make life a bit easier for the next clutch change if you fit the earthing strap in its correct place. Too often the strap is fixed via a bellhousing bolt, thus adding to the clutch change task. The correct and better place for the earth strap is to the bottom bolt on the engine lifting eye.

ASSEMBLING & FITTING THE ENGINE

The engine can easily be rebuilt at home, but Faversham Restorations (where the accompanying photograph sequence was taken), along with all our TR restoration specialists, is very experienced at rebuilding the Triumph 4 and 6-cylinder engines, and/or converting them to unleaded operation. You may consider it £1200/£1400 well spent to have that knowledge used on your engine, if only because responsibility and warranty for the engine rests with one reputable supplier. Cost out the parts and machining for a home rebuild; it's quite likely you won't be able to shave a significant sum from these figures for a full rebuild.

Since a picture speaks a thousand words, I have used a pictorial sequence with associated text starting at 7-23-1 to explain the fundamentals of assembly for those still anxious to do the work themselves, and have tried to cover the basics of fitting the engine in a photographic sequence stating at 7-24-1.

7-23-1. The basis of an engine rebuild must be regrinding the crankshaft. Here we see a TR6 main bearing being taken down to -0.020in or 20 thou undersize. The grinding subcontractor should supply new main and big end bearings to suit the new journal sizes as part of the deal. Opinions vary as to the necessity of subsequently having the crank Tuffrided (hardened and polished) and/or balanced. My preference would be to have the newly-ground journals – which will have lost most, if not all, of their hard surface during the grinding operation – Tuffrided to ensure longevity. However, a 6-cylinder engine crankshaft should not need to be balanced, although I have no doubt it will run that shade smoother if you do have it done. You will, of course, need to have the front pulley, flywheel and clutch balanced, too, to take full advantage of your crank expenditure, so the cost does not end at the crank balancing operation.

7-23-2. The second basis of a thorough engine overhaul is re-boring the cylinders. Here we see a six-pot crankcase (actually from a TR5) having number 3 cylinder re-bored. The equipment is very portable and accurate to 0.0005in. It clamps at the bottom of an adjacent cylinder bore. Again, the machining subcontractor should supply a set of new pistons and rings, to the appropriate oversize diameter, as part of the deal.

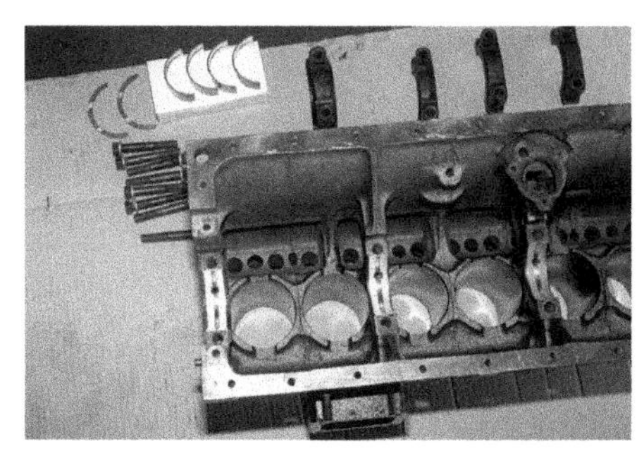

7-23-3. The first order of assembly will be to marry the crank to its refurbished cylinder block. However, a word about the importance of cleanliness first. Note the substantial area of fresh corrugated board on the floor, and that all components have been thoroughly washed, cleaned, and blown out well before being allowed anywhere near the assembly area. The photo shows that everything for the first step has been laid out in an orderly manner, and the top half main bearings have been carefully slotted home. You can see that the oil feed holes are clearly aligned with those in the block.

7-23-4. After liberally oiling the new half main bearings, the crank is offered to the block – a very satisfying moment! We are looking at the front of the engine.

7-23-5. Now it is time for the other half of the main bearings to be carefully slotted into the main bearing caps. There is no particular half bearing to marry to any particular cap, but you obviously must ensure each 'tang' enters its slot in each cap. Lots more clean engine oil (smooth the oil all over each bearing surface) on the half bearings before you drop each cap in place and loosely bolt it. You MUST use new spring washers under the bolt heads, and don't forget the half thrusts each side of the rear main cap. There's lots more information about these in the main text, but they should come pre-ground by the machine shop to give the crank the correct end float.

7-23-6. Lightly pull down the main bearing bolts and give the crank a spin; it should turn over with very little effort at this stage. If it does, get the torque wrench out as shown here. Somewhat to my surprise, the workshop manuals for long-back and short-back engines appear to call for different mains cap torques, but obviously it is important you use what is specified for your particular engine. I think it a good idea to spin the crank after you have torqued each cap – the engine should still turn over smoothly and fairly easily each time; investigate any tightness.

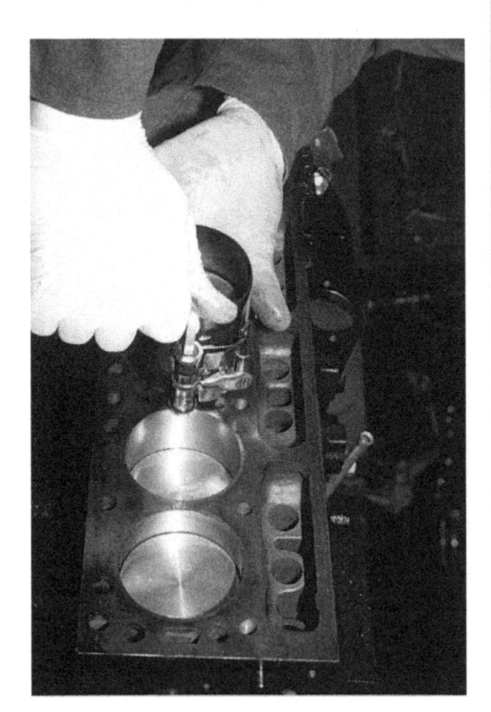

7-23-9. Apply the ring clamp after covering the top of the piston with lots of clean engine oil ...

7-23-7. The two alloy 'blocks' should be inserted next. The rear one contains the rear oil seal and, of course, goes in over the rear 'back', long or short. Ensure that the lip of the oil seal points inwards against the way the oil will try to escape, and that the lip is well oiled before assembly over the crank is attempted. The front one, pictured here, is interesting in two respects – firstly, it is no more than a bridge over the front main bearing cap, and, secondly, a piece of wood is driven into a recess at each end once it is securely in place. Use silicone sealant round the three sides of the rear alloy which come into contact with the crankcase, but none on the front one.

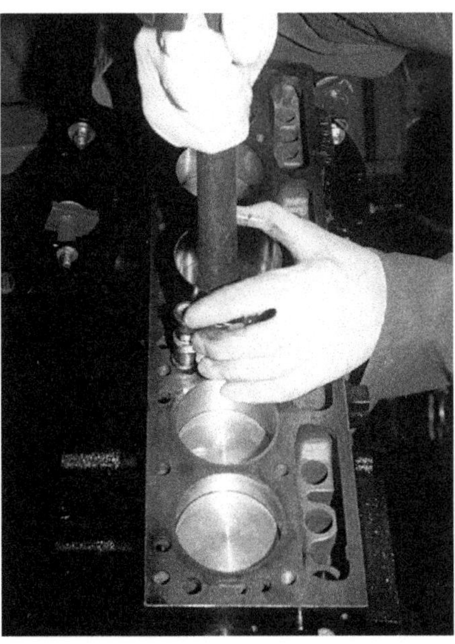

7-23-10. ... and tap each piston home ...

7-23-12. Tighten the big end and ensure that the crank revolves. Finally, torque the big end and check that the crank is not too tight, before moving on to the next piston/conrod assembly.

7-23-8. Pistons next ... or at least once you have them on the con-rods and put the rings in place. Follow the workshop manual when carrying out these operations, but it really is a good idea to try your first piston/con-rod (before the rings go on) in the engine. Do everything carefully, of course, but if you turn over the crank with this first assembly loosely in place you will find out before it is too late whether you have the con-rods and pistons assembled the right way round. If the crank turns over, you have the right con-rod-to-piston relationship; if the big ends just foul the crankcase, it is the wrong way round. Obviously, you then need to marry up all the pistons, con-rods and rings and then, I suggest, one-by-one put the big end shell in a con-rod and offer the piston assembly in its bore as per this picture.

7-23-11. ... guiding the big end down onto the crank (centre of picture).

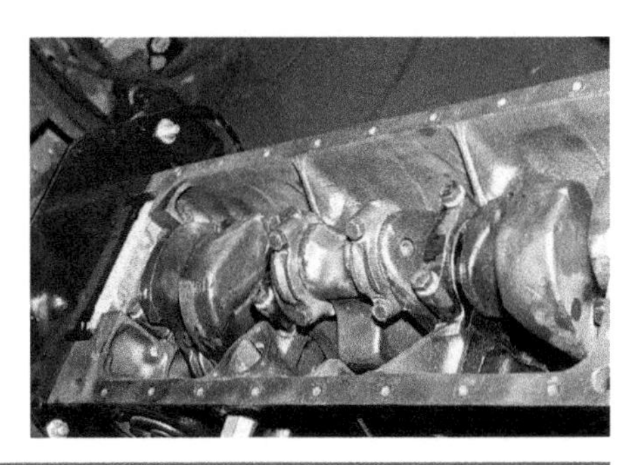

7-23-13. Now for the oil pump. Prepare the face with a little Hylomar; fit the pump.

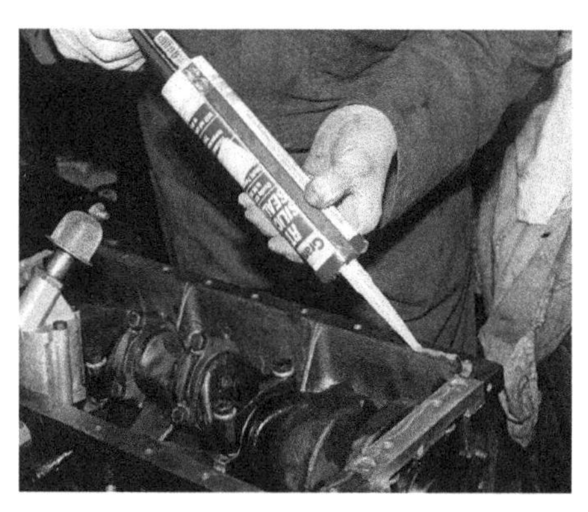

7-23-14. As you can see, the oil pump really was in place before fitting the sump was focused upon. A bead of silicone sealer is wise, provided you do not apply so much as to risk surplus bubbling into the inside of the sump (from where it can eventually get into and block the oilways). Fit the sump gasket, using another small bead of silicone sealer.

7-23-15. Then position the sump. Do not forget to use new spring washers on the sump bolts, and that the rear 4 bolts (where the sump is reinforced with a second strip of steel) are longer than the majority of bolts. When they are all in place, torque the sump bolts.

7-23-16. Make a start on the front by finding the thinner of the two gaskets that come in most gasket sets. It is never a bad idea to check that the holes line up.

7-23-17. Fit the front plate to the block after applying a little silicone sealer. No gasket required just yet.

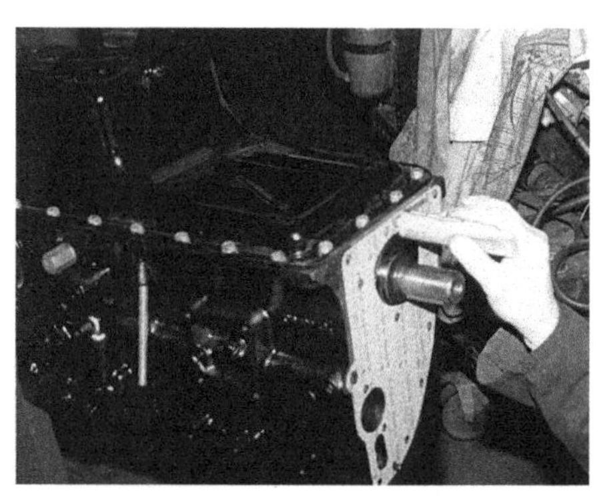

7-23-18. Now the front gasket goes on with a little Hylomar ready to seal the timing cover that will go on in due course.

7-23-19. Put the front oil seal in the timing cover, and check that the timing chain tensioner is not unacceptably worn. Current replacement tensioners are prone to wear very quickly, so if the original one is in good shape, use it. Temporarily put the cover on the engine.

7-23-20. Place the crank so that number one piston is at Top Dead Centre (use a DTI if necessary, although that is somewhat over-cautious) and offer up the front pulley temporarily. Mark the 'TDC' line. Remove the pulley and timing cover.

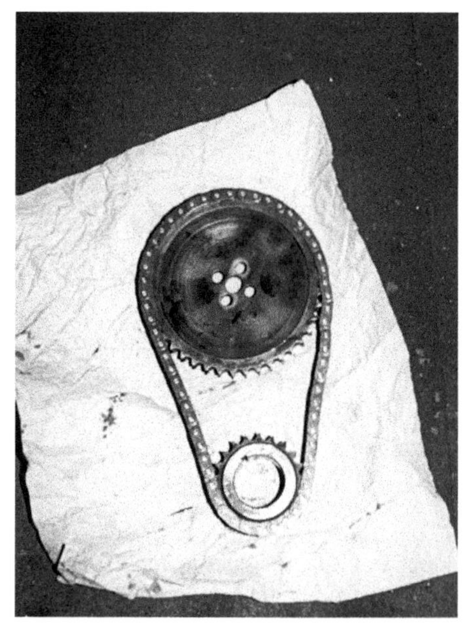

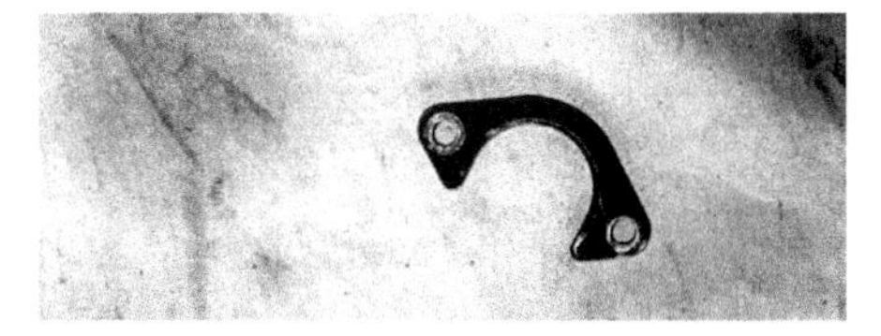

7-23-22. Offer the curved cut-out edge of the camshaft retaining plate to the camshaft front slot. It should not be tight, but also must not slop about. If all seems well, carefully put the camshaft into the block and secure with the retaining plate.

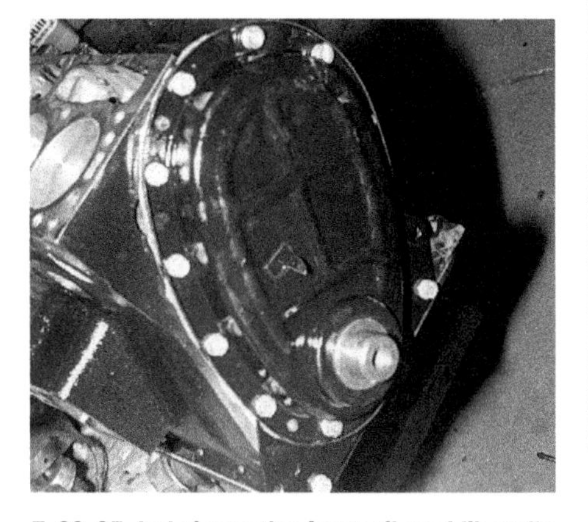

7-23-25. Lubricate the front oil seal liberally with fresh engine oil. Place the timing cover in position by coming in from the side to allow the tensioner to take up the inevitable slack in the chain. Take care to get the various cover fastenings in the correct places. A few pre-dismantling notes will pay off handsomely at this point. Finally, using lots more engine oil as pre-assembly lubrication, push the crank spacer home through the front oil seal.

Opposite. 7-23-21. Check that the centre popped timing marks are visible on the large/camshaft gear and on the smaller/ crank gear. Here, the old gears are in super condition and will be re-used with a new chain. However, were you fitting new gears, too, it would be necessary to transpose the timing marks from the old to the new gears. It's therefore very important that you hang onto the old gears when you strip the engine, even if they are obviously beyond re-use. One further advantage of re-using the old cam gear is that you will almost certainly find an imprint of the large and small slots (that are let into the front of the camshaft) on the rear of the cam gear, which further helps orientation of cam before you 'time' the engine by coupling crank and cam together with the timing chain. Do not be too vigorous when cleaning the rear of the old gear if you plan to re-use it.

7-23-23. Whilst the standard camshaft gear wheel is perfectly satisfactory for a standard camshaft, if you plan a 'fast road' or 'hotter' camshaft, then the precision with which you should adjust the timing may make a 'vernier' adjustment not only worthwhile, but essential to get maximum potential from your camshaft.

7-23-26. Slightly out of sequence in the photographic record, it is now that I would put the cam followers in place, again, using lots of clean engine oil or cam lubricant. Note that the block is of the earlier 'plain-bore' type – i.e. there are no machined recesses at the top of each cylinder.

7-23-24. Offer up the timing gear/chain set as one assembly, and keep the timing marks in line. Run a straightedge over the timing marks to ensure they are in line before tightening the two cam gear fastenings and folding over the lock tabs. Put the timing cover gasket in place, and don't forget the oil thrower ring on the crank.

7-23-27. Head studs are next. Look at photograph 7-23-31 to see the double nut method used to screw in most studs. Here, to guard against oil leaks, we see a little extra silicone sealer being applied to the thin area that runs down the pushrod side of the engine.

7-23-28. The cylinder head has been pre-assembled with unleaded fuel inserts in the exhaust valve seats, new valves, and new double valve springs. Here it is ready to drop onto its block after the correct 'plain-bore' head gasket has been put in place.

7-23-29. Putting the head on actually takes just a few moments, but what a leap forward the engine assembly seems to take!

7-23-30. Following the workshop manual, and applying both the correct cylinder head nut torque and tightening sequence is vital. Here we are getting towards the end of fitting the head to the block.

7-23-31. By tightening two nuts together as shown here, you can usually screw most studs in place with the minimum of fuss. After screwing two nuts together very tightly, use the top nut as your drive for the box spanner to fit the stud. As a matter of interest, although not strictly relevant here, when undoing a stud use the same two-nut technique, but use a ring spanner on the bottom nut.

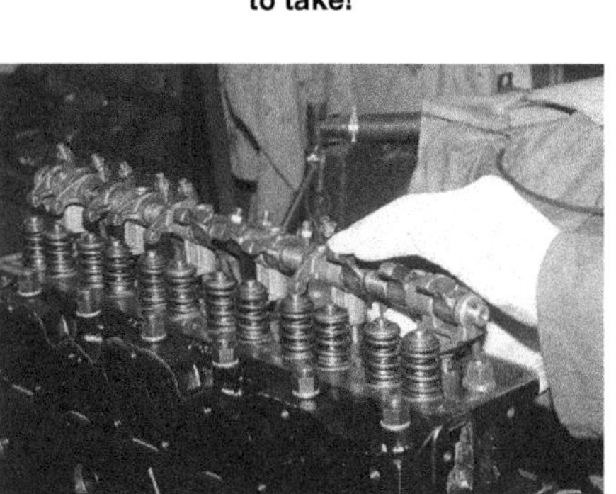

7-23-32. Drop the pushrods into the 12 cam followers and the rocker assembly over its studs. Torque the nuts, as it really is easy to over-tighten the rocker assembly securing nuts. Set the rocker clearances.

7-23-33. Water pump housing is next ...

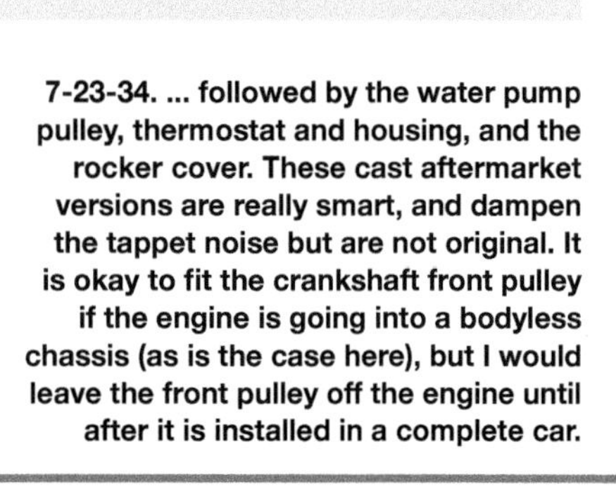

7-23-34. ... followed by the water pump pulley, thermostat and housing, and the rocker cover. These cast aftermarket versions are really smart, and dampen the tappet noise but are not original. It is okay to fit the crankshaft front pulley if the engine is going into a bodyless chassis (as is the case here), but I would leave the front pulley off the engine until after it is installed in a complete car.

7-23-35. Finally, fit the oil filter (in this case a non-original but very sensible 'spin-on' filter adapter), pedestal (for metering unit and distributor) coil, and an extra oil feed for the rocker shaft. Some engines fitted with this feed (which couples into the oil feed gallery serving the rockers) have been found to burn oil unacceptably. I'd like all my oil pressure at the crankshaft, so it's best you take advice on this particular extra.

7-24-1. Putting an engine into, or should I say onto, a bare chassis needs little additional explanation. However, popping the complete engine into a body-on chassis may deserve our attention for a few minutes. Although completely out of sequence, this picture is valuable, since it shows the extra support that the front lifting lug benefits from. Omit the front manifold stud and bolt a spacer (which you may have to have turned up, or you may find something suitable in your scrap bin) to give the lug something to lean against. It is less obvious, but while discussing preparatory steps, it's as well to mention that the engine needs to slope slightly down at the back – I would guess about 10-15 degrees.

7-24-2. Still out of sequence, strictly speaking, but this picture illustrates the additional preparatory work that is essential when dropping an engine into a more-or-less complete car. If you have removed the engine you will have already discovered that you need to unbolt both the steering rack and the front suspension crossmember, to allow them to slide forward. These steps will be equally necessary if you are about to fit an engine into your body-on chassis. You can see that the steering column has been disengaged from the rack (in a restoration project, you probably would not want to do more than temporarily secure the rack until the engine is in place). The front crossmember between turrets has been unbolted and, while still offering some support, slid forward by about two-thirds of its mounting (arrowed).

7-24-3. Here we are ready to go for it, although you'd be better off with the front pulley removed and the engine slopping down slightly at the back, certainly more than is shown here. You will, of course, have noted the absence of a bonnet on the car – another important preparatory step.

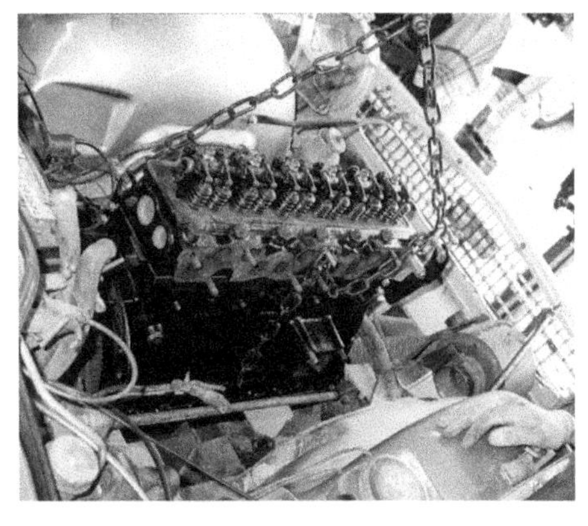

7-24-4. Getting the top (three) studs into the bellhousing and the clutch splines to align with the clutch plate all adds to the interest – and satisfaction – when the engine and gearbox do finally come together sufficiently to get some fastenings in place, and allow you to take a breather.

7-24-5. With at least some fastenings in place the next step is to get the engine mountings bolted both to the block and the chassis. I would still support the rear of the engine with a trolley jack until all engine to bellhousing fastenings are in place and tight. The next really satisfying step is to remove the (heavy duty) rope or chain you used to lift the engine into the car.

Chapter 8

Modern oils & fuels

RUNNING-IN/BREAKING-IN & ENGINE OILS

One area where modern developments can work against owners trying to improve their engines relates to the lubricating oil they initially select. When you're running-in (breaking-in) the engine in your TR, you can actually harm your doubtless expensively rebuilt power unit by over-pampering it, initially, with synthetic engine oils. Strange but true.

Synthetic and semi-synthetic oils have superb anti-wear characteristics, and it's worth considering them (after a while), although views on the suitability of synthetic oils for classic car engines are very polarised, both for and against. However, everyone agrees they must be avoided for at least the first 6000 miles, and possibly the first 9000 miles, since they'll delay the bedding-in of cylinders, rings and bores. Used too early, synthetic oils are likely to promote bore-glazing before the cylinder walls are smooth, and prevent pistons and rings achieving their best gas-tight seals – because, as strange as it seems, they are too slippery and prevent the running-in process from taking place effectively! So, by using synthetic oil during the running-in period you will end up with an under-performing engine that may burn more oil than you would expect from a newly reconditioned engine, however carefully you have driven it during this period.

Let a good purpose-made running-in mineral oil from Castrol or Penrite do its job during the initial 500 miles, and allow the bores to be smoothed off and the rings to seat. Change the oil and filter at 500 miles. Most of the rest of the running-in process occurs within the next 2500 miles. Therefore, to give your engine the best start in life, change the oil and filter at 3000 miles and then again, I suggest, at 6000 miles, and use a very good mineral oil for each refill. Castrol Classic 20W/50 (Castrolite to us old 'uns) or Penrite Classic 20W/50 are ideal after the 500 mile mark.

The concern of those opposed to the use of synthetic oils in classic engines usually relates to the vigorous detergents used in many synthetics. Certainly, there is a real danger that a well-coked engine with many thousands of miles under the flywheel can lose compression, start burning huge quantities of oil and lose performance if introduced to a synthetic oil high in vigorous detergents. However, it is my opinion that a classic engine which has just completed, say, 6000/9000 miles since a complete rebuild, filled with a low detergent synthetic oil does bear consideration and I have followed this policy with my own cars.

While talking about running-in a newly reconditioned 'classic' engine, let's just touch on the best rpm band to use during this critical period in your engine's life. In short, keep rpm between 2500 and 3000 for the first 500 miles and 2500 to 3500 for the next 1500 miles or so, and do ensure (to state the obvious) that the fuel mixture is correct and not causing bore wash through over-richness.

OIL THERMOSTATS

Mineral oils are formulated to perform best at between 90 and 110 degrees C (194-230 F). When they become too hot they thin, and thicken when too cold. At both extremes they give less than optimum lubricating performance.

If you have elected not to have an oil cooler fitted, then you need read no further, since this section does not apply to your car. However, for those with, or thinking about, an oil cooler it seems a good idea to also consider an oil 'stat. These go in the line to your oil cooler and will make the oil bypass the cooler when it's not up to temperature. Obviously, this will accelerate warming up of the oil and assist initial lubrication, particularly in cold climatic conditions. Oil 'stats are widely available with push-on or standard 5/8in threaded unions, and will enable you to have an oil cooler in circuit all the time and available when oil temperature rises as the result of hard work or hot weather. At a stroke, overheating and overcooling your oil are eliminated. Picture 8-7 shows any who are unfamiliar with oil 'stats what to expect.

USING UNLEADED FUEL

By now most UK TR owners will have become used to filling up with unleaded fuel of one type or another, and have, no doubt, found it necessary to retard the timing by a couple of degrees or so. Probably most have given some thought to having hardened inserts fitted into the cylinder head exhaust valve seats. It's unlikely that TR engines will run indefinitely on unleaded fuels without some action, although this may well be necessary later rather than sooner, and a few may never require new valve seats at all. So, unless you have a TR5 or 6 with petrol injection, procrastinate! If you have a PI TR you do need to take action straight away, and you should read chapter 10 before deciding on your action plan.

For readers with carburettors, let's look at the situation in a bit more detail, starting with the undeniable fact that you will probably have to retard the timing slightly immediately you make the change from leaded to unleaded fuel. The exhaust valve seats are, however, much less clear cut and longevity will depend on a number of factors, but largely the vigour with which you drive the car. They were originally machined straight into the cast iron head and are therefore made from material that will not withstand continuous vigorous driving using unleaded fuel, but don't panic. If you've been using leaded fuel in your engine for a number of years a thin but very worthwhile protective lead deposit will have built up on the working surfaces. This will, depending upon how you drive your car, last for maybe thousands of miles. Obviously, if you do a de-coke, drive flat out for hours, or tow a caravan, the 'lead effect' is dramatically foreshortened, but, with light driving at, say, 50-55mph, it could last for several years.

In due course you will have to adjust the tappets with increasing frequency, signalling the start of valve seat recession. But so what; tappet adjustment takes but an hour from start to finish and, if you're covering only some 5000 miles per annum, you could go many years with just an annual tappet adjustment. Do not worry that you are doing your engine irreparable damage, and running the risk of needing a new and very expensive cylinder head. Eventually, when the time comes to have the head modified, the valve seats that have been recessing due to the use of unleaded fuels will be machined out and replaced with a hardened insert. By adopting a "wait and see" policy you are doing your engine no damage whatsoever.

Then there are numerous additives for unleaded fuel that further delay valve seat recession, although no additive will protect or perform as well as leaded fuel. These additives (the full name is Anti-Wear Additives or AWAs) will also be most effective in light to normal driving conditions. In the US, leaded additives have been readily available at every autopart store for some 15 years or so, and a distribution network and agents are becoming established all over the UK even as I write. Sodium, phosphorous, potassium and manganese components form the basis of most of these additives, which will, unless you drive fairly hard, add years to the life of your cylinder head. Use a reputable make and always stick to the same brand. Never mix additive with leaded or lead replacement (more in a moment) fuels. A number of additive products underwent comparative tests at MIRA in late 1998, and four were initially declared as having met the pre-set valve seat recession criteria. Since then, additional products have been submitted for tests and several products have been added to the list of those proven to delay the onset of valve set recession. At the time of writing the full list is as follows:

Additives with octane boosters (treatment costs about 8 pence per imperial gallon of fuel)

- Millers VSP-Plus
- Castrol Valvemaster Plus
- Nitrox 4-Lead

Additives without octane boosters (treatment costs about 3-4 pence per imperial gallon of fuel)

- Red Line Lead Substitute
- Superblend Zero Lead 2000
- Valvemaster
- Nitrox 4-Star

On the face of it, it seems there's little point buying the more expensive "with octane booster" product. However, if your car needs the higher octane rating of super unleaded fuels (see section on anti-run-on valves), you may find a cost saving is possible by using standard unleaded 95 RON fuel with the octane boosting additive. Its makers claim the octane booster increases ratings by 2 or 3 points, which brings the resulting combination back to leaded fuel octane. You would do well to study all the respective manufacturer's claims and the chemical constituents of each product. As I understand the information released to date, Superblend showed the lowest valve seat recession during tests and uses potassium as the basis of its protection. Red Line's product is based on sodium and has been in use in the USA since the eighties. Valvemaster bases its product on the chemical properties of phosphorous. Send off for more information from a short-list of suppliers. Once you've started to use a lead substitute product, it's wise to regularly check tappet clearances, and keep a record of any adjustments necessary until you are confident that VSR (valve seat recession) is minimal.

Then there's lead replacement petrol – LRP – which replaced the vast majority of UK four star leaded fuel pumps from autumn 1999. LRP fuel will probably be satisfactory for light driving, but cars used at high speed could experience some recession, and will still need to have the cylinder head attended to – eventually.

Obviously, the situation changes if and when you come to an engine rebuild for other reasons; clearly, it makes no sense to refurbish the bottom end and clamp the old head back in place. No, this is when the time for procrastination is over and an unleaded head conversion called for. Many automotive machine shops are capable of creating the cavity, and inserting a hardened valve seat insert at each exhaust valve position in your cylinder head, but I think it makes sense to entrust the machining work to one of our premier TR specialists. The original BL valves are of very high quality, although some non-OE replacements supplied in more recent times are not so good. When the head work takes place a specialist's advice may be required about whether to replace or retain your valves. Most will elect to do the job properly and have new OE valves, or special 'unleaded' valves with hard chrome flashed stems fitted. In any event you should have phosphor bronze valve guides fitted, but the message is, don't rush to spend your money too soon. Wait until it is essential, or you are inside the engine for other reasons.

'UNLEADED' CYLINDER HEAD

So, how do you modify a cylinder head so that it is suitable for hard or sustained fast driving using unleaded fuel? The process is the same for aluminium heads and cast iron ones, and involves machining a recess for the hard-sintered valve seat insert used for each exhaust valve and shown in photograph 8-1. This is definitely not a DIY job, but, since it's a topic that will be relevant to every reader at some time, I thought it would be of interest to establish the work and cost involved.

The machine shop needs to select the appropriate insert for your head, which means not only choosing the correct grade of sinter, but the appropriate sizes (bore and outside diameter), too. The size issue is relatively self-explanatory: you need a bore that is compatible with the throat size of the particular cylinder head you are about to machine, and an outside diameter that will accommodate the exhaust valve in question. There are hundreds to choose from! What most of us will not, perhaps, have appreciated is that there are also at least 3 grades of sintered insert to choose from, too. For unleaded petrol/gas, it is usual to use a mid-range insert, but if one were preparing a cylinder head for, say, propane, LPG or natural gas, it would be necessary to use a harder grade of insert. The 'HM' (high-machinability) grade consists of a blend of finely dispersed Tungsten-carbide tool-steel, and alloys of iron which the makers say provides machining characteristics comparable to cast iron, yet the insert is hard enough and therefore suitable for naturally aspirated and turbocharged engines.

8-1. These are the sintered insert rings as received by the cylinder head reconditioner. A huge range is available, but this is how the hard basic ring looks before being pressed into a machined recess formed where each exhaust valve sits. They are more complex, metallurgically-speaking, than they look.

The first step in the process is to remove the old valve guides and fit new ones. This is important since virtually every subsequent operation will rely on a mandrel that will be positioned in each valve guide to ensure the accuracy of each machining operation. The machining can be carried out by numerous pieces of 'kit'. Where high volume/repetition

8-2. Whatever machine and holding fixture the restoration company uses, a high degree of accuracy is required to ensure the true vertical movement of the cutter while the head always stays absolutely horizontal. This machine has been especially modified to ensure these requirements are met, and that the speed of the cutter is compatible with the machining of both cylinder head recesses, and the subsequent work required in the sintered exhaust valve seat inserts.

8-3. The cutter can be seen forming number 2 cylinder's exhaust valve recess, while, in the foreground, the result of number 1 machining can be seen.

8-4. Here, the insert can be seen standing proud of its recess for a few seconds. The hammer blow required to overcome the interference between insert and bored recess is quite considerable. Fortunately, the insert has a special chamfer on one end to aid entry into the recess.

8-5. In the foreground the insert has been machined flush with the combustion chamber, while the next insert is being machined to the profile of the exhaust valves provided by this TR owner. You can establish whether your valves seem likely to be up to the rigours of unleaded fuel by a quick magnetic check. If the valve attracts a magnet it will be too soft for use with unleaded fuels, and must be replaced with one that contains more stainless/nimonic materials, which will not attract a magnet. Not a lot of people know that!

production is involved, the machines can be very sophisticated indeed, and correspondingly expensive. Photograph 8-2 shows how it is more likely to be carried out in a low-volume 'classic' car-focused restoration business such as Bailey and Liddle (see Appendix for details). The old valve seat is machined away next, as can be seen in photograph 8-3, but with great care, for it is important to ensure the outside diameter of the hole is absolutely correct for the selected insert. In fact, the hole needs to be a prescribed amount less than the insert outside size, and, to ensure this, Bailey and Liddle actually bore the hole twice. The first cut is never quite to size because the cutter warms up and expands. Consequently, the boring tool is slightly adjusted after the first holes have been bored and before the second, much finer dead-size cut. The depth of the bore is also important and is regulated to a shade less than the depth of the insert; once the insert is in place, it will be slightly proud of the combustion chamber. More on this in a moment.

Next up is driving the inserts into their recessed bores as per photograph 8-4. The picture is slightly deceptive in that the mandrel mentioned earlier is in place to ensure the insert can only be driven home square to the valve guide. Although already an interference fit, the insert plays its part, too, by expanding into the bored recess, thus providing a dependable valve seat that will not move even at the elevated temperatures in a combustion chamber.

You may be surprised to learn that there are still two machining operations to carry out. The first, shown in photograph 8-5, involves the same set-up (including the all-important but unseen guide mandrel) and brings the top of the sintered insert down to the level of each combustion chamber. The second requires selecting a cutter that matches the profile of the exhaust valve to be used, and machining out the inside of each insert until the valve sits properly in the head/insert/combustion chamber – nothing to it, really! The final operation is carried out by hand to blend-in, where required, the base of each insert with the respective exhaust passage.

The cost of fitting new valve guides and unleaded compatible exhaust valve seats to a 4-cylinder TR head will range from about £120 for just fitting inserts to your stripped and pre-cleaned cylinder head, to about £300 for the comprehensive service of fully stripping, fitting inserts, supplying all parts and re-assembling your 6-cylinder head ready to bolt to the block.

ANTI-RUN-ON VALVES

I am anxious not to cause any alarm, but it might be wise to mention that, as the fuel octane rating you use in your TR reduces, engines with a compression ratio of about 9.5, possibly 10:1, can start to experience 'running-on'. That is to say, the engine continues to turn over, usually very roughly, after the ignition switch is turned off. Sometimes the running-on period is a matter of a few seconds, but occasionally can be longer.

This never happens with modern EFI (electronic fuel injected) cars for the injectors cease to open once the

8-6. The anti-run-on valve; two, in fact! This TR6 must be running on carburettors and, on unleaded fuel, the engine will run-on after the ignition has been switched off.

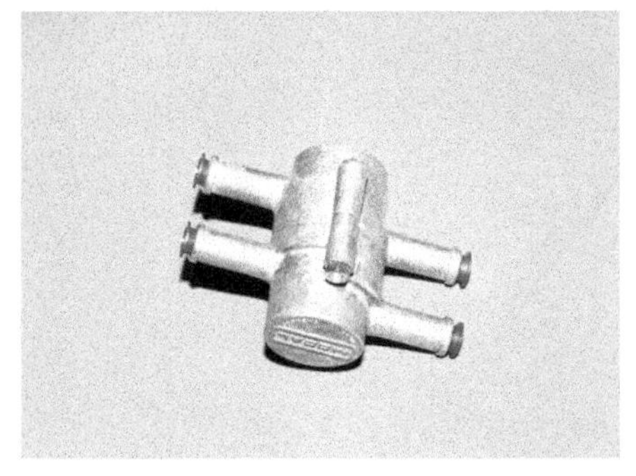

8-7. A typical oil thermostat which allows hot oil to circulate through an oil cooler when necessary, but to bypass the cooler when the oil temperature is insufficient to warrant cooling. This particular example has four ends that accept plain hoses fixed by jubilee-like clips. Some 'stats have screwed end couplings that necessitate corresponding couplings on the ends of the oil lines. I think 'plain end' 'stats are easier to fit (particularly retrospectively) than those with screwed ends.

ignition switch is 'off'. TRs with PI are also very unlikely to suffer from run-on since the fuel pump stops pressurising the fuel and opening the injectors. However, carburettor cars have a reservoir of fuel in the float chambers, and air drawn through the carburettors by the engine revolutions will continue to carry fuel to the cylinders – for quite some time!

A reduction in fuel octane may not be the only cause of run-on. Your cylinder head may be in need of a de-coke and have carbon deposits that glow and act as a substitute for the spark plugs. However, in the context of a change from 98 RON (leaded fuel octane rating) to 95 RON (unleaded octane), if you start to experience running-on your first step should be to start adding an octane boosting additive to your unleaded fuel. If running-on problems persist, try switching to super unleaded fuel, which has an octane rating closer to that of leaded fuels.

If running-on still persists, you should try fitting anti-run-on valves. These electromagnetic valves 'close' when the ignition is 'on' and are ineffective in that mode. However, they are coupled to the inlet manifold and, as soon as you switch 'off', the valve(s) will automatically open and allow air into the inlet manifold. This dilutes the fuel/air mixture, hopefully, to the point where your engine has insufficient fuel to run-on. Photograph 8-6 shows an anti-run-on valve.

The valves are available from Rover dealers and most TR specialists at about £35 each, and the part number is ADU9535. You should initially try one, which will usually do the trick, but resort to two, as this owner had to, if your first valve does not work. They work best when mounted vertically. If you still have problems, look for other reasons, like compression ratios in excess of 10.5:1.

Chapter 9

Gearbox & overdrive

GEARBOXES

Stripping and re-assembly of your gearbox is a job for an experienced expert, and outside the scope of this book. If you doubt this, consider that even the majority of professional restorers will not usually do more than basic gearbox operational repairs, but will send the box to a specialist who is repairing them all day every day.

That said, there's still a lot to consider, starting, I suggest, with a review of which gearbox should be fitted to each car, and whether you have the correct box for your car.

Triumph used a prefix numbering system that tells us whether the gearbox was originally assembled to a TR, and what type of gearbox it was. Your car's commission number will not be reflected in the gearbox number, which comprises a prefix and number, and is stamped in various positions on the left side of the gearbox. However, as we will discover, there are lots of good reasons why the case of your particular box may show it to be from another Triumph model.

As parts become both more expensive and more difficult to find, it is increasingly common practice to utilise components from sister cars to repair TR gearboxes. Consequently, the case,

9-1. The clutch fork is all-important on all TRs, but particularly crucial on the 6-cylinder models where the securing pin is prone to fracture. On this example the owner, Mark Price, drilled an extra 0.25in (6mm) diameter hole through the fork and its cross shaft and fitted a high tensile bolt and (just visible) new nyloc nut. He will have ensured that the plain shank of the bolt came almost through to the outside of the fork. As we can see, he correctly lock-wired the, no doubt new, pin. Most professional TR restorers carry out a similar modification to improve the reliability of this vulnerable part of the TR clutch. However, the hole is rarely more than 3mm or 4mm in diameter, and securing is by a driven-in roll pin. The larger than average hole in the bellhousing flange directly to the right of the cross shaft is unused in 6-cylinder cars, as it was originally intended for a dowel pin in 4-cylinder applications.

or some of the internal parts within your 'box, may well be non-original. But don't panic; few will know and even less will care. What does it matter if, say, your first gear countershaft came from a saloon gearbox? If your thin flanged bellhousing has cracked, surely the most important consideration is to get the car back on the road, and not that the case is now an ex-saloon, thick flanged one, particularly as the thicker flange became a Triumph gearbox improvement right across the whole model range?

I do understand the issue of originality prohibits such variations for some owners, but, for the majority, the message is do not be overly concerned if the gearbox case numbers are non-TR in origin. 4-cylinder, 3-synchro 'boxes were prefixed TS and UF, 4-cylinder, 4-synchro ones CT, LE, LF, MD, ME, MG, MK, VA and VF, while the 6-cylinder 'boxes carry the prefix CC, CD, CF or CP. Factory reconditioned units carry a GR prefix.

While the majority of this chapter focuses on the gearbox and overdrive problems, we will spend just a few minutes addressing the problems that some owner's experience with the clutch generally, and the bell housing cross shaft and clutch fork in particular. If you are experiencing clutch drag or gear engagement difficulties, before the gearbox is removed, do ensure that the clutch slave cylinder pushrod extends by 0.625in (16mm) when the clutch pedal is depressed. Clearly, any shortfall needs to be corrected and worn forks and/or clevis pins replaced. Photograph 9-1 shows the clutch actuation assembly *in situ* and with the gearbox separated from the engine, you need to check and replace as required the two plain bearings within the bell housing in which the cross shaft runs. Always fit a new hardened pin to secure the clutch actuation fork (photograph 9-2) to the cross shaft each time the gearbox is removed, and always replace the slave pushrod fork and clevis pin in the middle hole (of 3 options) in the side lever/crank shown in photograph 9-3. The bottom hole will result in insufficient clutch separation, while the top hole would increase clutch travel but with significantly increased pedal pressure.

TR5 and TR6 owners in particular are likely at some time to have to handle a broken pin that secures the clutch fork to its cross shaft. The pin in question, unfortunately, usually shears at about a 45 degree angle, thus making it impossible to just remove the cross shaft. If the cross shaft is worn and needs replacing anyway, then there's no doubt that the easiest and quickest solution is to cut the shaft adjacent to the fork on the external lever side of the fork and remove the sub-assembly in two parts, salvage the fork and fit a new cross shaft, part number 136354.

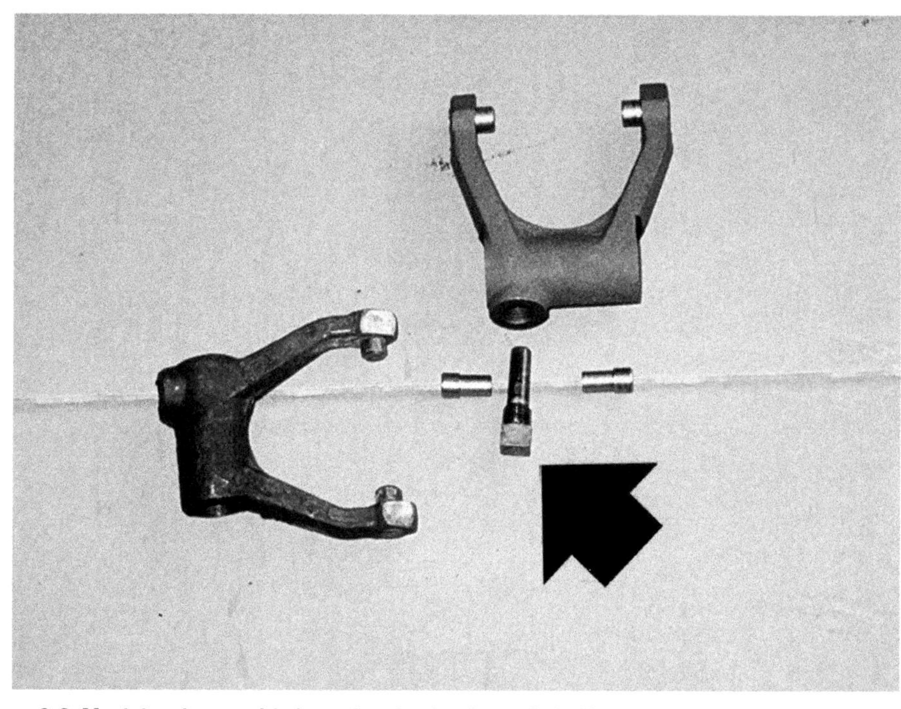

9-2. Much has been said about the clutch release fork. Here are two examples, along with the hardened locating pin (arrowed) that should be replaced every time you are inside the bellhousing. The picture also shows two replacement examples of the pins that engage the sliding sleeve. While you cannot see the wear in this photograph, these pins do become 'flatted' and should also be replaced in the fork (since the flats reduce the movement of the sleeve).

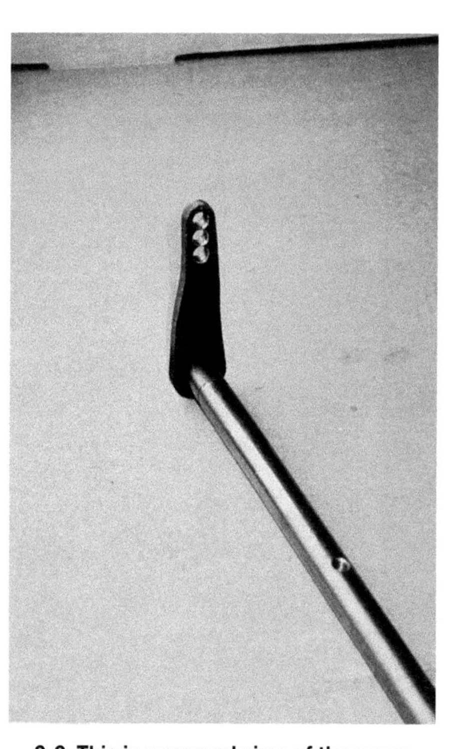

9-3. This is a general view of the cross shaft included to clearly show the three holes in the actuation arm. For all TRs the central hole is used.

However, if the shaft is not worn, we should try to salvage both shaft and fork by drilling the pin out, except the pin is hardened and a 'spark-erosion' operation will cost more than the replacement parts! So, for a salvage attempt on the basis that you have nothing to lose, start by removing the square head and a short section of shank of the pin. Put a small piece of thin tube in the resultant hole to protect the thread in the fork (or drill a hole through the centre of a suitable screw) and, using a (hardened) pin punch, try to break up the protruding part of the original pin. If that fails to allow the shaft to exit its fork, you could try drilling a

3 or 4mm diameter hole into the other side of the fork diametrically opposite the threaded boss, and use a pin punch to break up the original pin from the other end.

There are some very bad quality components on the market that may not be spotted by the home re-builder, but will be all too obvious after hours of work, once the box is in the car and put to use. In this context see my notes under 'Rear axles' about service exchange units being sourced preferably from a reputable dealer. You can, of course, fit complete ex-saloon gearboxes with perhaps 4-synchro and/or overdrive, and I will give you more information on this with some pictorial help in just a moment. However, if you are of a mind to delve into your gearbox, do not do so thinking you may be able to convert a 3-synchro gearbox to a four synchromesh 'box. You, or perhaps more likely an expert, can add overdrive to either type of gearbox, provided you have the correct overdrive assembly and the tail shaft and adapter plate for your gearbox. More on this, too, in a moment.

Before we get into the unusual, let's explore some of the much more usual gearbox and overdrive issues that you can at least confirm at home, some of which may be following up suspicions formed when we first inspected the car prior to purchase:

- Drain the oil and inspect it closely. A slight brassy look is normal. If the oil is very grey in colour, or if grey coloured steel chips are evident, remove the top cover casting and look for damaged gears. You may need to revolve a number of them, but it's likely you will find missing or at least chipped teeth. Check second gear for play. If present, the thrust washer is probably broken, which will have caused further damage to both the gear and the main shaft. A thick, grey sludge means the layshaft and/or gears have disintegrated (we will probably have noticed the tell-tale hiss on our test drive). Brass pieces signal broken synchro rings or bushes, which will also have been obvious on our test, but any or all of the above imply you need a major overhaul, or an exchange or replacement 'box.
- When refilling the repaired, exchange or replacement gearbox, the quantity of oil for a normal gearbox is 1.5 pints, or 0.85 litre, but a 'box with overdrive will need 3.5 pints or 2 litres. Use ST90 gear oil for all types of standard gearboxes, but when filling the overdrive gearboxes do allow plenty of time for oil to fill the overdrive. It's best not to use synthetic oils in an overdrive gearbox.

OVERDRIVE MATTERS
Problems and solutions common to 'A' and 'J' types

It is not the objective of this book to act as a workshop manual, or to cover all eventualities and problems, but there are some frequent overdrive problems that we should explore.

First and foremost watch for the classic overdrive problem where the overdrive switch on the steering column rotates and earths out, thus permanently engaging overdrive, with potentially expensive consequences ... So if your column switch is loose, fix it – now! As a preventative step, whatever you do before re-assembling a car you are re-building or re-clutching, do treble-check that the overdrive inhibitor switches atop your gearbox are functioning reliably and consistently. You could have one switch (if your overdrive functions on 3rd and 4th gears), but those gearboxes that also enjoy overdrive on 2nd gear will have two switches. Use a battery and light bulb or a multi-meter to check for electrical continuity when the requisite gear is engaged and, equally important, no continuity when non-overdrive gears are engaged. It is particularly important that there is no continuity when reverse gear is engaged.

While you are there with the relevant equipment, take a moment to check that the reverse light switch is working.

A sudden and complete non-operating overdrive probably signals an electrical problem, but one that most owners of classic cars should be able to cope with. At first many will think this sounds like really bad news but, in fact, the problem could be fixed quite quickly and cheaply, and it really is worth exploring the situation yourself before going to a specialist.

The earlier, 'A' type overdrives are notorious for the way the solenoid valve operates; the design tends to allow them to gum up with water, dirt and oil. This certainly is a problem that virtually every owner can check for and rectify. You will need to take the gearbox cover off the 'box and, using your operations manual and/or the following photographs, identify the solenoid and its through shafts. Jack up the back of the car and safely secure it, mindful that, eventually, the rear wheels may be rotating and changing rpm, too, either or both of which can cause the car to jerk off conventional axle stands. You might therefore want to support the chassis with ramps, or find some alternative, very solid method of getting the rear wheels off the ground.

However, for this first check, we need only to have the ignition switched on so there's no need to start the engine, although you will need to engage, say, third gear and flick the overdrive switch in and out. Obviously, you should see the solenoid and its mechanism operate each time you alter the switch position. If nothing appears to be happening, try the same test in, say, top gear and then use your multi-meter (on a low volt setting) or a 12v bulb on an earth lead to detect whether power is being presented to the solenoid. If things do not improve, you need to work backwards until you resolve the electrical problem. More than likely you will find power is reaching the solenoid when it should, so you will need to free-off a seized/frozen mechanism with penetrating oil, or possibly replace the solenoid. If you feel that the problem is going to take more than a couple of minutes to resolve, pull (and wrap a little tape around the termination), a low tension lead from the coil to save the ignition system from prolonged use.

If the solenoid mechanism is working, we need to look downstream for the problem. Again, with the car still safely on its very stable stands, and the engine running a shade above tickover rpm, loosen the operating valve plug. Once third gear is engaged, oil should bleed past the loose plug. Allow a few moments for any trapped air to bleed, but if no oil is forthcoming, the pump is probably not working, so re-tighten the bleed plug. Let things cool off somewhat and drain the gearbox, looking carefully at the oil. Refresh your mind to the points made in chapter 2 on listening to the gearbox before purchase, and (earlier in this chapter) what the colour of the oil tells us.

Now consider that the same oil circulates around both gearbox and overdrive, which means that a gearbox problem can and will circulate contaminated oil through the overdrive, and damage the overdrive unit, too. If the oil does not look too bad, remove, inspect and clean the overdrive filter. If you now remove the pump body plug, the base of the pump will be exposed. Tap it gently. Put the 'box in neutral and have a helper turn a rear wheel so that the propshaft turns and the pump moves up and down freely. If it doesn't, or if the pump sticks again, you need to talk to your friendly TR specialist, overdrive repairer (see Appendix) – and bank manager!

A running car could experience a couple of overdrive difficulties that are rather puzzling to those not *au fait* with this type of mechanism. The first is the very slow uptake of the overdrive. We'll discuss the exceptionally fast and average engagement of overdrives in a moment, but when you have to wait several seconds before the overdrive unit cuts in you have a very slow unit. This is almost certainly due to a worn pump. It is worth checking first for clogged filters (refer to your workshop manual), but thereafter you'll have to decide if, how and when to resolve the problem of a worn pump, particularly if the overdrive stays 'in' once it engages.

You can usually take your time and fix this problem at a convenient moment, provided the overdrive will disengage when required. Do not drive the car anywhere if overdrive is not disengaging. If overdrive is eventually there when you want it, although the pump might be so worn that the return springs cannot be compressed and overdrive fails to engage, or hold 'in', you can at least take your time fixing it.

There is another reason why the overdrive take-up can vary. It may cut in okay sometimes; you may feel it is trying but never really makes it; sometimes it feels as if it has actually dropped out, but hasn't: this almost certainly signals worn brake ring linings in the overdrive unit. Conical brake ring linings wear and eventually fail to properly hold the conical clutch when it is pushed backwards by the pump. Consequently, the clutch does not lock and drive is not directed through the sun wheels. You might feel you can tackle the former problem, but the latter is really not a job for the inexperienced. Furthermore, you are likely to lose overdrive completely in the very near future, so a call to a specialist is appropriate.

9-4. You will get few clearer views of an 'A' overdrive. Note the vertical actuating solenoid on the left of the unit and slightly angled (downward) speedo drive right centre.

9-5. For comparison, here are two 'A' type gearboxes with overdrives – identified at a glance by the large brass sump nuts. The one on the left is the variant from a Saloon (2000 and/or 2500), or a Mark 1 Stag, as revealed by the horizontally-mounted solenoid and vertical mounting studs. Were this gearbox from a TR, the solenoid would be mounted vertically, as per the right side unit. You will have noted the flat rear platform for the rubber Metalastic mounting pad on the right side unit, identifying this as being of TR origin. Note the solenoid lever and through-shaft and how vulnerable it is.

The 'A' type overdrive

The 'A' type overdrive in photograph 9-4 shows a genuine TR 'A' type overdrive *in situ*. One of the most frequent 'A' type faults is that the valve in the right side of the overdrive becomes blocked for want of servicing, which holds

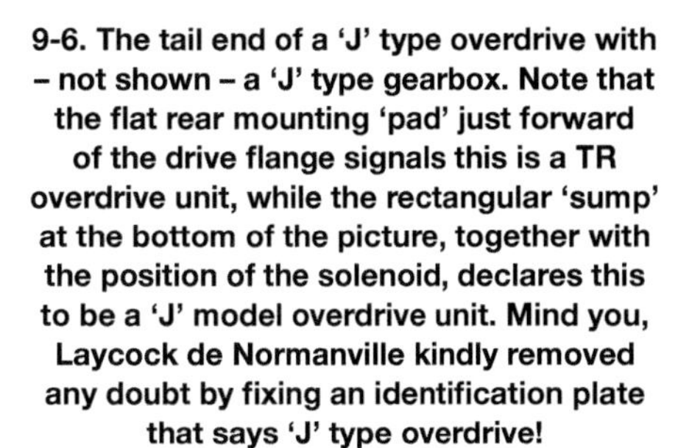

9-6. The tail end of a 'J' type overdrive with – not shown – a 'J' type gearbox. Note that the flat rear mounting 'pad' just forward of the drive flange signals this is a TR overdrive unit, while the rectangular 'sump' at the bottom of the picture, together with the position of the solenoid, declares this to be a 'J' model overdrive unit. Mind you, Laycock de Normanville kindly removed any doubt by fixing an identification plate that says 'J' type overdrive!

9-7.

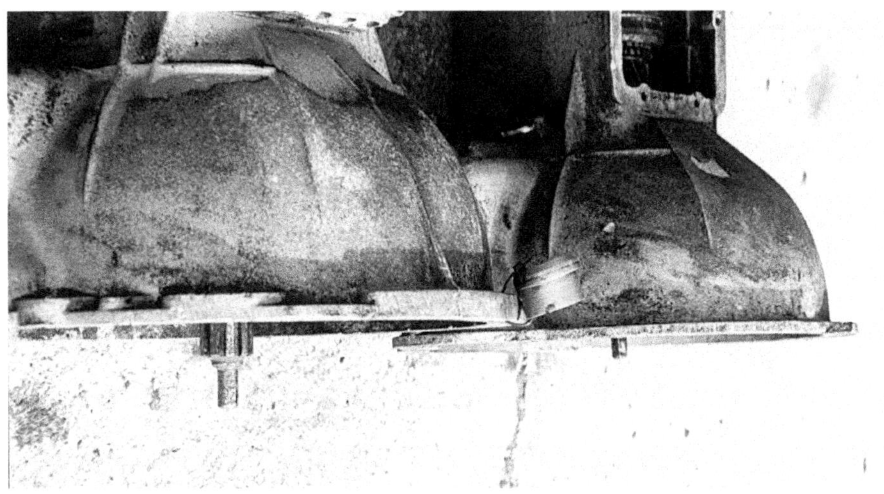

9-7 and 9-8 (above). Comparative views of two bellhousings and first motion shafts. On the left side of both pictures is a Triumph Stag gearbox that is similar in many ways to the TR gearbox shown on the right. Although the splines are the same, the spigot for the Stag can readily be seen extending about 2in (50mm) too far for immediate TR use. A specialist can correct this by substituting first motion shafts, which puts you in business. I wonder if you have spotted a less obvious but important detail? The right-hand TR bellhousing flange is only about half the thickness of the left-hand 'box. In fact, the right-hand bellhousing flange is actually broken for about 45 degrees along the top edge, reinforcing the wisdom of choosing, whenever possible, the later, thicker flanged, stronger Triumph gearbox, whatever your intended application. All Triumph applications use the same interchangeable casings, and you'd be well advised to swap your thin flanged TR casing for a thick – say, ex-Saloon – one if you plan even modest increases in performance.

the overdrive 'in'. Not a tremendous problem (cleaning the filter will usually restore normal service) provided you notice that overdrive is not disengaging. However, if you fail to notice that the car suddenly has very high gearing all of the time, and engage reverse gear, you'll find that the back of the car will rise up (as if the handbrake has been left on). The 'J' type overdrive is protected as it will slip before anything breaks, but the 'A' type will break and require a professional rebuild.

If you have your 'A' type rebuilt, always have an improved/modified uni-directional gear fitted, as it is much more robust. Furthermore, the 'A' type is not strong enough to withstand the torque generated by a hard-driven TR in second gear. The factory realised this and, late in the 'A' type's life, stopped putting overdrive inhibitor switches on 'A' type second gears. In fact, overdrive on second gear was not re-introduced until the CR/CF TR6s. If you do have overdrive on second gear and an 'A' overdrive, do treat the gearbox/overdrive unit with care or, for

9-9 and 9-10 (right). Three comparative gearboxes, all with overdrives fitted. On the far side of the first photograph we have the correct TR2 to early '6 'A' type gearbox and overdrive. The central gearbox is a 'J' type correct for later TR6s. Note its extra support for the gearchange casing. In the right photo the vertical mounting (nearest the camera) tells us this is a Saloon gearbox, confirmed by the fact that the speedo drive exits the 'box high on the casing and horizontally. The TR speedo drive will be low down and angled at about 30 degrees to miss the tunnel cover, as can be seen in both other units.

9.10.

9-11. It may sound strange when you learn for the first time that the exhaust system you order is dictated by your gearbox/overdrive. This picture shows the mid-point mounting for a TR6 exhaust system, and perhaps explains why you need to sort out your gearbox details before you order the exhaust system. Perhaps more important in the context of a chapter on gearbox matters is the detail of this 'A' overdrive speedo drive point. It is angled down at about 30 degrees so we know that this is a real TR installation. The flat 'A' rear mountings are pretty clear, too.

insurance, remove the second gear inhibitor switch next time you have the gearbox cover off. Conversely, if you have a 'J' type overdrive, you can use overdrive in second gear, and may even wish to fit a second gear inhibitor switch if you do not already one.

There were actually two types of TR 'A' type overdrive units and, as the saying goes, not a lot of people know that! Photograph 9-5 does show two different types of 'A' overdrive but, unfortunately, only one is a TR 'A' type. The large brass nut located below what is the accumulator protrusion clearly identifies all 'A' units.

The picture helps to illustrate my next point: there were two sizes of accumulator. The very early units had what turned out to be a larger oil accumulator, which 'bangs' the overdrive 'in' pretty fiercely. After some time the protrusion into which the sump drain nut is screwed was reduced to the size shown in this picture – thus reducing the size of the accumulator and softening the take-up of overdrive.

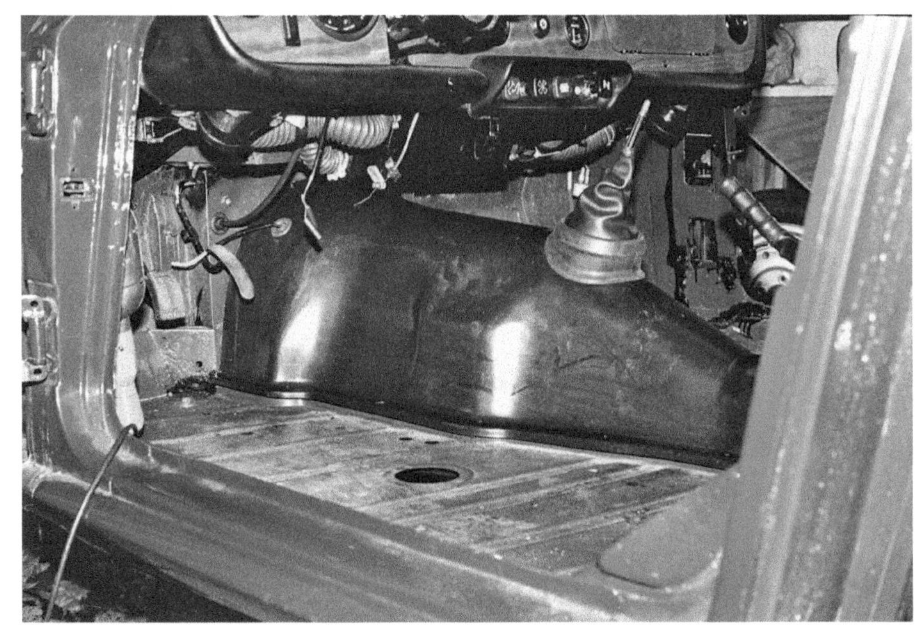

9-12.

9-12 and 9-13 (above). The first picture shows the non-original but much improved PVC moulded gearbox cover in place with, in the second shot, an extra access door. In this instance the extra opening and cover have been cut to ease the servicing of an 'A'-type overdrive but you would need to revise the location of your opening(s) if your car is fitted with a 'J' type overdrive. Normally these extra access 'doors' are positioned to allow you to service the speedo-drive and overdrive solenoid. Mild steel would be the choice of cover material affixed with self-tappers. This cover is actually a special fibreglass moulding made in place before the hole was cut in the base PVC.

Obviously, this is not applicable if you plan any track racing, since I doubt you would want either the extra weight or the reduced revs an overdrive unit brings. However, if you intend to use your 'A' overdrive in competition/ rallying, you might want to be selective by using an early casing with a 3/4ins (20mm) deeper protrusion above the brass nut rather than the shallow protrusion shown in photograph 9-5. The casings with the deeper protrusion are usually more robust. Furthermore, if the brass sump-plug sits on a 3/4in (20mm) protrusion, there's a chance it is the earlier, stronger, harder, quicker unit with the larger accumulator. I am sorry to say that an expert will have to get inside the overdrive unit to establish that it has the larger capacity accumulator. However, the latest 'A' overdrive units with a relatively short protrusion (maybe 5mm) will certainly not have the large accumulator, so will not be as desirable for competition use.

Photograph 9-5 serves another purpose, too, in that, because of the considerable amount of highly specialised work involved in converting a saloon gearbox/overdrive to TR specification, there is very little – if any – saving in using the non-TR 'A' unit shown in the picture. While it can be done – and you will not know the difference at the end of the exercise – it's as well to pay the apparently substantial sum for the real thing and be done with it!

An ex-saloon gearbox is not without its uses since if your TR 'A' unit has internal or casing damage, then a good ex-saloon 'box will be a much cheaper way of acquiring the replacement parts you need; you could transfer the lot and if the saloon 'box is in good shape that's not such a bad idea.

The 'J' type overdrive

The 'J' type gearbox and overdrive were introduced late 1972 for the 'CR' UK cars, and an example of the overdrive is shown in photograph 9-6. Few US 'CF' series cars actually had overdrive fitted, but it was nevertheless the same gearbox.

If your car does not have overdrive fitted then this 'J' type is the easiest and most reliable unit to have, though you will need the 'J' type gearbox to go with it. The 'J' overdrive was fitted to several Triumph cars of that era (Saloons and Stags, as well as TRs) and one or two examples appear in photographs 9-7 to 9-10 to help identification.

One further advantage of the 'J' type overdrive unit is that the speedo drive gear on the rear driveshaft of the gearbox can be changed, thereby adjusting the speedo drive ratio to suit the TR's 15 inch wheels. The other Triumph models all had smaller wheel sizes and, without the ability to change speedo gear ratios, would have given some strange speedo readings! The earlier 'A' type gearbox – and more particularly the 'A' type overdrive units – did not have this flexibility. Consequently, using an 'A' gearbox and overdrive unit from, say, a 14 inch wheeled Stag in your TR will result in either a very expensive and specialist weld/re-hobbing operation, or a recalibrated and non-original speedo. The latter change can be accomplished for about 10 per cent of the cost of the re-hobbing, so will be the most popular choice.

If you are using a non-TR gearbox/ overdrive, you will have to consider the drive flange, and either have the one you get with the non-TR 'box altered, or get an ex-TR rear drive flange. The alteration is not particularly difficult: a slight skim is taken from the non-TR flange recess, the flange is turned through 45 degrees and the holes redrilled on a slightly bigger PCD.

If you wanted to fit a non-TR 'J' type gearbox and overdrive unit to a pre-1972 TR, this can be done quite successfully, but requires an adapter kit to allow the rear gearbox mounting to sit securely and comfortably in your TR. Some welding to the chassis of suitable mounting plates may be required; these are, of course, non-original in respect to both gearbox and chassis.

So, as a potential purchaser you do need to reassure yourself that the overdrive fitted to your prospective purchase is as original as you want it to be, and that any chassis modifications have been carried out in a professional manner.

It is worth making the point that the 'J' type gearbox and overdrive is a much better unit than was its predecessor. First gear is a little taller, to advantage, and the overdrive solenoid is more reliable for two reasons. Firstly, the overdrive on the earlier 'A' type is operated by a genuine solenoid, whilst the later 'J' type is more of an integral electromagnetic valve. Secondly, the earlier unit operates through an external lever/bellcrank that is, along with its driveshafts, somewhat exposed to the elements. You can just see the mechanism on the far gearbox/ overdrive shown in photograph 9-9.

It is unlikely to be a major influence on your decision, but you should be aware that the 'A' type unit is notoriously susceptible to oil leaks. The overdrive units run at 400/450psi and so need to be designed to retain high pressure oil over a prolonged period. The 'J' type wins comfortably in this respect – although it would be better still if it had been designed with horizontal joints for superior oil retention.

A very useful tip from Alan Wadley concerns the servicing of 'J' units. He tells me that, in his experience, many owners do not appreciate that the 'J' overdrive has two filters, both of which need cleaning from time to time. Most owners know about and attend to the first (suction) gauze located under a ribbed 'sump'. However, there is a large plug above the suction filter with a second (pressure) filter above that. The plug is the largest of those in the base of the 'J' and, of the two filters, this plug covers what is, in fact, the more important filter. However, it rarely gets the attention it deserves and should perhaps be added to your annual 'super service'.

In conclusion, a couple of relatively minor but helpful details are shown in photograph 9-11 (how the exhaust system mountings are affected by the overdrive you choose), and photographs 9-12 and 9-13, which illustrate the improved gearbox covers that are available these days.

Photograph 9-14 is included to demonstrate why you are better not delving into the 'box without the necessary experience.

9-14. In closing, I thought you might like a look at what is inside a gearbox. You can take the top off (not shown in this shot) and peak at the contents without getting yourself into too much trouble. The overdrive is also missing from this picture, which is really only included to re-emphasise that gearboxes – and overdrives, for that matter – are a speciality in themselves, and unless you have experience, or the close assistance of an experienced specialist, are best left to the experts.

Chapter 10

Petrol injection (PI) system

PETROL INJECTION REPUTATION

I am pleased to report that an increasing number of owners have started to understand their PI (petrol injection) system and that, today, it does not have to be unreliable.

True, initially, PI systems were prone to problems; the pump overheating with resultant fuel vaporisation and starvation being the main one. The second most frequent problem tended to creep up unnoticed. As the Lucas fuel pump gears wore, resulting in less fuel being pumped, the primary method of cooling the pump (i.e. the fuel) gradually lost efficiency. However, our classics do not now, in the main, do the tens of thousands of miles required to bring about this degree of wear, and, in any event, there are alternative pumps available.

A well-maintained engine and PI system gives a significant improvement in engine and vehicle performance, so 'PI' is a highly desirable feature of the 'sixes'. Its problems are well understood by many owners and a number of TR and PI specialists – and solutions are available. We will explore the main problems and solutions in the pages that follow, but I am indebted to one specialist, Malcolm Jones of PDI (see Appendix), for checking my initial draft and making a number of very helpful suggestions.

The complete PI system is spread around the car: photographs 10-1 to 10-5 give an overview of the main constituents.

FAULTS MISTAKENLY ATTRIBUTED TO PI

Perhaps because of its reputation, the PI system is wrongly blamed for some faults.

- First and foremost the engine wears, over many thousands of miles,

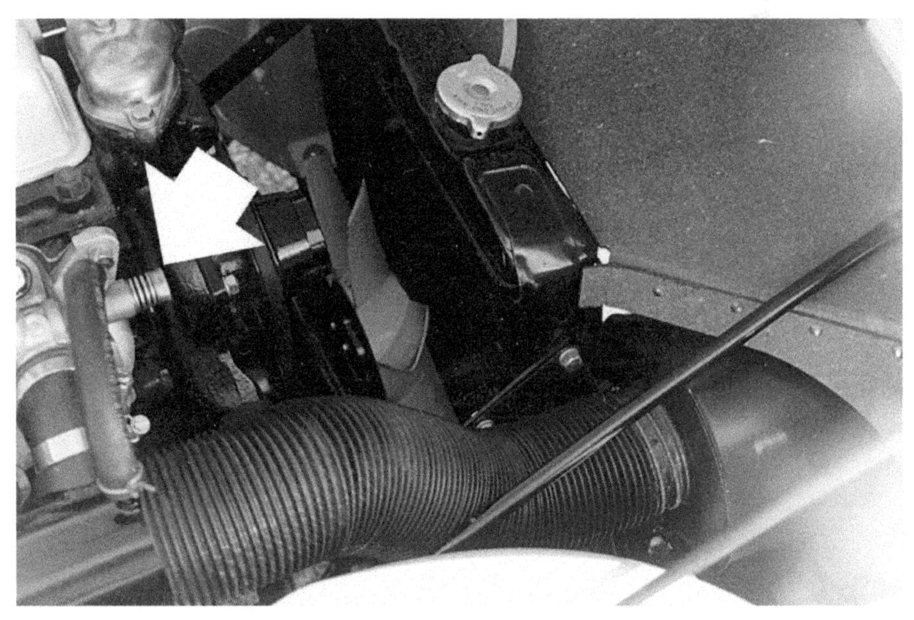

10-1. The air filter casing feeds through an appropriate hole in the radiator duct and delivers the coolest air possible to the front of the air manifold – which can just be seen in the bottom left corner of this picture. Note the air control valve centre left of the picture.

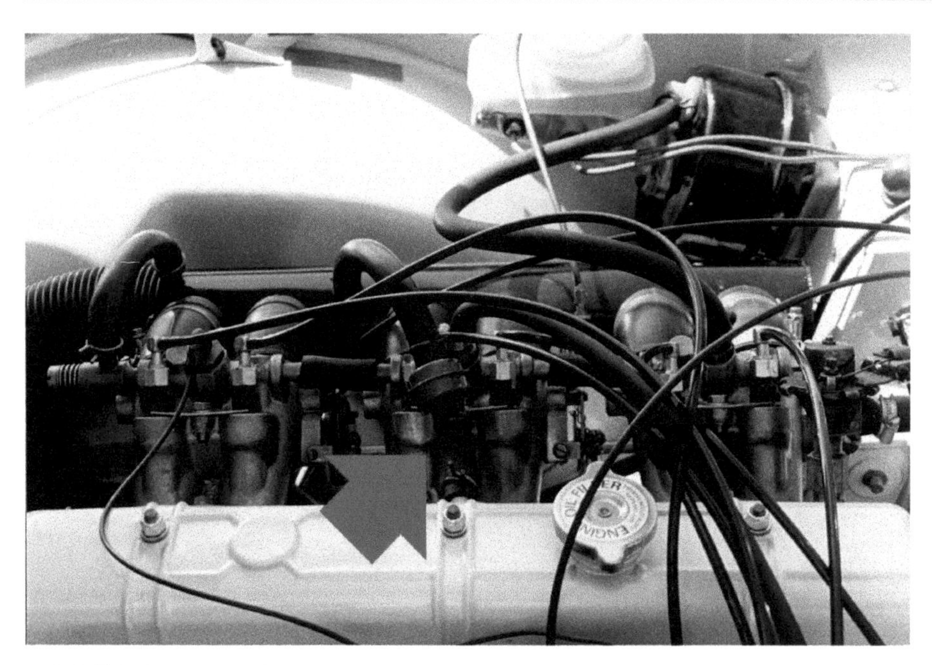

10-2. This is the induction arrangement from an early PI system, showing the '150' inlet manifolds (identified by their single balance pipes between the 3 inlet manifolds), and the air supply, feeding from the left. Note the air control valve centre left of the manifold balance pipes, and its looped air feed from the air supply manifold.

10-3. There are many important parts to the TR petrol injection system, but none more so than those shown in this photograph. The distributor is obvious, but, compared to most early 1970s contemporary cars, unusual in standard form in that there should be no vacuum advance. This early TR6 has, in fact, a vacuum advance diaphragm, probably from a Saloon, but retains the standard TR cable-driven tachometer. The fuel metering unit is just right of centre with two of the six injector pipes visible exiting the top of the main body. The fuel control unit has a domed top and is standing upright towards the right edge of the picture.

of course. The manifold vacuum (gradually) falls and the metering unit steadily 'compensates' by enriching the mixture. For the Lucas PI system to perform well, the engine needs to be in very good order, and a vacuum gauge on the inlet manifold of a CP (150bhp) car should pull a shade over 6in HG (six inches of Mercury) at tickover (850rpm). A CR (125bhp) car should pull 7ins HG.

- Engine wear also results in poor cylinder compression, which, in turn, means low vacuum readings, an inconsistent metering unit with unpredictable performance, and an engine that is most unlikely to run properly. If your vacuum reading is low, check all six cylinder compressions. If they are less than 135psi after a couple of strokes, or vary more than 10psi cylinder-to-cylinder, you will need to find out why, and may even need to rebuild the engine. There may be absolutely nothing wrong with your PI system.
- The battery must be well charged and in good enough condition to give a full 12 volts.
- The electrical system generally, and the feed wire to the pump in particular, must be in excellent order. Early cars had an inadequate electrical feed cable to the pump that dropped the voltage available at the pump. The cable in later cars was upgraded, but check yours and, if necessary, upgrade it to at least 28 strand x 0.3. Ensure it is supplied from a good source of power – the positive battery terminal, if in doubt. Simultaneously, run a separate earth cable of similar capacity either from the negative battery terminal, or from the earth strap that connects to the rear engine lifting lug (which should be looped straight off the negative battery terminal).
- If you have not already got a relay fitted it would be a very good idea to fit one and, as per the foregoing suggestion, ensure the secondary feed to the relay is of good capacity and is from an adequate (not overloaded) power source. Still follow the extra earth lead suggestion.
- Fuel pumps generally – but Bosch conversions in particular – need a very good earth. If you are reluctant to run an additional earth lead from the engine compartment, do at least ignore the pump earth wire within the harness and fit a second hefty earth lead straight to a convenient earth point in the boot/trunk. It then becomes equally important to take a close look in the engine compartment to ensure that the earth strap from the battery is

10-4. It is surprisingly difficult to find an absolutely original PI fuel pump/filter in a TR5 or 6 boot/trunk. This is pretty close to original in that only the detail is different. The picture certainly achieves its objective of illustrating an original PI fuel pump and CAV filter in situ in the left-hand corner of the boot. For some reason, the owner has added a fuel filter to the leak drain – something I would not suggest you copy. The second modification is rather more common and perfectly acceptable. The original pipe from the tank to pump has been replaced with a copper one, which has been routed round the back of the filter. You will note that the Lucas pump is mounted above the bottom of the fuel tank – something that you definitely cannot successfully emulate with the Bosch pumps most PI enthusiasts fit.

10-5. The fuel pump and filter are normally unobtrusive behind a specially-shaped boot/trunk front panel, although the CAV filter is just in view dead centre of this shot.

properly connected to both engine and chassis. Omit this apparently remote check and you run the risk of a dubious earth, with its attendant poor fuel pump performance.

- The pump cables carry a significant current that will reveal any weaknesses in the terminations at both ends of the cable. Consequently, you should change any that are corroded or not in pristine condition, and ensure that the terminations on the pump/cut-out/relay are bright and clean. I would advise soldering any crimped cable terminations as added insurance.
- The grey inertia cut-out button (designed to cut off the fuel pump in the event of an accident) needs routine cleaning to ensure reliability. The internal terminations can corrode, set up an electrical resistance that can stop the fuel pump or, in extreme cases, perhaps exacerbated by the extra current required by a Bosch pump, melt the grey switch! Pushing it up and down a couple of times once a month will help, but annually, with a small screwdriver, undo the end where the terminations exit, clean the contacts and ball with fine emery paper and reassemble the switch. To further guard against meltdown you should ensure the inertia switch is wired into the primary circuit of your fuel pump relay, although you'll still need to exercise the contacts from time to time. Jaguar XJ6s and XJSs have switches that provide similar accident cut-off protection, and these are reputed to be very reliable. They may be non-original but worth consideration if your inertia switch has given you aggravation.
- Be careful that the grey inertia switch does not 'pop' when you slam the door, or if your car gets a nudge while parked. If the car does not fire up with its usual speed, check the inertia switch before you flatten the battery. As a complete aside, you can fit a hidden switch in your fuel pump power line as a security measure, but make absolutely sure that the switch you select is more than capable of carrying the current involved. The fuel pump's direct electrical feed will carry about 10 amps so you need an above-average switch to handle that current consistently. A more modest switch becomes practical if you fit it in the primary circuit of the fuel pump relay.
- Although the change to a Bosch

pump is an 'improvement', and space dictates that we look at the detail in a later volume, this is so common a modification these days that I must make some reference to two very important installation details. If you plan to fit a Bosch fuel pump, you must mount it lower than the original Lucas pump, as demonstrated by photograph 10-12. The Bosch pump pressurises the fuel superbly but it cannot suck fuel as well as the original Lucas unit, so must be mounted lower than the Lucas pump. You should also upgrade the capacity of your alternator, preferably simultaneously, but, at worst, very shortly afterwards. The extra amperage required by the Bosch pump will so load the original 16 ACR alternator that it will burn out within a few months. A larger capacity 18 ACR is essential for long-term reliability.

- If you have not replaced the air filter in the last 10,000 miles, do so without further delay.

PI SYSTEM PROBLEMS & SOLUTIONS

- If you cannot get the engine to run at all, first check to ensure that fuel is reaching the metering unit at a minimum of 85psi. While things may not be to your satisfaction, the shuttle should operate once fuel pressure reaches 80psi, and the engine should at least run if you are getting that fuel pressure – provided the shuttle has not seized or partially seized. If minimum fuel pressure is present at the metering unit, remove all six injectors (take care to put the loose bolts well out of harm's way) and crank the car. If some injectors operate then you need to look further down the page for a solution, but if no fuel is present at the injectors you should remove the metering unit at the pedestal. This is best accomplished by removing it in one assembly, along with the distributor and pedestal, as shown in photograph 10-6. Take care as you remove the pedestal since you run the risk of upsetting the ignition timing.

The distributor is driven by a bevel gear which will rotate through, say, 20 degrees as you withdraw the pedestal. Mark not only the orientation of the rotor arm before you start to remove the pedestal (the 'timing' position), but also its position just as it clears the engine block. I am sure you are ahead of me here, but clearly you need to start re-assembly with the arm in its 'exit' orientation, and to watch that it gradually assumes its true 'timing' position as you push the pedestal fully home.

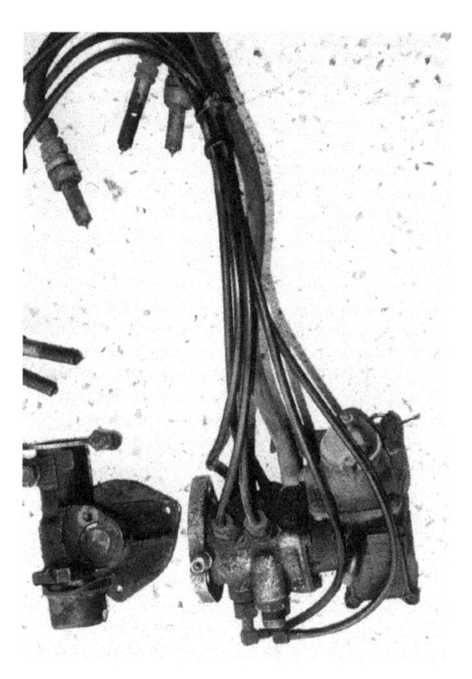

10-6. Although the pedestal has since been unbolted from the metering unit, this picture nevertheless gives you a very good idea of how to initially remove the metering unit from your engine – in one assembly. The main body of the unit can be seen with the six injector lines still attached. Note the different fittings connecting injector line to body. Note, too, the dome-shaped mixture control assembly at the right-hand end of the unit.

With the unit off the car, check that the drive dog is okay and that the rotor turns. If both these items seem not to be the problem, then there's bad news and good news. The bad is that it's most likely your shuttle has partially or completely seized and an exchange metering unit is required. The good (or at least not-so-bad news) is that, these days, there's no such thing as a scrap metering unit; they can all be repaired. More on this later.

- If the engine is 'missing' in one or more cylinders but the plug looks perfectly ok, you should start the car and feel the 'pulse' of each injector fuel line at a point about 1.5ins (40mm) short of each injector. Grasp each plastic injection feed pipe in turn between thumb and forefinger and feel for the pulse that should be evident as fuel pressure is raised to open the injector. If you find a feed pipe without a pulse, it is possible there's an air bubble in that feed pipe. This acts as a shock absorber or cushion and prevents fuel pressure from getting through to the injector to open it. Remove the injector retaining plate with, I suggest, the engine stopped. Take great care to place the plate and its retaining bolt well out of harm's way. If you are concerned about more that one injector then it may be prudent to remove all three retaining plates and their respective bolts before you start the engine. You would not be the first to decide you need to remove an injector with the engine running, only to have the engine ingest a retaining plate bolt!

Remove and point the injector high up into the air and keep the feed pipe as straight as possible in order to allow the air bubble a route upwards to the injector and, hopefully, escape. Try the injector again back in the engine but be prepared to repeat this exercise three, or even four, times.

- If the engine is 'missing' on one cylinder and, on investigation, one particular plug is firing well but keeps becoming fouled with fuel, replace the plug and try again. If the problem persists, remove the injector retention plate(s) that hold the injector(s) in place. Start the engine and carefully remove the injector. Unfortunately, the petrol injectors and exhaust manifold are on the same side of the six-cylinder Triumph engine, so you must take great care to protect the (hot) exhaust manifold from the spray of fuel, and to always point the injector spray away from the engine. A cardboard or plastic (be sure the material is resistant to fuel) box or jam jar is quite a good idea for a few moments, but in any event, do be absolutely sure not to allow fuel to spray on or even near the exhaust manifold.

The fuel should spray in a fine conical-shaped mist from the injector, and any large droplets in the spray pattern signal trouble. If you can see overly-large droplets, switch off the engine, allow things to cool for a few minutes and, still ensuring you don't allow fuel to fall on the exhaust manifold, with the injector out, switch the ignition on again without starting the engine. There should be little fuel passing the injector: one drip every 10 or 12 seconds is quite normal but more frequent drips signal trouble, certainly if they are as frequent as one per second.

In a car that has been laid up for some time, the bung seal in the metering unit is the most likely cause. Most likely it will be passing fuel into the fuel line(s), which will necessitate removing the metering unit, having the seal replaced and the metering unit re-calibrated by a PI specialist such as PDI. If your car has not been laid up recently, any weeping, or even a small jet from the injector, will tell you the end seal on that injector probably needs replacing. Switch off the ignition. It might be worth removing that one injector and its supply line at the metering unit and using a proper connection, passing filtered compressed air at about 80psi through the line. This should open the end seal and dislodge any particles of dirt trapped under the seal. After disconnecting the airline, refit the fuel line to the metering unit, bleed as explained below and see whether there is an improvement. If the seal still leaks, dribbles or whatever, then a service exchange injector looks favourite. It makes sense to check the remaining 5 injectors and, of course, to replace the lot if more than one is a problem. Remember that any leakage allows the respective fuel line(s) to empty after a run. As a double-check can you recall whether you have experienced any cold starting delays of late?

- Most injectors can be replaced on a service exchange basis and, since there is a bit more to it than just replacing the 'O' ring, I suggest service exchanges are your best route to trouble-free injectors. However, there are some important details to note when you do come to fit more than one replacement injector to a running car (i.e. a car that has not been completely stripped, or has to be started for the first time).

You should never change all the injectors at once – do them one at a time. For each injector, start the engine to get it reasonably warm and, after turning off the engine, replace the first injector. Restart the engine and get that injector operating before you follow the same routine with the next injector. Using this method you are, in effect, using the engine to bleed each injector.

- When starting an engine from scratch, after, say, a complete rebuild, a different technique is required. Remove all 6 spark plugs and all 6 injectors. Place the ends of the injectors in one large glass jar, or perhaps two smaller (perhaps jam) jars, and crank the engine in ten second busts. Longer cranking runs can damage the starter motor. Tap the injectors on the side of the jar until 4 have 'bled' – at which point you should replace the 4 working injectors and their respective spark plugs and start the engine. Thereafter, use the engine to bleed the remaining pair of injectors.
- The fuel pipe threads should not be too tight nor too loose, and you must be alert to the possibility that insufficient threads have engaged. Too few interlocked threads and there's a possibility that your 100psi fuel line could separate, spraying large amounts of fuel over a hot exhaust system. More probably you will strip the threads when tightening together male and female unions – which is very much less serious but equally avoidable if you check the threads in the first place.
- Ensure you change the flexible fuel hoses, particularly the supply pipe to the fuel pump, well before they wear due to a combination of age and the 100/110psi most are subjected to. If yours are 10 years old, change them. I recommend you fit Aeroquip stainless steel braided, available from TR Bitz or PI specialists such as PDI.
- While carrying out this routine – or, indeed, any – servicing on a PI system, remember that the original seals and 'O' rings are not compatible with unleaded fuels, and will, in any event, have hardened as a consequence of age, temperature and exposure to fuel. As a result, any sealing rings removed or disturbed during an investigation, service or test are unlikely to provide an efficient seal and should be replaced with new ones.
- In fact, all delivery valves in the metering unit, the vacuum diaphragm and the injector seals should be replaced with seals made from unleaded compatible material as a matter of course. This is particularly important if you are likely to use LRP or unleaded fuel. The seals are available from your favourite TR or PI specialist. A few lines back we discussed some difficulties that might cause you to consider service exchange injectors. Your car may be running perfectly well but you should also change all six injectors for unleaded compatible sealed ones on a service exchange basis if you are switching to lead-free (or LRP) fuel. Note, however, that I have not suggested that unleaded compatible seals should be fitted to Lucas fuel pumps (photograph 10-7). More on this very shortly.

THE PI SYSTEM IN OUTLINE

In order to ensure that the detail which

10-7. A picture of the much maligned Lucas fuel pump and a pressure relief valve. This pump is actually from a Triumph PI Saloon and differs from TR installations by virtue of the special Saloon mounting bracket, which mounts the pump in the car horizontally instead of the TR's vertical installation without a bracket. The Saloons suffered just as much from fuel vaporisation, but the horizontally mounted pump in the Saloons did extend the life of the pump bearings.

follows is considered in context, it seems best to take a simple overview of how the PI system controls fuel volumes.

The key to both regulating mixture strength and distributing the fuel is the metering unit seen in photographs 10-6 and 10-11. The regulating or mixture control mechanism is located at one end of the metering unit distributor, and controls the amount of fuel delivered to the engine for each injection, whose needs are signalled by the throttle opening and manifold vacuum. The amount of vacuum/throttle position varies the length a piston, officially called a 'shuttle', may travel by altering the position of a diaphragm within the control mechanism of the metering unit. The control mechanism has two links affixed to the diaphragm, two control springs, a fuel cam and a cam follower. There is an enriching mechanism, or 'choke', that can increase the mixture by up to 300 per cent, but I think it more important to focus upon the principals of fuel distribution in normal running conditions.

The main body of the metering unit contains one fuel inlet connection accepting fuel from the fuel pump/ pressure relief valve at 105-110psi. The fuel inlet feed can be seen pointing vertically downwards in photograph 10-11. The metering unit also has six outlets spaced in pairs at 120 degrees. One pair can also be seen in picture 10-11. A return fuel line connects to the metering unit to route that fuel which leaks past the shuttle back to the fuel tank, and this is the horizontal pipe that loops out of sight in photograph 10-11. A rotor within the main body of the metering unit has two outlet/radial ports, and rotates at half engine speed via a driveshaft from the side of the distributor. The inside of the main body is constantly full of pressurised fuel. As the rotor turns, every 120 degrees one of the inlet ports in the rotor registers with the inlet port in the main body, allowing pressurised fuel to enter the centre of the rotor, simultaneously pushing the piston/shuttle towards an end stop. When the rotor turns through a further 120 degrees, another outlet aligns with a port in the rotor and pressurised fuel again enters the rotor centre bore, driving the shuttle towards the other stop.

As a matter of interest, the movement of the shuttle up and down the metering unit can sometimes be heard, particularly when the engine is idling. As the metering unit directs 105/110psi fuel into each of the six fuel lines that lead to the injectors, the pressure rises in each line in turn and the injector end opens for as long as the fuel pressure in that line exceeds 60psi.

You hardly need me to tell you that the fuel pump has a vital role in that it must supply fuel at a constant volume and pressure. The original PI system used a Lucas pump but, as we will discuss in greater detail later, the majority of PI systems today are pressurised by a more modern replacement – the Bosch fuel pump. Both pumps supply fuel at high pressure, but, conversely, do not suck fuels all that well – particularly the Bosch pump. Therefore, if you are experiencing troubles, start by ensuring that the fuel filter is clean, the lines are not kinked or collapsed, and the Bosch pump, if fitted, is mounted below the bottom of the fuel tank as shown in picture 10-12. A few lines on fuel filters are appropriate, as some can reduce the effectiveness of the pump. We will talk in more detail shortly, but whatever filter you are using it must obviously be serviceably clean.

The filter element you are using can also reduce effectiveness of the fuel pump, so before you dash off to your local auto-factor, we had better touch on a couple of points. The correct filter element is vital to the effectiveness of the original Lucas pump and the CAV filter system, but becomes even more critical should you decide to fit a Bosch fuel pump. Opinions and advice vary, and several experts will assure you that the original CAV filter arrangement will work perfectly satisfactory with a Bosch fuel pump, provided you use the correct filter element and change it at the specified service intervals. Conversely there are those who feel that the Bosch Pump must have a Bosch filter to work effectively.

A third opinion, which I understand originates from Bosch and, frankly, rings true to me, is that the Bosch pump will work satisfactorily with a CAV filter, but the life expectancy of the pump will be reduced. Clearly, you will have to make up your own mind but note that no-one suggests the Bosch pump and an in-line EFI filter combination is other than first-class, and I therefore feel this to be your safest option in the longer term. If cost is a discouraging factor, I strongly suggest you shop around.

Moving on to the next step in our outline of the PI system, the Lucas pump pressurises the fuel to about 160psi (Bosch pump, 130psi) and sends it to a pressure relief valve, also seen in picture 10-7, which maintains a main line pressure of at least 100psi. Excess fuel is routed back to the tank, which sounds an unnecessary complication but, in fact, this constant flow of fuel provides the main cooling for the Lucas fuel pump and is therefore essential to the long-term reliability of this pump. The main fuel line carries fuel to the metering unit that we have already established is mounted on a pedestal beneath the ignition distributor.

THROTTLE MECHANISM

Continuing with the elimination of easier-to-solve problem areas before moving on to the 'heavy' stuff, the different throttle mechanisms within the three inlet manifolds are not without respective synchronising difficulties. Since there are three different mechanisms, it seems best to review each in chronological order.

The very first TR5s using petrol injection did not incorporate an air bleed adjusting screw. The air for engine idle was supplied by cracking open the throttle butterflies, just as one would with a carburettor. However, with three pairs of butterflies necessary in a six-cylinder PI system, it provided impossible to set up these very early systems to idle consistently and smoothly. In fact, Lucas/Triumph quickly issued a retrofit kit that incorporated a bleed valve, and the vast majority of cars will long since have been modified. If yours hasn't been modified fit a bleed valve to the front (currently blanked-off) balance pipe, or fit a set of early, ex-2500 PI Saloon inlet manifolds with the air valve already in place. This will put your system on a par with the middle and best of the three PI arrangements. This second system was fitted to late TR5s and early TR6s – in short, the 'CP' TR models made up until about November 1972 that had one balance pipe between each manifold, and can be seen on the left of photograph 10-8.

These throttle mechanisms allowed the butterflies to be set closed at idle,

10-8. This picture would never apply in a real situation since it is a mixture of two different injection manifolds. However, I assembled them for this photo to illustrate the differences between the earlier '150' manifold on the left, and the '125' manifold on the right with its extra balance pipe at the rear (top of the picture).

10-9. Although space dictates no more than a glimpse of this desirable PI improvement, most readers will be interested to see a picture of Revington TR's neat Bosch 110psi fuel pump painted black at the bottom of this close-coupled kit. The Bosch pump is available from all premier TR specialists, and it's worth phoning around as prices vary considerably. The top and slightly larger unit in this picture is the in-line Bosch filter that is also an improvement, but not absolutely essential over the original Lucas/CAV filter. The unit to the left of the main assembly is a gauze pre-filter in-line between the pump and a tap screwed directly into the petrol tank. Note the anti-vibration rubber mountings included in this Revington TR package. Most TR and PI specialists do a similar package.

since idling air was supplied through the air valve situated at the front of the front throttle body, as we have already discussed. You can confirm identification of these systems by a single long throttle cross shaft that runs from front to back just below the throttle bodies. Three (adjustable) rods activate the throttle spindles.

The third PI system was introduced for the CR models post-November 1972 and reverted to a difficult-to-set-smoothly arrangement. Not, I hasten to add, the very early TR5 one, but a system I can best identify by saying that it had two balance pipes between each manifold and three throttle spindles that were (loosely) linked one into the next. One of these (three per car) bodies is shown on the right of photograph 10-8. The throttle drive was from the front pair, which drove the centre pair that, in turn, activated the spindle within the rear throttle body.

The last – ironically latest – system can be dramatically improved if you can bring yourself to carry out a completely reversible modification. The change involves installing the idle air control in its earlier 'CP' position in order to improve uniformity of the system's fuel/air mixture. Remove the manifolds, linkages, etc. Remove and reverse the air valve arrowed in photograph 10-1 and the blanking plug (closing the other balance pipe). You will have to take a couple of inches (50mm) off the pipe from your repositioned air valve to the air box. You can also install a much improved, special throttle linkage that runs across the top of the throttle bodies. This type of linkage is available from all premier TR and PI specialists. All facilitate much more accurate individual throttle butterfly settings, but photograph 10-13 shows PDI's product as an example.

CHECKING & CORRECTING FUEL PRESSURE

The first signs of fuel vaporisation in the Lucas pump will most likely occur during a long run on a warm day when the engine will stop without warning. You may notice that the pump appears to be running much faster than usual, sometimes also slowly changing in pitch. If this occurs or has happened to you recently, the chances are that, at this point, pump temperature has risen sufficiently to vaporise the fuel in the gears, giving them nothing to drive and depressurising the fuel system. This is even more likely if you are using unleaded fuel, as we will see in a moment.

You will have heard stories of owners putting a bag of frozen peas on the Lucas pump to cool it down, and this really is a good idea and works well in the short-term. However, the pump/motor shaft seal can become damaged, either by the excessive heat or unleaded fuels, in which case fuel will seep past the seal. You should first check for this. If all is well, thereafter drive with at least half a tank of fuel in an effort to keep the temperature of your fuel as low as possible, and to provide as much 'head' of fuel above the pump as possible.

However this is no more than a temporary remedy as you will need to establish and rectify the fundamental cause of the problem as soon as possible. As we will detail shortly, this will almost certainly necessitate a change to a Bosch fuel pump similar to that shown in photograph 10-9.

We have already discussed the electrical supply to the pump and the earth and the need to ensure they are adequate. Your next consideration should be whether there is undue friction within the pump, and what output pressure the pump is achieving. If your pump squeaks, particularly when you first start up in the morning, the bearings are probably running dry, and it's a pretty safe assumption that there is indeed frictional heating within the pump. The Lucas pump was always marginal, even with the leaded 98 octane fuels it was designed for. Today's 95 octane unleaded fuels vaporise at much lower temperatures than the fuels of yesteryear so, unless you can buy 97 octane leaded fuel, or are building a 'concours' car, the only truly practical remedy is to fit a Bosch fuel pump.

For those keen to try stripping, cleaning and re-assembly of the Lucas pump we will very briefly explore this exercise soon. Basically, if your pump is squeaking you need to fix it or replace it.

The next step that can be accomplished without a pump overhaul is to check the fuel pressure via a 'tee' piece fitted into the high pressure fuel line just in front of the metering unit. Clearly, the open 'base' of the tee needs to be capped off when not in use, and should be opened with great care only when the system is not pressurised.

To check the system's pressure, connect a 0–150psi pressure gauge to your tee piece. A steady reading of 100 to 110psi should be established within a few seconds of switching on the ignition (do not start the engine) and tells you all is well. If you are outside the specified band, adjust the pressure control valve located in the boot/trunk. If it refuses to respond, remove the valve, strip, clean and retry, replacing the valve if you still cannot get within the 100/110psi band.

If you are still unable to get sufficient pressure, it will be necessary to step back a pace and check the delivery capability of the pump. Start by removing the car's fuel return line from the control valve, and replacing it

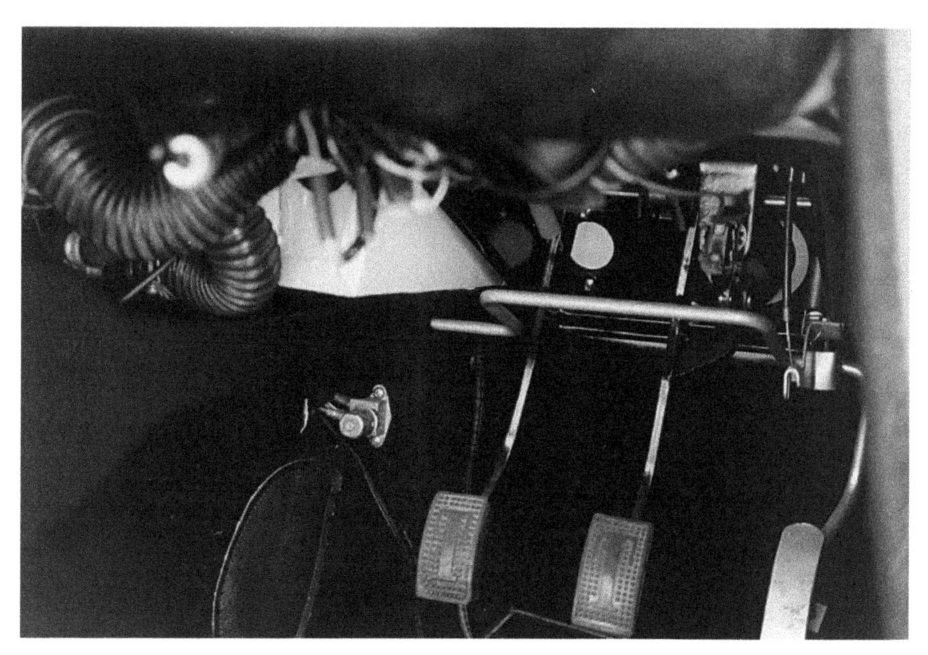

10-10. Probably of no more than academic interest to our friends Stateside, this is a view of a RHD car pedal arrangement.

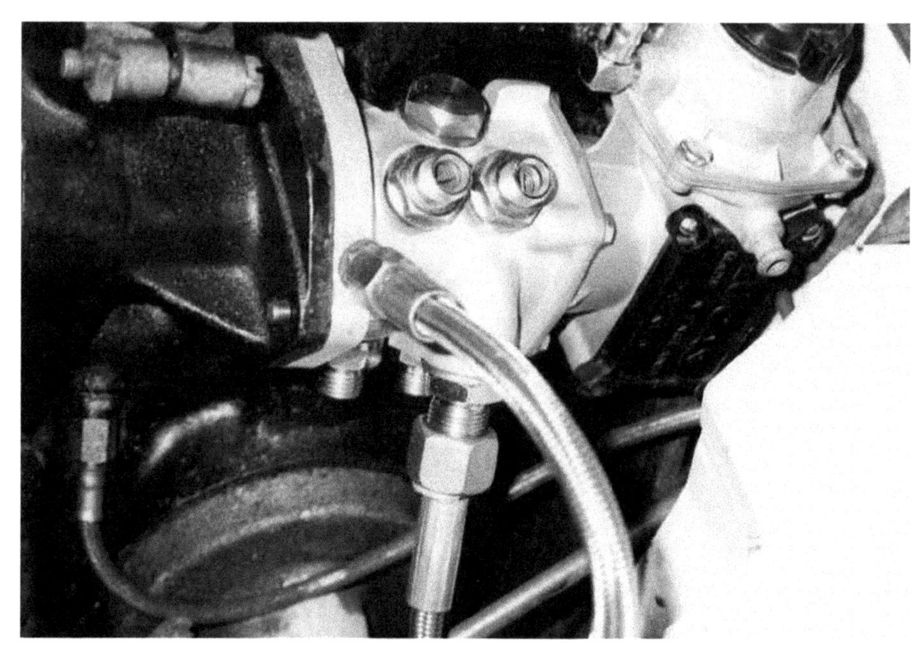

10-11. A slightly more close up shot of the fuel feed and return pipes coupled to the metering unit. The injector pipes have been omitted for clarity. The metering unit is a fairly early one that would normally have a push-on fuel return pipe. However, the later screw-on return pipe provides for a more fuel-tight connection; this unit has very sensibly been altered to the later connection we see here.

10-12. The now very popular Bosch fuel pump under an in-line Bosch filter. Note that it is tucked neatly into the front left corner of this TR6 spare wheelwell. The objective is to get the pump as low as possible because, for all its virtues, the Bosch pump does not draw fuel at all well. The installation is, you will note, pushed forward as far as it will go and this is aided by the front pipework (that loops from pump to filter) which uses 90 degree banjo unions. You should follow this example even if you continue to use the CAV filter in that the pump needs to be in the wheelwell and as far forward on the left wall as possible. The spare wheel still goes in, of course.

with a length of fuel tubing leading to a demi-john or the like. Switch on the ignition and note the time it takes to deliver a litre (two pints) of fuel. It should be approximately two minutes. If longer, move to the second pump check which involves fixing your demi-john fuel line to the direct output from the pump. This time you should fill a litre container in about 1 minute, with a slower fill time calling into question the supply of fuel to the pump and, initially, the fuel filter.

THE ORIGINAL CAV FUEL FILTER

A few paragraphs ago I mentioned that opinions vary about whether a CAV type filter will be satisfactory in the long-term with replacement Bosch fuel pumps. All agree, however, that most of the current 'original' style replacement elements for the CAV filter system are not manufactured with a sufficient area of filtration paper around the core, and that this is a crucial factor in determining the effectiveness of the CAV system, whether it is filtering fuel for a Lucas or a Bosch pump.

All filters of this type are manufactured so that the filtering media (paper for fuel, felt for oil) is zigzagged to provide a deceptively large area for the fuel to flow through. The larger the area the easier the flow of fuel and the lower the pressure drop. In response to a number of owner difficulties, Revington TR has carried out some research in order to establish what current replacement elements are suitable for an original CAV filter, and I am indebted to Neil Revington for providing me with the conclusions. Fundamentally, Revington believes there is but one element/cartridge (which has to be imported) that will provide for satisfactorily fuel flow, and it is important that you use this cartridge if you want reliable performance from your PI TR. It may be available from premier TR and PI specialists, but Revington sells it under part number GFE5296CAV. If you plan to continue with a CAV filter, regardless of pump, with one exception that we detail in a moment, you should accept no substitute elements from anyone other than one of the premier TR/PI specialists listed in this book – and only then if you are assured of its suitability.

An easy and sure alternative available from your local Lucas agent is a genuine Lucas/CAV replacement element, part number CAV296! This element is suited to leaded and unleaded fuels and only needs to be replaced with another Lucas product on or before the appropriate service interval to ensure the satisfactory continued use of your original CAV filter arrangement.

If you have tried changing filter elements but are still experiencing slow fills, then your tank outlets could be (partially) blocked, which, of course, requires that you remove the inlet pipe, free the blockage and clean the bottom of the tank.

By this point you will have either solved the problem and re-assembled the pipe work for another try at adjusting the fuel pressure, or will be looking at your pump in a very suspicious way. One last check/change may be worthwhile before you focus on the pump – consider fitting a modern EFI fuel filter (yet another one of my small if non-original modifications). Note that these are metal cased and usually about 2.5in (60mm) in diameter, and do not have to be of Bosch manufacture.

THE FUEL PUMP & MOTOR

The almost universal use of unleaded fuels and the availability of Bosch pumps makes servicing the Lucas fuel pump questionable. I am sorry to say that unleaded fuel additives or LRP (Lead Replacement Petrol) will not make any difference to the need for pump seals made from Viton, or the other unleaded compatible materials essential to keeping the original Lucas pump operating properly.

Contrary to popular belief, these seals are now available, so for 'concours' enthusiasts, those intent on using the Lucas pump, and those few owners who can still regularly obtain leaded fuel, here's how to service the drive motor of the Lucas pump.

Start by checking that there are scribe lines on the pump and motor housing and that they are aligned. Remove the two screws that hold the motor together and remove the motor cover. The armature will come out as you remove the cover as the result of the pull of the stator (a permanent magnet), so be prepared and ensure you remove the armature as slowly as possible so as not to lose the brush springs or the thrust washer(s). Clean the motor with a dry soft cloth and check that the stator magnets are not cracked. If they are cracked it's time to

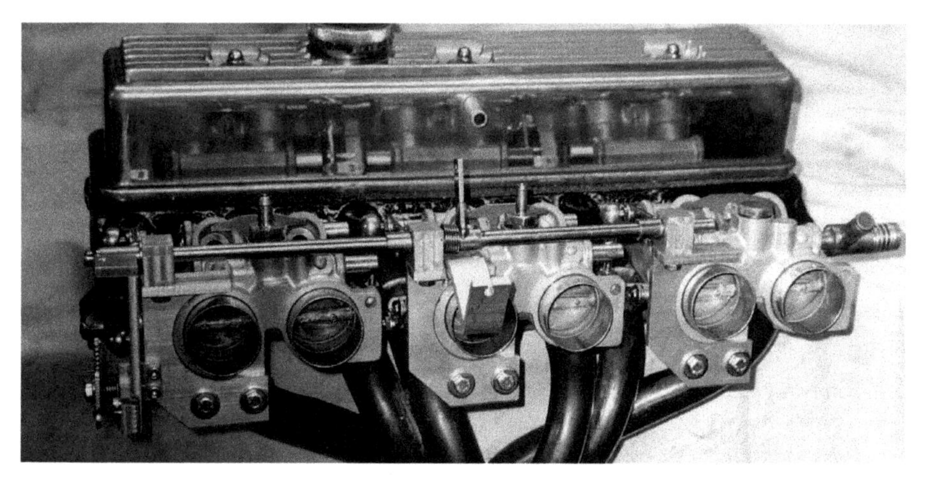

10-13. An improved throttle linkage from PDI. Malcolm Jones tells me that it makes for much easier balancing and adjustment.

switch to a Bosch pump without more ado.

If all is well with the stator, turn your attention to the armature. Ask your local Lucas agent to check it for continuity and shorts and, if all is well, to polish the commutator and supply new brushes. You will need a Gato type seal (part number 517419) and 3 'O' rings (part number 517413) made from unleaded compatible materials. While I am sure they are available elsewhere also, Revington TR can definitely supply them in Viton, and Prestige Developments and Injection can supply the seals in a material that is definitely unleaded compatible.

Re-assembly starts with applying a few drops of STP (engine oil additive) to the shaft ends and housing bearings. You might find a plastic knitting needle helps with the second operation but, in any event, avoid getting lubricant on the brush housings. Slide the armature into its housing and release the brushes on to the commutator. Ensure that the pump and motor scribe lines are correctly positioned and fit the end cap screws. Refit the pump and fuel lines and recheck the fuel pressure. If it is close to your target, start the engine and, after allowing the system to warm up (say, 5 minutes), set the thrust bearings in the motor end cap. You can do this at home by screwing the thrust bearing in until the motor just starts to slow down, then back off a half turn and lock it.

Check and adjust the fuel pressure and, if you are still in trouble, you can be fairly certain it is the pump that is unserviceable and the whole motor/pump assembly needs replacing, probably with a non-original Bosch unit.

I appreciate we are trying to maintain and rebuild your TR to its original specification but, for those able to use the original Lucas pump for some time yet, there is one very easily reversible fuel pump modification that I think you would be prudent to adopt, particularly if you are running on unleaded fuel. Back in the '70s, Lucas introduced a round tubular cooling coil to fit over the motor body. That's good enough for me, for increased reliability makes for more enjoyment and less hassle, and I'm all for that! What am I talking about? A square section tubular cooling coil made by Revington TR that brings more cooling coil surface into contact with the motor body. What does the cooling job? Petrol that has been directed back to the tank by the pressure relief valve. You will note, however, that it will have little cooling effect if your fuel pump is so ineffective that it's pumping low volumes of fuel in the first place. It is no solution for a poorly performing overheating fuel pump, but is valuable additional insurance, aiding cooling of a well maintained, properly performing Lucas pump.

THE METERING UNIT

The PI metering units (photographs 10-3 and 10-6) have a much longer working life than reputation suggests. However, it is not generally appreciated that they do have a shelf life (5 years maximum), so when offered that very rare 'new' unit at an auto-jumble, do remember that it will require refurbishing for two reasons, even if it has genuinely never been on a car: it will require unleaded compatible seals, and will have certainly passed its 'sell-by' date.

You can reasonably expect to get 75,000 miles from a new unit, not that there are many about these days, but a reconditioned unit with a new rotor that re-establishes the original Lucas tolerances should have the same life expectancy. You should budget for about half that life expectancy from a properly reconditioned and re-calibrated one that does not have a new rotor fitted.

Often the units are replaced because they are over-fuelling the engine, but, as we touched upon right at the beginning of this chapter, this is not an automatic reason to assume your unit has reached the end of its life. However, it is true to say that, as the rotor, datum track and pivot pins all wear, each tend to richen the mixture. A period of inactivity can allow the formation of rust on the metering unit's internal surfaces, which can wreck a metering unit in minutes, as can inexperienced hands investigating seized rotors and the like.

So this is an area in which to tread very carefully; if in doubt, leave it to the experts. As Moss said in its TR5/6 catalogue: "disassembling your petrol injection unit can seriously damage your wallet" ...

At this juncture I should point out that time and technology have advanced in the sense that, at one time, a seized rotor almost certainly meant that that unit was scrap. This is no longer the case since PDI can fit a new rotor, so the number of scrap metering units has been dramatically reduced. If you are told your unit is scrap, get in touch with PDI (details in the Appendix) before you do much else. Yours may have miles and miles of life in it still, and since there are some areas that can be reasonably safely explored, provided care is exercised, I think we should start by establishing the various types of metering unit you might encounter.

Three basic types of metering unit were manufactured. Whilst almost identical, they were 'tuned' to suit the different inlet manifold depression characteristics of the applicable six-cylinder engine. In fact, the only mechanical differences were the actual springs and spring covers, so although there were no external markings, it is relatively easy to ensure you have the correct unit for your TR, and involves removing four screws and the metal dome adjacent to the bulkhead. If you considering changing the unit set-up, please remember that there are some simple but very important adjustments that can only be done accurately with a fuel-flow measuring rig. Since there are only three of these in existence, it's worth asking yourself whether it would be just as well to trade your complete unit for a service exchange unit that will not only have the correct springs, but will also have been properly adjusted.

However, if you are removing the covers to check for compatibility, do take care to prevent the springs beneath the cap from flying about! On the TR5 and 'CP' TR6 there should be two pink springs, one large and one small, while blue springs tell you it is an early 2.5PI Saloon metering unit. The 'CR' '6s actually had 3 black springs and a different, non-interchangeable spring barrel. The early 2.5 PI and TR domes are identical, so if you have the wrong springs, you can put the correct ones in, although you will be missing the re-calibration that is the product of experience and very specialised equipment.

It's worth adding here a few words that I hope will dispel a fairly frequent misconception regarding upgrading the performance of a PI car. Often owners think that they can improve performance by changing the metering unit from, say, 125 to 150bhp specification. This is not possible, or at least it is not possible in isolation, for the metering unit and camshaft work in (close) correlation and must be matched. This becomes obvious if you bear in mind my references to the importance of the inlet manifold vacuum to the operation of the PI system. It will become clearer still if you turn back to my respective vacuum figures for the 125 and 150bhp systems; there really is quite a difference. That said, if you are prepared to change your camshaft and the metering unit to compatible 150 units, you will certainly increase the performance of a '125' engine.

It follows that one of the checks you should make when dissatisfied with the performance of your PI system is to establish exactly which camshaft you have. If you cannot be sure, change the cam for something you know the pedigree of, for you will now appreciate that a 125 metering unit will not work at its most efficient with a 150 camshaft, and vice-versa.

Moving on to the areas you can and should check if your PI system is giving you grief, the most obvious is to be absolutely sure that the choke cable and actuating arm really are fully home on the metering unit when the choke knob is 'in'. The adjusting screw must have circa 0.025ins (a little less than 0.5mm) clearance with its cam. If the car has been laid up for some time, you must also check that the datum track is not sticking and partly enriching the mixture. Remove the rectangular plastic cover and look for a broken datum track spring, which will be obvious and require replacing. Liberal use of a freeing agent is required if the movement is at all suspect, after which apply lots of engine oil to the mechanism.

Beneath the domed plastic cap is a depression chamber with a diaphragm/ seal that must be in good shape for the unit to operate properly. You can test the diaphragm by removing the pipe (from the manifold end) that runs from the inlet manifold to the top cover of the metering unit. Suck as hard as you can and, like many a distributor's diaphragm, the vacuum should hold the linkage's upward movement. If all seems well, replace the rectangular cover over the datum track. No vacuum, or inconsistent movement signals the need for further investigation, and you will almost certainly find a new diaphragm will be necessary.

You could also find that the thread on the linkage holder (which passes through the centre of the diaphragm) has stripped, allowing the link to move, sometimes falling to the bottom of the metering unit. Make notes of the assembly sequence and take some close-up photographs before you go too far and cannot remember which way around something goes.

Do avoid one mistake frequently made when assembling the metering unit pedestal to the block: assuming that the workshop manual 'as required' quantity for gasket 104939 means 'one off'. It probably would have been better called a shim, for you could need 3 or 4 – even 5 is not unheard of – for it adjusts the height of the pedestal as dictated by the driveshaft. If your pedestal has been leaking oil from its joint with the block, chances are that your driveshaft has bottomed out and the, no doubt, single gasket has never been allowed to carry out its dual role of sealing and spacing. Fit a couple more gaskets!

Take care not to jump to the wrong conclusion, though, for an oil leak in that area could come from an adjacent but quite separate problem. It is unusual for owners of PI cars to give enough consideration to the other two seals that are fitted to the pedestal. Part numbered 145720, they fit back-to-back on the metering unit driveshaft to stop fuel from the metering unit escaping forwards and oil from the engine escaping backwards. There is a 3mm drain hole beneath and between the two seals just forward of the metering unit flange, and any leakage of either fluid will drip from here and signal trouble.

When you do choose to or have to replace the seals, make a note of their respective positions and which way each seal faces before you push them from their housings. For the record, the oil seal re-assembles first and faces the oil that is trying to escape the engine. There then needs to be a gap over the drain hole (stick a drill shank in the hole to help you) before the fuel seal (facing the other way to meet the fuel pressure) is eased into the housing.

Whenever you have the pedestal off the car it really is worthwhile carrying out some preventative maintenance, not only replacing these two seals but also replacing the two roll pins that secure the drive gears. They are actually called 'mills-pins' in my parts book, but the part numbers will be the same and are 500974 and 500975. They can wear and become sloppy so, as I say, are best replaced.

If the car still runs poorly or uses excessive fuel, then the metering unit could be in need of re-calibration. As I explained earlier, accurate re-calibration requires very specialised equipment, and only a few very experienced specialists will contemplate altering the calibration of your unit on the car. It is certainly not a task most readers should contemplate, so, sadly, we're back to purchasing a pre-calibrated service exchange unit.

Chapter 11

Carburettors

IDENTIFYING YOUR CARBURETTOR

When SU carburettors were originally manufactured, a thin triangular aluminium tag stamped with the part number easily identified them. The tags were secured by a float bowl cover screw. Most tags have been lost over the years; one or two may have been discarded as otherwise identical carburettors from another vehicle (which would carry a different identification number) replaced the original TR carburettor. My own TR6 is running on HS6 SUs which came with Volvo identity tags, which I hope you will understand were discarded!

If your early carburettors have been lost or irreparably damaged, don't despair. Your chances are slim of finding an actual pair of original carburettors, but there are still a large number of identical carburettors about, and a visit to a salvage yard that has some cars from the 1950s and 1960s – particularly BMC/BL models – is likely to resolve your problem. Alternatively, look in the relevant BMC/BL motor club advertisement columns. You will, of course, need to change the jets and needles but, since you will almost certainly be refurbishing the whole unit(s), this won't be a hardship.

Fortunately, it's still possible to obtain virtually every TR SU service component. Supply of components to overhaul Zenith Stromberg carburettors is also pretty good, and you should be able to rebuild your original carburettors with few problems. Like the SU situation, there are alternatives for any major Stromberg breakage or omissions, and I would suggest Stags and early Saloons as potential sources.

The TR2 used twin 1 1/2 inch (measured at the inlet manifold end of the carburettor body) H4 SUs. These are easily identified by there being only two carburettor to inlet manifold securing holes. To the best of my knowledge, these were the only (TR) SUs to use only two studs. The correct identification tag number is AUC721. The TR3 increased the size of the choke to 1 3/4 ins which automatically gives it the SU designation 'H6'. The twin configuration was retained but these carburettors were secured by four mounting studs, and designated AUC786 up to 1958. To be absolutely correct for these early TR3s the fuel inlet pipe is secured by a screwed fitting.

The TR3A, post-1958, and the TR4 (up to CT21470) used a very similar carburettor, but the fuel supply pipe was pushed over a short male pipe cast into the float chamber top. This should carry the tag number AUC878.

A significant change occurred at TR4 CT21470 for, after what I am sure must have been much soul searching, Triumph designed its own carburettors and introduced a change in carburettor manufacturer – Zenith-Stromberg. Easily identified by the much 'dumpier' body shown in picture 11-1, the change was clearly brought about because of stricter emission regulations. Many TR owners consider the Zenith-Stromberg to be inferior to the equivalent SU. It was, perhaps, never as easy for the home enthusiast to adjust and tune, possibly by design, for it was pre-set at the factory. In fact, it should be regarded as providing equal power to the SU but fewer emissions. Furthermore, TR250/6 owners who have switched to SUs from the original Zenith-Stromberg (without further engine modifications) have found it impossible to detect a difference in their car's performance. Having said that it is interesting to note that Triumph actually used Strombergs up to 1973 for the 2000 Saloons but switched to SUs for the larger capacity six-cylinder Saloons thereafter.

So, on balance, the HS6 SU is

11-1. The 1.75in Zenith-Stromberg as fitted to all TR250s and TR6s destined for the USA. It's probably underrated but, in view of the relative difficulty in setting and tuning the Stromberg, it's best replaced by the SU.

11-2. A pair of 1.75in HS6 SU carburettors fitted to a TR6 and demonstrating some interesting features. You will note the much less restrictive 'K&N' KN56-1400 pancake air filters which have replaced the standard paper element and casing. These filters require less maintenance, too, since they need only be cleaned and oiled perhaps once a year. Note the twin choke cables, one to each carburettor, which means that this set-up is using a PI choke cable. The rear heat shield is non-standard, too, but also serves to provide a location for throttle and choke return springs. The original late TR6 inlet manifold (out of shot) used here to carry the mixture to the inlet ports has no provision for return springs. An ex-2500 Saloon/Sedan carburettor and inlet manifold would have provision for return springs. This heat shield would be more effective were it to have self-adhesive aluminium reflective foil fixed to the rear face.

probably the carburettor of choice – although it's best to avoid SUs with waxstats on the bottom if at all possible, particularly on the TR, where the carb and engine/exhaust are in close proximity. The consequential heat on the 'stat can incorrectly adjust the SU!

Whatever your SU, it's a good idea to fit a heat shield between carburettor, engine and the exhaust as per photograph 11-2, particularly when unleaded fuel is in use since this boils at a much lower temperature than leaded fuel. Those TR250/6 owners switching to 1.75 inch SUs could well have used ex-Saloon (2500TC or 2500S, post-1974), which used a pair of HS6s with the identity tag 'AHD678' and, incidentally the best inlet manifold for the 'sixes' (with UK inlet ports, that is) depicted in photograph 11-3. If you have the US ported head that is discussed in some detail in chapter 7, you are more or less stuck with the slightly less efficient 'squarer' manifold shown in photograph 11-4.

PRELIMINARY OPERATIONAL CHECKS

The optimum performance of each carburettor, and therefore your car, depends so much upon condition, very fine adjustment and balance. All of these requirements are difficult to explain, and even more difficult to illustrate, so read this chapter ever-mindful of the close mechanical tolerances required throughout an efficient pair of carburettors. A workshop manual will be a particularly invaluable aid to carburettor rebuilding.

The very first check to carry out is to establish that the almost inevitable throttle spindle wear has not reached an unacceptable level. You may need to disconnect the various return springs to make a true evaluation. The test is simple: do the throttle spindles move up, down, in or out within their housings? The smallest movement is possibly acceptable, but if you can see a gap opening and closing, you may as well skip the next few sections and cut straight to refurbishing the throttle body. If spindle wear looks acceptable, but you are unhappy with the idle, performance or fuel economy of your car, read on.

Before you resign yourself to rebuilding your SU or Zenith-

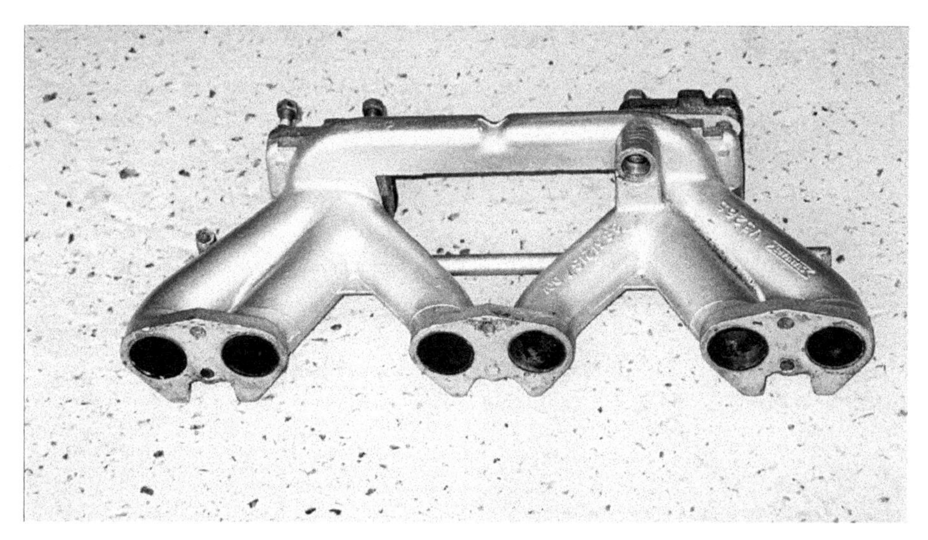

11-3. Regardless of whether you are fitting Strombergs or SUs, this is the later Stanpart inlet manifold of choice because of its more sweeping air flow. Both types of carb bolt to the manifold. Manifolds from either the 2500TC or 2500S Saloons have return spring attachment points cast on the top, telling us that this is an ex-TR6 example. If your car has an early US cylinder head, take care – the shape of this manifold may mean it will not fit your inlet port widths.

11-4. A pair of 1.75in Strombergs on a TR250. Points of note are the left side steering column, all the bellhousing bolts are correctly orientated, the 6-into-one exhaust manifold, and the early (and not ideal) inlet manifold, which you should compare to the rather more efficient and slightly later one shown in the previous picture.

Stromberg, we need to consider the simpler carburettor adjustments and diagnostics first. As the sort of carburettors we are discussing are simple in concept and detail, there are even earlier basics we should consider first before adjusting the carburettors.

The initial step is to look for and correct any obvious mechanical defects, starting with the condition, adjustment and operation of the choke, fast idle and throttle linkages. Not so obvious is the smooth operation of the SU pistons in their housings, but they are definitely worth checking as dirt or corrosion can adversely affect operation to the serious detriment of carburettor performance. Remove the (3) top cover securing screws and correct any obvious contamination by the gentlest means possible. If there is nothing too obvious, or you have tried the car after what you thought was an adequate cleaning of piston and bores, remove the cover(s) again. This time look for signs of (slight) interference between piston and bore; it may only be a small single 'high spot' but could be enough to slow or even hold piston movement, resulting in some very strange performance characteristics. You do need to polish the bore at the point where the piston 'binds', but this must be done carefully with either very fine emery paper or metal polish. Remove only just enough material local to the high spot, remembering that this needs a slow "do a bit and try it" approach.

If you are still uncomfortable with what you think is a carburettor problem, give thought to servicing matters in general, and condition of the ignition system in particular. It's far more likely that ignition components will need attention than, say, your carburettor balance. Do you have a filter fitted in the fuel line to the carbs and, if so, has it been replaced in recent history? If you do not have a fuel filter in line, first clean the two carburettor float chambers of the inevitable dirt that will have collected. Check that the fuel cut-off (needle) valves in the float chamber tops are opening and closing fully when they should, and then fit an in-line fuel filter.

If your needle valves are not working to your complete satisfaction, or you have been experiencing flooding, consider not only replacing your needle valve/seat assembly, but upgrading them via the use of 'Grose Jet' valves. They are not, strictly speaking, original, but are much more effective and reliable than the original valves. The Moss part number for H and HS Type SU carbs is GAC9201X, and the Zenith Stromberg carbs carry part number GAC9200X.

With or without Grose-Jets you need to check that fuel levels within the float chambers are set at a sensible level, and are more or less equal in each carburettor. Use the Repair Operations workshop manual to check float levels. Alternatively, with the piston/needle assemblies removed again, and aided by a battery-powered torch/flashlight (do not use a mains powered lead light as it gets too hot to use in close proximity to fuel), look down into the jets at the fuel level. You may (almost certainly will) need to ruffle the surface of the fuel by blowing gently. The fuel level should be approximately 5-6mm (1/4 in) below the top of the jet. If it isn't it means that the floats, float pivots (you'll be amazed how much they can wear) and needle valves will require some attention.

Are the tappets correctly adjusted? When significant engine wear occurs the inlet vacuum will reduce and the induction requirements of the engine will change. It is possible to tweak the carbs to partly compensate for this, but the car's performance will be below par and, in any case, proper carb adjustment is hardly possible when engine wear is advanced.

As the name implies, the dash pots in both SU and Zenith-Stromberg carburettors are designed to dampen/slow the piston as it moves up and down in the carburettor bore in response to the inlet vacuum created by the engine and driving conditions. As you increase throttle opening, the engine demands more air, which increases the vacuum within the inlet manifold and the carburettors in particular. Consequently, the piston rises, presenting a narrower section on the control needle and allowing more fuel to be sucked into the incoming airflow.

As far as the damping effect is concerned, I am sure you will appreciate that the thinner the oil in the damper/dash pot, the faster the piston will rise and the quicker the injection of fuel. So thin oil sounds just the business? Well, yes, provided you are not already over-fuelling the engine. It's usually a matter of trial and error, so start with 20W (SAE20), or somewhere in the middle of the range, and subjectively decide what seems best for your engine. Any subsequent significant carb, needle, ignition or engine changes may require you to re-trial the damper oil. ATF fluid is the thinnest oil and would be categorised as having a 5W (SAE5) viscosity. It will allow your piston to rise in the shortest possible time. You will find a 3-in-1 oil in a red can has a 10W viscosity rating, while 3-in-1 in a blue can is 20W. In the UK Hermetite offers a special carburettor damper oil at SAE 20 viscosity. I dislike engine oils as the viscosity thickens as the temperature increases. However, many successfully use 15W-40 or 20W-50 engine oils at the thickest (slowest rise) end of the range. Do not overfill the dampers. About 12mm (0.5in) above the top of the piston is sufficient.

ADJUSTING IDLE SPEED

Do not try to get your TR to tickover at too low an rpm. Obviously, it depends upon the level of tune and the car in question, but, in general terms, consider 700/900rpm as the most suited idle/tickover speed.

If this is difficult to achieve look for the obvious mechanical causes. The jet/choke linkage position is the first, while obviously you should have a small clearance between the fast idle adjusting screw and the fast idle cam. If all is well here, or if you experience difficulty in achieving a consistent idle, check the induction system for vacuum leaks.

Start with a close visual examination of the carburettors and inlet manifold; do not forget the servo valve and/or vacuum piping, if fitted. If there's nothing obvious the next step is to start the engine. Set it at a fast idle/tickover and spray the various joints and spindles with an aerosol of carburettor cleaner (Moss part number MRD1023). Have a partner watch the exhaust for a changed (usually darker) emission whilst you listen for a slight change in engine tone. A systematic approach is usually helpful, and I would start with the manifold to head joint and spray test each possible source of a vacuum leak progressively nearer the carburettor spindle(s). Do not spray carburettor cleaner into Zenith-Stromberg carburettors without first removing the diaphragms, as the carburettor cleaner will quickly render the diaphragms useless. This does not apply to SU carbs, although you're not likely to reveal a vacuum leak from the inside of the carburettor.

Chances are that your vacuum leak test will reveal at least one worn throttle spindle, which will need repair before you progress to carburettor tuning and balancing. We will explore how to carry out this repair shortly, but now it's time to check and adjust carburettor balance.

BALANCING TWIN CARBURETTORS

This procedure is explained in detail in the Triumph Repair Operations manuals. It is largely the same for both types of carburettor, and revolves around the changes in engine speed that occur as the mixture is altered.

The air cleaners should be removed, the inter-carb throttle linkages loosened and the tickover/idle set at about 800-900rpm. Using either a small screwdriver or the carb's in-built piston lift button, raise one piston about 3 or 4 mm (0.15in). The engine will respond in one of 3 ways:

- An immediate increase in engine revolutions will signal that the carburettor is adjusted to give too rich a mixture. Expect the increase to be about 50-100rpm, so, although a partner watching the tachometer could be useful, your clearest sign will be a change in engine note. Close the main hexagon by one flat and try again.
- A reduction in engine revs signals that the carb is currently presenting too weak a mixture; the adjustment hexagon should be lowered by one flat and the test repeated.
- A slight change in engine rpm (a momentary hesitation), quickly followed by a return to the previous rpm, indicates that the carburettor mixture is correct.

Adjusting SU carburettors is relatively straightforward, but it's as well to mention that, if you have Zenith-Stromberg carburettors, you should buy or borrow a special tool that passes down through the top of the carburettor. This tool stops the piston from turning (which will damage the diaphragm) and, via a hexagon key, adjusts the mixture. These procedures are effective, but you cannot better a modern gas analysis tune-up, which will check CO, CO2 and unburned hydrocarbons and balance the carbs, too.

If following the DIY route, you need now to adjust the butterfly settings to provide equal or balanced air flows through each venturi. You can buy a proprietary balancing tool, but quite satisfactory results can be achieved by using a 12 inch length of 0.25 inch bore rubber or plastic tube as a stethoscope. You are trying to achieve an identical sound (hiss) from each carb, so listen to the (same) inlet area of each carb and adjust the butterflies/throttle spindles until they sound the same.

Correct adjustment of the 'choke' or fuel enrichment mechanism of any carburettor is of great importance, and may be tackled at this point. The later Zenith-Stromberg uses a conventional 'choke' or flaps which pivot to restrict airflow through the carb venturi, thus increasing the flow of fuel. The earlier SUs do not use the conventional method. Instead, depending upon the type of SU, they either draw the lower jet downward to increase the area between needle and jet, or employ a bar that pivots up under the air pistons, restricting airflow and increasing fuel flow. Before choke adjustment can be made accurately, the choke cable should be disconnected from the carburettors and the instructions in the workshop manual followed.

REPLACING THE THROTTLE SPINDLE

Well, having tried everything else, there will come a time when you decide that the throttle spindle(s) need replacing. We will look at this operation to establish if it is necessary for you to bush the body of your carburettors.

It's important to understand the reason for this refurbishment work in order to help you decide whether it is really necessary, and to ensure that it will provide the remedy you are seeking. Wear between body and shaft results in 'play', revealed by simply wiggling the shaft in its body. As unlikely as it sounds, this wear will result in a vacuum leak, which will cause an inconsistent (often rough) tickover, poor performance and poor fuel economy. The greater the wear the more pronounced these consequences. Some wear is to be expected, but you will be unable to adjust the carburettor to compensate once the 'play' becomes 0.005in.

As explained earlier, carburettor cleaner sprayed (both sides) where the shaft exits the body will tell you what you do not want to hear. If the tickover changes for a moment, your shafts and/ or bodies have unacceptable vacuum leaks and need rectification.

There are some excellent workshop manuals about that make detailed instructions superfluous here, but let us explore a few tips and the general principals of what might be involved. If wear is not too bad, the solution may be as simple as replacing the throttle shaft(s). Identify the top of the butterfly plate(s) with felt tip pen or typewriter correction fluid, before closing the splayed ends of each of the two butterfly grub screws and removing the screws, butterfly(s) and shaft(s). Offer an unworn length of shaft to each side of the carburettor body and test for body wear by moving the shaft up, down, in and out. Any movement over and above the very minimum will almost certainly mean body and spindle wear. If movement is minimal, a new throttle shaft of standard diameter should solve the problem satisfactorily.

If movement is clearly present but judged not excessive, consider using an oversize throttle shaft. These take up slight wear in the carburettor body, but are best fitted by your local engineering shop, as each body needs to be through-reamed to a specific size that is very slightly larger than the diameter of the new shaft. The most drastic solution is required when the carburettor body is too worn for oversized shafts to be effective. In these circumstances, the body will have to be cross-drilled, reamed and fitted with a pair of phosphor bronze bushes, or service exchanged for a re-bushed body. If your carburettor bodies have got to this state I recommend you go the service exchange route as re-bushing to achieve a subsequent smooth-operating, standard-sized spindle is best left to the professionals.

When reassembling your carburettor(s), remember to refit the butterfly the correct way up, and to splay the butterfly securing screws as soon as you are sure the butterfly fully closes the carburettor venturi.

Fuels are getting more and more volatile, which means they boil at lower and lower temperatures, which is bad news for our classics, particularly when the carburettor(s) and fuel-lines are above the exhaust manifold (as is the case with all Triumph TRs). Consequently, many older cars experience fuel vaporisation either in the fuel line or in the carburettor itself. This results in (when hot) running on, pinking, poor performance, lumpy running in traffic, cutting out (sometimes with no restarts for 20 minutes), and a variety of other running issues, and it is probably not your carburation or ignition that is faulty. However, you can minimise the problems by insulating the carburettors and siting/insulating fuel lines as you re-assemble the car.

Picture 11-2 showed an additional rear heat shield in place between the carburettors and engine, and suggests further thermal improvement using reflective/thermal insulation. Picture 11-5 also shows some interesting insulating details too, and highlights the need to reduce radiated heat from the exhaust.

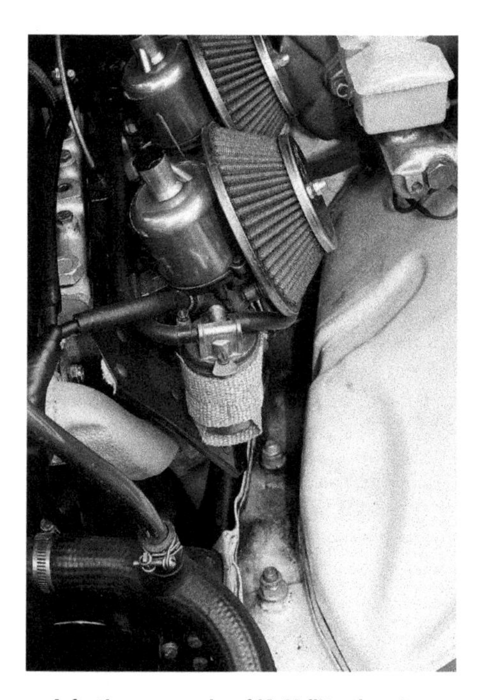

A further example of K+N filtration, but the primary purpose of the picture is to illustrate thermal insulation methods. This cast-iron manifold/header has (wisely) only been partially wrapped to reduce radiated heat, while the same silicone tape has been wrapped around the float bowl to reduce heat percolation and fuel vaporisation. You should not fully wrap a cast-iron nor a mild steel fabricated exhaust header, because cracking or distortion respectively can take place. A thermal blanket would have been a better option to more extensively cover the manifold, and self-adhesive reflective foil/ insulation might have been an alternative and perhaps more effective method on the float bowls.

Chapter 12

Front suspension & steering

Numerous important changes/improvements have been established by the TR fraternity over the years. Certainly, when the car is undergoing a complete restoration, the opportunity should be taken to not only restore the car to its original status, but also weld a number of additional strengthening brackets and gussets to the front suspension. Such structural improvements can mostly be accomplished when the chassis and body are united, of course, though few would disagree that they are easier with the body removed from the chassis.

I consider these welded changes to be chassis related, so have covered them in the chapter devoted to chassis restoration and improvements.

The main focus of this chapter is on matters to do with maintenance, improvement and/or repair of standard cars. Owners of IRS cars should read it in conjunction with chapter 5, for it might be necessary to also carry out some of the structural changes.

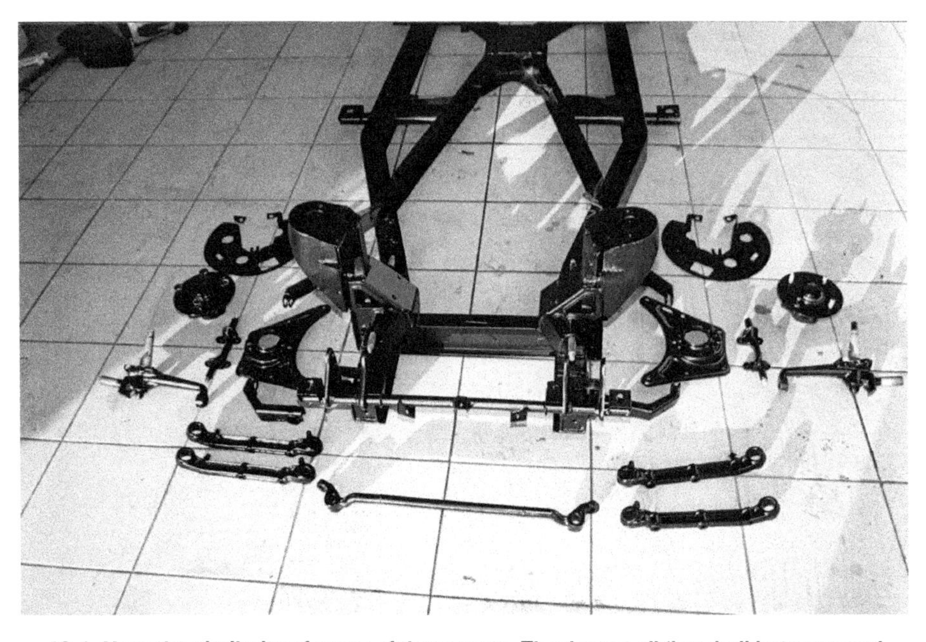

12-1. Note the similarity of many of these parts. They're not all 'handed' but you need only assemble one wrongly for it to mean that there will be a correspondingly incorrect one on the other side of the car in the not-too-distant future. Note the holes in the lower wishbone spring pans – they're ideal for removing and replacing the telescopic shock absorbers.

REASONS FOR A SUSPENSION REBUILD

There are several reasons why you may feel that the front suspension or steering needs some attention over and above routine maintenance. Stiffness in the steering, or mechanical noises in the front suspension when changing direction and/or when braking hard and/or reversing are not unknown, and all must certainly receive immediate attention. In this context, the springs, shock absorbers, wheel bearings, hub, ball joint, trunnion bolt and bushes, and top and bottom wishbone bushes all deserve close investigation, and there are four particularly important details to be aware of since you need to:

- Check the lower chassis and suspension mountings for evidence of fatigue or accident damage.
- Ensure all suspension mounting bolts are of adequate length so that none are relying on the threaded part of the bolt as a bearing surface.
- Check that the steering rack mounting pads on later cars have been pushed sufficiently outwards on assembly to prevent rack movement.
- If your steering is stiffening, establish where the fault lies. Any stiffness in the steering should be taken very seriously; consider the situation particularly carefully if either of the bottom trunnions are suspected of being at fault. They are known to snap after first exhibiting symptoms of stiffening.

There's more on each of these matters as this chapter progresses, but first let's consider the front suspension and steering on a less critical level.

Overhauling TR suspension, front or rear, is not, generally speaking, difficult, so I do not intend to duplicate the information given in the operations manual. There are a few pitfalls with the front suspension, and we will explore these in some detail in just a moment.

Guard against getting handed parts mixed up; they look very similar when off the car, as photograph 12-1 shows. It may be sensible to tackle one half of the front suspension at a time. I have bought several 'incomplete restoration projects' from inexperienced owners who have stripped everything in sight, piled it all into one huge box and then realised they have lost all reference to what goes where and in what order. Mind you, it's very hard to finish the front suspension before the chassis is fixed in a complete restoration, when all the various sub-assemblies are likely to have to come off the chassis before it can be repaired. In this event take lots of photographs of each sub-assembly *before* you start removing bits of car! Identify the parts by name, number or position and tie a label to each, as per photograph 12-2. Make some sketches of things you cannot photograph well and keep the information in separate folders for easy reference.

Buy a copy (it does not have to be new) of the parts manual for your car. When taking a sub-assembly off the car (and certainly when you take the sub-assembly apart), take a photocopy of the relevant page in your parts manual and mark the copy up with additional information, such as assembly sequences, angles and spacers, and even which way round the bolt-head/nut goes. The parts manual will usually have the assembly sequence information shown in an 'exploded' illustration, but 6 or 12 months later that may be insufficient. If you do not plan to refurbish a sub-assembly – say, the front suspension – immediately, try and remove it from the chassis in the largest complete piece you can, and only strip it to component parts when you are ready to refurbish it. As you strip major assemblies from the car, try and replace fastenings in their respective places.

12-2. It is obvious what parts we are looking at, but note and follow this example of identifying and labelling each component. Front suspension parts are particularly easy to mix up, confuse or get the sides muddled, which makes this job very worthwhile. Some components are handed, so mark the 'top' or even 'this side up' on each part. Note, too, that the various fastenings have been temporarily replaced: this really is a good idea.

Unfortunately, the foregoing advice is the exact opposite of that elsewhere in the book, when refinishing components are dealt with 'in-bulk' for maximum cost effectiveness. You'll appreciate, however, that there's no point in saving money by bulk refinishing components if you have no idea what goes where!

IMPROVING HANDLING

TRs are noted for having good handling characteristics, so, assuming your car is operational but not handling to your satisfaction, let us first look at some of the most common reasons why before we go into the detail of a complete front suspension rebuild.

Inevitably, there will be some reference to rear suspension details, as poor handling cannot be solely down to the front suspension. Since it is hardly practical to separate them in this context, I hope you will forgive a slight overlap. If your front suspension work involves removing and/or replacing the front springs, do read the relevant operations manual and take note of the pointers that follow later in this chapter.

The TR2, 3, 3A and 4

The complete standard sidescreen suspension can be seen in photograph 12-3. There are all sorts of chassis/steering improvements that can be contemplated but, in general, we will focus here on the more mundane standard details.

First up, the lower wishbone arms on all ladder chassis cars (TR2/3/3A/4) are not shimmable, as photograph 12-4 demonstrates. If your chassis is in any way out of true, then the car is unlikely to handle well, and the only solution is to get the chassis repaired; likely to be a body-off job.

The upper fulcrum pin (part

12-3. A good view of a Sidescreen TR front suspension assembly. Note the fully tubular top wishbone and compare it and its top ball joint to the later example shown in photographs 12-6 and 12-7. Sharp eyes may possibly spot the brake caliper differences – with this earlier car's bridge pipe connecting the two halves of the caliper. Later calipers were internally drilled to carry the fluid to the outer piston without a bridge pipe.

12-4. A nice view of a beautifully repaired and powder coated TR2, 3 or 3A chassis, identified by the non-shimmable 'through' lower wishbone pivots and steering box mounting.

12-5. The two ways of assembling the top fulcrum pin. It may be hard to see, but can you spot the worn 'necking' (arrowed) on the right side of the top one, making it suitable for the scrap bin only?

number 200659, photograph 12-5) on all cars deserves your close attention during assembly since the same pin can be fitted two ways, depending upon the model. If you are rebuilding your car, get it the right way round before you fit the bodyshell as it is ten times as difficult to correct with the body in place. If your car is complete and operational but does not feel right, or if you have trouble adjusting the camber angle, probably the best thing to check first is that the upper fulcrum pin is assembled the right way round. More details follow shortly.

While discussing putting the front suspension together the correct way round, note that the word 'Top' is stamped on the upper rear wishbone arms. Not only does this mean the arms fit to the top wishbone assembly, but it is also intended to ensure that they are fitted 'top way up'. Put another way, the word 'top' should be visible when looking down on the assembled front suspension. The arms are handed and, if you follow this suggestion, you will fit the right arm on the correct side of the car. The front arm is not handed and can be fitted either side.

The TR4A/TR250/TR6

The front suspension can be seen in photographs 12-6 and 12-7. From the introduction of the TR4A, Triumph revised the mounting of all lower wishbones to the car's chassis. Two shimmed brackets, each with a removable bolt passing through a rubber-to-metal-bonded ('metalastic') bush, were introduced. This made adjustment, stripping and re-assembly a much easier process, although the resultant bottom suspension mounting is less rugged than the original design.

It is in these lower wishbone mounting bracket areas that most of the chassis modifications, mentioned earlier, emanate from. If your IRS car

12-6. A TR250 front suspension differs very little from that of earlier cars; were it not for the 6-cylinder inlet manifold, it would be hard to identify this car. The turret brace is well in evidence, and you will note two body mounting holes in the brace. Note the grease nipple to lubricate the top ball joint. There is a second nipple each side in the bottom trunnion, and it is most important that this is regularly greased. All four grease nipples should receive attention every 3000 miles.

12-7. The front suspension from an IRS car with rack and pinion steering is clearly quite different from the Sidescreened car's arrangement shown in photograph 12-3. The two-piece channel section top wishbones are also very clear.

handles poorly, first off have a four-wheel alignment check carried out. This is a job for your professional TR restorer since there are various shims to adjust. If all is still not well, you need to rectify the car's general tendency to rear end steer by fitting polyurethane bushes to the rear trailing arms and better rear shock absorbers. Superflex bushes are recommended in the former case and Koni or Spax telescopic shocks in the latter. The shock absorbers will require a pair of special mounting brackets, but these changes will transform the overall handling, and the rear end in particular. We will look at these in more detail in chapter 13.

Still not happy with your car? Then it's the front suspension we focus upon next, starting with the bottom wishbone pivot bushes, which are the sturdier 'Metalastic' type with a steel inner sleeve. You will have to press them out in the vice and, whilst direct replacements will be appropriate after years of service, there should be no need to upgrade the bushes. However, the top wishbone bushes come in two halves and will probably push out without too much effort. They are much less satisfactory and polyurethane/Superflex bushes in the top wishbones are strongly recommended as they will improve handling and last longer, with minimal increase in road noise.

While you have the front suspension apart I suggest you fit a pair of slightly uprated lower springs. The TR4A, 250 and 5 did not have a front anti-roll/sway bar, and you should fit one of the sort used in the '6. This tightens up the front end dramatically and is completely reversible if you subsequently need originality restored. We will discuss solid steering rack mounts shortly, but, if you are still concerned, it is time now to look at the upper fulcrum pins in more detail.

The upper fulcrum pin

The two ways that this pin can be fitted are shown in photograph 12-5. This is a frequently made mistake, so it's worth checking yours, even if you aren't unhappy with your car's handling or steering.

On early cars (TR2 to TR3A), the pin is positioned as shown in the workshop manual. You can take off a wheel and feel the fulcrum, which should be fitted with the curve radiusing inwards, away from you. To help further, the curvature of the pin seems the correct shape relative to the turret, as it seems to be following the rear contour of the turret. On TR4A to TR6 cars it looks as if the pin has been assembled the wrong way round as it seems to run against the turret's curvature when correctly fitted. It should curve outwards towards you.

The pin on a TR4 could go either way! In order to change the front geometry, Triumph switched the position of this pin in 1962, and changed several other details simultaneously at TR4 chassis number CT6343 (wire wheels) and CT6390 (steel wheels). If, therefore, your car is about that commission number, you could do worse than take professional advice from a TR specialist if you have doubts about handling, or before you re-assemble the front suspension.

To expand on the changes; note that early TRs had no caster angle built into the suspension geometry. From the above chassis numbers a three degree castor angle was introduced, with the result that the top of the vertical link leans backwards slightly, and is nearer the rear of the car than the bottom. The situation is actually helped by the simultaneous changes that Triumph introduced. These will give you a clue to which design of suspension your car is intended to have, as an improved/ different top ball joint was introduced and the trunnion became handed.

Be aware, however, that the one component that does cause more problems than any other is the upper fulcrum pin.

With the above suggestions in place, you will probably have transformed the handling of your TR. If, however, a more rigorous refurbishment is required, read on ...

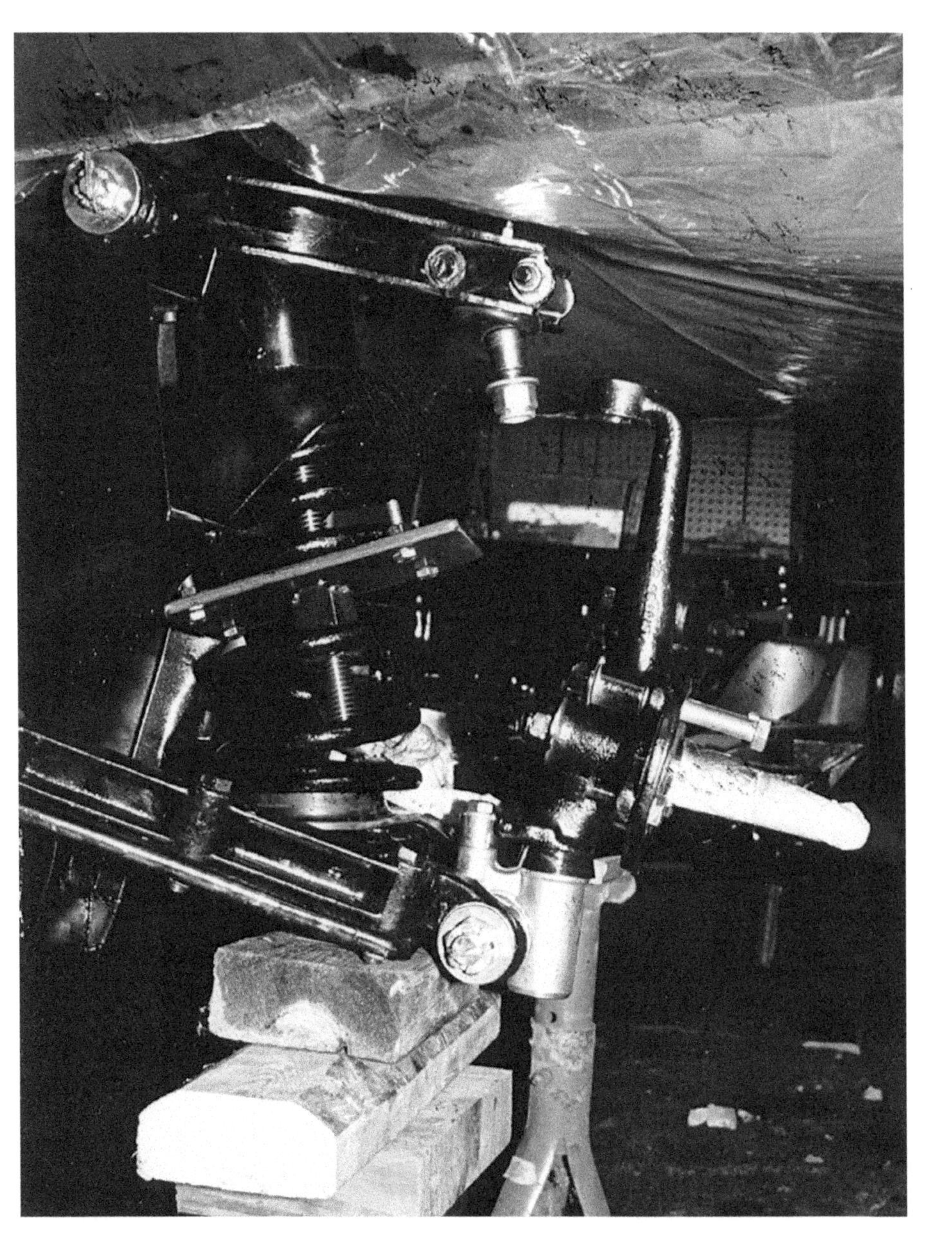

12-8. This picture shows the front swivel pin with an internal front spring compressor in action. The vertical link is free to move as required, and the top ball joint is positioned ready to be coupled up when everything is nicely located. The trolley jack at the bottom of the picture will lift the assembly whilst the top ball joint is connected.

IMPROVING THE STEERING

Regardless of which model you have, take great care if the steering is, or becomes, tight in any way. While this is not a very frequent problem, the consequences are such that it is worth alerting you to it.

The potential consequence of ignoring stiffening steering is that one vertical post, seen in photograph 12-8, snaps clean off at the top of the threaded portion. In the first instance, suspect one or both of the lower swivel pins, or trunnions, as they are officially known. There are 3 types which vary with caster angle, but all do the same job of allowing the steering to swivel where the lower wishbones pivot. The best initial test is to jack up both front wheels and take hold of either road wheel and swivel it. It should move from full lock to full lock without stiffness or binding. If it is hard going, take one track rod end off and try to deduce which swivel pin, if either, is causing the trouble, and drop that one for individual examination. With the suspect isolated you should strip the trunnion and establish that there is no untoward wear. Use liberal quantities of lubricant (more on this shortly) to prevent re-occurrence during re-assembly. Fully tighten the cross bolt/nut and liberally grease the trunnions at frequent intervals in future. The workshop manual will tell you that you should oil the bottom trunnion, but the oil drains out. So you're better advised to use grease, but frequently.

If you have up and down wear/play in the steering wheel of your TR2/3/3A, you need to replace the top steering column felt bush. This play occurs as a result of the up and down movement of the inner column, but the replacement felt should be soaked in oil for at least a week prior to fitting. The oil impregnates the felt and delays it drying out, which will reintroduce the play. This is, of course, a non-original improvement, but you can substitute the later, rubber-based bushes from, say, a TR6. We will look at other "improvements" in a later book.

In order to improve the actual steering on the sidescreened cars you have several options. Steering can be dramatically improved by the use of a properly rebuilt steering box. Rebuilding a box can be an expensive operation, but can make an unbelievable difference, so if your current box has play and/or the car's steering feels like that of a tractor, ask a TR specialist about the cost of an exchange box. Revington TR may have a satisfactory and relatively low cost solution via its special spring-loaded steering box replacement top, which is designed to not only reduce play, but be self-adjusting thereafter.

However, before getting too deeply into rebuilding or replacing the steering box on a sidescreened TR, do take stock. Space here dictates that I no more than touch on the alternative, which is to replace the box with a steering rack. This is a non-original improvement and will receive our full attention in a later book. The improvement is so dramatic – even compared to a rebuilt steering box – that I could not fail to mention it here.

While a rack and pinion kit is currently available from three TR specialists, I must also mention that there are differences in the offerings. Revington TR's modification is undoubtedly best for competition use, but involves welding brackets to the chassis of the car which, of course, are subsequently not always easy to remove. TR Bitz and Proteck offer bolted conversions that are well suited to fast road application, are easier for home installation, and certainly much easier to revert to original should that ever become necessary. I would also recommend this route to owners contemplating a LH to RH drive conversion.

TR4s used the more direct steering 'rack' as standard. These are easily replaced and are not expensive, so I suggest you exchange your rack almost as a matter of course when rebuilding your later car, or if you are in any doubt whatsoever about the smoothness and operation of your rack.

I mentioned earlier the truly dramatic change that a steering rack can bring to a sidescreened car. The TR4 took that further with the introduction of steering via rack and pinion, which improved directness and feel very considerably. As photograph 12-9 shows, however, the early TR4 racks were mounted on two tall ear-like brackets which, not to put too fine a point on it, were very Mickey Mouse-like. The rack was very securely mounted to the floppy 'ears' by commendable solid aluminium blocks and U-bolts.

At CT20063 (LHD) and CT20265 (RHD) the rack movement allowed by these vertical eared brackets was improved by using a much more rigid

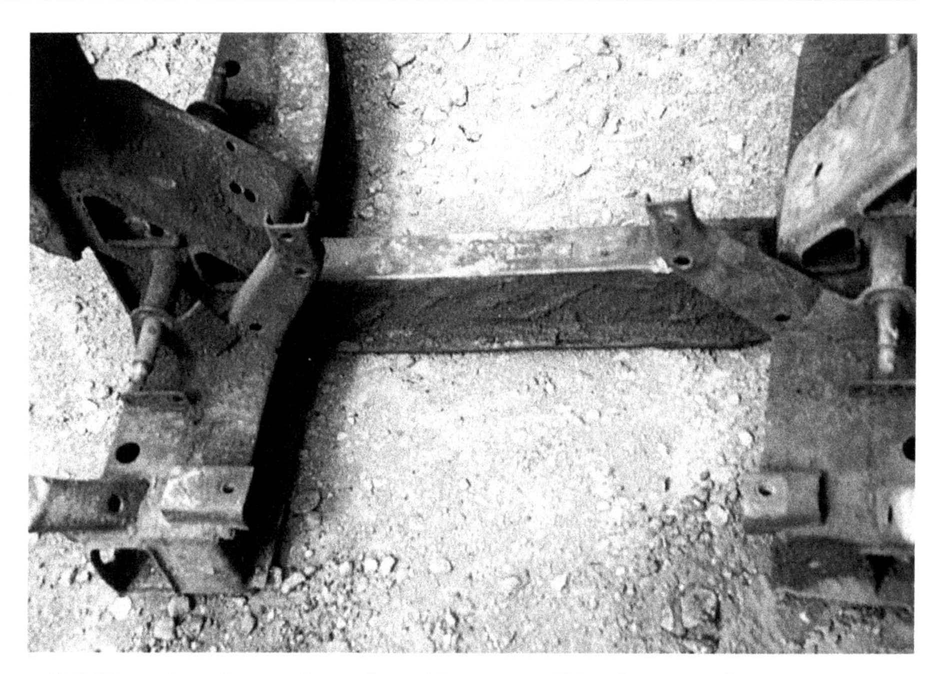

12-9. The early rack mountings allowed for a very solid rack to mounting arrangement; trouble was the 'Mickey Mouse' brackets were not as rigid as they should have been and were replaced in later TRs.

12-10. The later steering racks are mounted on rubber bushes, but all too rarely are these bushes installed correctly. They do need to be spread outwards prior to the two 'U'-bolt fastenings being tightened in order to avoid the (slight) sideways rack movement that can otherwise occur. There is a proper tool that works like a turnbuckle on both rack clamps simultaneously, which is shown in this picture. However, the trade normally uses a pair of wide-nosed mole grip-like welding clamps to pull each 'U'-shaped clamp outwards before individually tightening each.

rack mounting via a horizontally-mounted chassis plate. The overall length of the system, ball joint to ball joint, remained the same as on earlier cars but, simultaneously, both track control arms increased in length so the rack had to be shortened. This is one point to note, should you ever go looking for a TR4 steering rack, for there are two non-interchangeable types!

However, all was still not absolutely wonderful as the original solid aluminium mountings were replaced by rubber bushes held captive by a U-shaped clamp, and the later/ rubber bushed rack can move within the mounting rubbers after a couple of years. This movement can be improved and/or delayed by ensuring the U-clamps are pressed apart before the nuts are tightened. A turnbuckle tool is helpful (see photograph 12-10), or pre-tensioning can be done by weld clamps.

Many individuals fit a pair of the original aluminium/solid mountings to later cars. This modification gives a very direct feel to the steering at the expense of increased harshness, but is completely reversible if not to your liking, or you wish to restore originality. The same modification is also applicable to TR5/250/6s.

Significant steering column play at the wheel in a TR4-TR6 is probably due to the securing bracket on the inside of the bulkhead/firewall having fatigued and broken. It is a one-piece pressing made from circa 18swg material, available as part number 815834SB from all our premier TR restorers. As an aside, we will touch upon this very same bracket when we talk about left-hand to right-hand drive conversions later in the book.

If you are rebuilding your car for Concours d'Elegance purposes then you have little choice but to make a welded repair to the original bracket, or replace it. Either way, a great deal of work is involved in creating sufficient space to carry out the necessary welding if repairing a running car. If a total refurbishment is being done, and you are not planning to enter concours competitions, consider making up a bracket with a much stronger 'bridge'. My original bracket had more or less disintegrated, and I fabricated a completely new bracket, mostly from 16swg material, but with welded in 1in x 0.125in steel for the all-important column bridge/support. Furthermore, I supported the column with a second 0.125in (3mm) cross packer to ensure that however tight I pulled up the 'U' clamp nuts, there was no chance of distorting the new bulkhead bracket.

Alternatively, for those reluctant to strip or weld beneath the assembled dashboard, I guess a new bracket can be fabricated from a section of aluminium or steel angle and bolted in place. However, I think for something as important as this bracket I would consider this to be no more than a temporary measure to get me through the fun months, stripping the dash and welding a stronger-than-original replacement during the winter months.

ADDITIONAL TIPS AND SUGGESTIONS

- Check your vertical links carefully as, believe it or not, it is possible to bend them in an accident. If your wheel rim rubs on the top ball joint, a bent vertical link is the most likely cause.
- On early cars (TR2/3/4) the wishbone bushes need reaming after being pressed into the wishbones. Few home restorers will have a 5/8in reamer, so you will need to find a (probably small) local engineering business that can run a reamer through each of the bushes. If you pre-arrange the task, it can probably be carried out while you wait.
- Ball joints and wheel bearings should be replaced almost as a matter of routine. However, there is one very important detail to beware of when fitting a replacement wheel bearing kit – the felt oil seal is always too thick. The current felt is not to OE specification and needs to be cut with a razor or similar sharp blade to approximately half its thickness. If you fit the felt washer as supplied in the kit, you will find it impossible to tighten the wheel bearing nut properly. You may think you have done the job well and the torque wrench may 'click', but the over-thick felt will have stopped the bearing spacers seating as intended.
- The front wheel bearings need a certain amount of float, as the workshop manual calls for, to allow for the bearings to warm up and expand, but to do so without seizing/freezing. As a consequence of assuming all 'play' is due to this clearance, many home restorers fail to recognise a couple of mistakes when reassembling the front hubs. Firstly, your outer bearing ring (the 'A' ring) must be an interference fit in the hub casing. If it 'falls' in there will be an unacceptable degree of play in that front wheel, which no amount of bearing replacement or adjustment will eradicate. You can use the correct grade of Locktite to hold it in its housing, but personally I would find a replacement hub as soon as possible. The second fault that new TR-ers think is of little consequence is to overlook a worn bearing seat on the stub axle. The inner (or 'B') ring should, of course, be a sliding fit on the stub axle but, for a variety of reasons, a previous owner could have had an inner ring rotating for a while, and this will have worn or polished a witness mark, even a groove, on the stub axle. It's tempting to think that a new pair of properly adjusted bearings will resolve the resultant play in the front wheel, and you may be lucky, but in all probability the play will remain or return after a short while because the 'B' ring is moving ever so slightly on the shaft. Like the sloppy outer bearing ring in the hub, that movement is multiplied many times to give play in the front wheel.
- The trunnions that form the lower steering swivel are supposed to be oiled as part of the car's routine servicing. I mentioned a short while ago that you should liberally lubricate the trunnions upon re-assembly with trunnion oil, and frequently grease them thereafter. You may be interested in one professional 'dodge'; provided you liberally oil each trunnion with Hypoy EP90 oil on assembly, you can use a conventional grease during each service, secure in the knowledge that this oil will keep the grease soft.
- Earlier I recommended replacing the rear lever arm shock absorbers with Koni or Spax telescopic units. Whilst the improvement might not be as dramatic, many TR owners advocate a similar change at the front. The original front shocks are, of course, the telescopic type but are non-adjustable. However, both Koni and Spax are adjustable and consequently can be fine-tuned to suit your driving and car use. Spax are adjustable *in situ* but you will need to take the usually more expensive Koni's off the car to adjust them.
- The shank of a bolt is that plain

portion between the underside of the head and the point at which the threaded section of the bolt starts. There is a danger that, at some earlier date, a bolt or bolts with too short a shank length have been used to assemble the front and/or rear suspensions. The consequence of this is not immediately apparent but, over the years, a threaded area of the bolt bears against the bracket, wears prematurely and introduces play into the suspension mounting point(s). This is clearly very undesirable, but a more serious consequence is that some lower fulcrum bracket failures are attributed to the snatching loads imposed by the resultant play in these suspension components. Therefore, do ensure you purchase bolts of sufficient length to provide a complete unthreaded surface right through the assemblies, even if you have to add an extra plain washer to take up any surplus bolt shank.

- All rubber bushes throughout the front suspension will need replacing if you are carrying out a full rebuild. However, the OE upper fulcrum pin bushes in particular are – not to put too fine a point on it – useless. I would advise fitting – admittedly non-original – polyurethane bushes throughout the front suspension. If you do not wish to fit polyurethane bushes throughout the whole front suspension, you would be well advised to fit them to the top – unless you fancy a little job each year! If your car is not handling to your satisfaction this would likely be a worthwhile step, whatever model TR you own. Superflex front suspension bushes are recommended at the expense of slight additional road noise.
- Take care not to tighten the inner fastenings until the suspension is loaded and the car assumes its road going stance. If you tighten the bolts too soon you will eventually force the bushes to flex into areas they were never intended to and they will shear. On the other hand, make a careful note somewhere that the fastenings remain loose, because you must not forget to fully tighten them once the car is on the ground.
- It's not a great idea to leave your TR suspended on its new front suspension bushes for several years while you sort out the body work. You certainly need to stand IRS cars on their suspension while you sort out panel fit, so why not leave the old suspension in place until the car is nearing the end of its restoration journey, before fitting new rubber suspension bushes?

12-11. Three of the various spring compressors available. Top is a proprietary internal compressor ideal for use on TRs. Right side below is another internal/front unit that has been slightly modified. The original 'ears' were cut off and the end replaced by a thick piece of steel drilled with a clearance hole for the threaded rod. The four bolts are placed so as to centre the plate on the coil spring. This unit can be seen in action in photograph 12-8. Bottom left is one of a pair of external compressors; these must be used in pairs and can be seen in use in chapter 13 (rear suspension).

FRONT COIL SPRING COMPRESSOR

With the engine and body in place, as would be the case with a running car, you probably won't need a coil spring compressing tool, since the weight of the car/engine/body assembly should be sufficient to compress the road spring using a trolley-jack. Note the specific reference to a trolley-jack; only this type of jack has the stability to allow you to contemplate this task in reasonable safety. Do not try compressing a road spring with any other type of jack, and in any case only if your car is virtually completely assembled.

If your car is largely un-assembled, you will almost certainly need a coil spring compressor. Photograph 12-11 shows various types of spring compressor and in photograph 12-8 one can be seen in use. You can buy or make one, but an internal one is recommended in view of the space restrictions. If you wish to buy yours it should be from a TR specialist such as Revington TR. Spring compressors can be made at home from something like $^{1}/_{2}$ in UNC studding. However, appreciate that the forces within a coil spring under compression are considerable, and you must ensure that your compressor is made from material that is comfortably up to the job. Remember that a flying coil spring can cause much damage.

With an external compressor you will need to control and compress the lower wishbone in order to squeeze the spring up against the top spring/fixed chassis 'cup'. Better yet, you can safely unbolt the four fastenings at the bottom and one nut at the top to release and remove the shock absorbers. The shock absorber comes right out through the bottom wishbone, leaving plenty of room for an 'inside' [the spring] compressor.

The 'inside' compressor basically consists of thick plate steel with a hole in the centre for a threaded rod to pass through, with large nuts and washers top and bottom. Slowly, evenly and under control, release the spring tension after it has been removed from your car.

Chapter 13

Rear suspension and differential

TRIUMPH IRS SYSTEM

Most TR4As and all TR5 and 6s have a rear suspension set-up that allows the two rear wheels to move independently of each other, which should allow a more comfortable ride and better road holding.

With a live rear axle, the road surface will cause one wheel to move up and down, and this movement is transferred to the opposite wheel because of the beam-like construction of the axle. The angle of the second wheel to the road consequently varies, which can have an adverse effect on road holding capability, hence the desirability of Independent Rear Suspension (IRS), where this effect is minimised, if not completely eliminated. Furthermore, IRS offers the designer an opportunity to reduce the unsprung weight on the rear wheels by mounting the differential on the chassis, instead of in the rear axle. This also improves road holding at the back end.

The TRs in question have a type of IRS called 'trailing arm'; the rear wheels and hubs are kept in place by the two trailing arms shown in photographs 13-1 and 13-2. These arms are supported upon a pair of coil springs. Two shock absorbers control suspension movement, and one side can be seen in the general view provided by photograph 13-3.

13-1. A nice illustration of the trailing arms and the difference that can be achieved between before and after refurbishment. The top trailing arm is as it came off this TR250, whereas the lower arm has been thoroughly cleaned, rebushed and finished by being sprayed with a clear plasticoat.

On each side of the differential is the IRS equivalent of a halfshaft (photograph 13-4), officially called the 'outer axle shaft assembly'. An outer universal joint terminates in an outer axle shaft that passes through the rear hub housing. The hub(s) are mounted to the trailing arm via six studs, as shown in photograph 13-5. An inner universal joint terminates in a flange that is bolted

13-2. An excellent view of the rear suspension trailing arm, its mounting brackets, and the all-important chassis mounting member. Note that both mounting brackets have identification notches on the top edges. The trailing arm bushes are polyurethane. Did you correctly identify the flexible brake hose mounting on the top of the trailing arm?

13-3. The complete IRS rear suspension. The top bump stop is clearly evident, as is the coil spring and shock absorber. Can you spot the outer handbrake cable attachment point?

13-4. This may look familiar at first but, in fact, it's not quite what it seems. This new driveshaft and hub are one way you can uprate the rear driveshafts using Revington TR's rolling 'splines', which provide four sliding ball bearing-type interfaces down the length of the driveshaft. Clearly, this eliminates friction and any tendency for the splines to lock, even under higher-than-standard torque conditions. The assembly fits through the TR hub in the usual way.

to the rear axle drive flange on the differential.

With the differential fixed to the chassis in effect, and the rear hubs moving as road conditions dictate, both rear driveshafts need to vary (slightly) in length and angle; hence the provision of a sliding spline in the middle of each shaft, and a universal joint at each end. This arrangement is shown in photograph 13-6 and these features are discussed in more detail a little later.

A further anomaly that is difficult to resolve is the now infamous low rear end, particularly noticeable under hard acceleration (sometimes termed 'squatting'). Triumph engineers must have been faced with a dilemma: fit softish rear springs to make the ride as comfortable as was decent for a sports car (with lots of attendant rear end droop), or go for hard springs to prevent rear end dip under acceleration, but with attendant rock-like suspension. Opinion today is that, initially too soft a spring rating – at about 280lb/in – was chosen as the standard for the TR4A. This is borne out by the fact that, by the time the TR6 was introduced, the rating had been increased by some 25 per cent to around 350lb/in. Most TR4A owners upgrade the rear springs to at least the TR6 spring rating. However, the '6 was/is still well known for rear end squat under acceleration, and many IRS owners fit 390lb/in rating, which does chatter the teeth at times but keeps tail drag to a minimum.

TRAILING ARMS

You have almost certainly read the chapter on chassis repair to IRS mountings, so I propose to do no more here than very briefly re-emphasise some of the most important points regarding the mountings for the trailing arms. I make no apologies for raising certain topics again, since we are talking about matters that could have

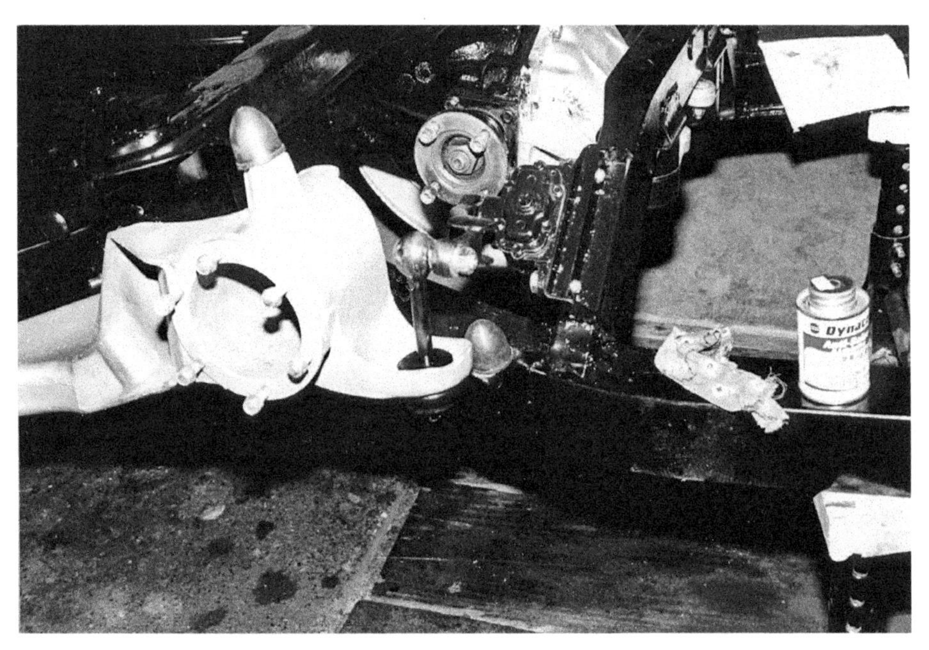

13-5. The trailing arm in situ, nicely contrasting the rest of the rear suspension. Note the original shock absorber and linkage, but as yet no driveshaft.

13-7. Yes, we have looked at a very similar picture in the chapter on chassis repairs. Nevertheless, since this is one of the most vulnerable yet crucial parts of an IRS chassis, I make no apologies for including this second view as a reminder of the importance of these two structural members (and particularly their internal stiffeners/spacers) to the integrity of an IRS chassis.

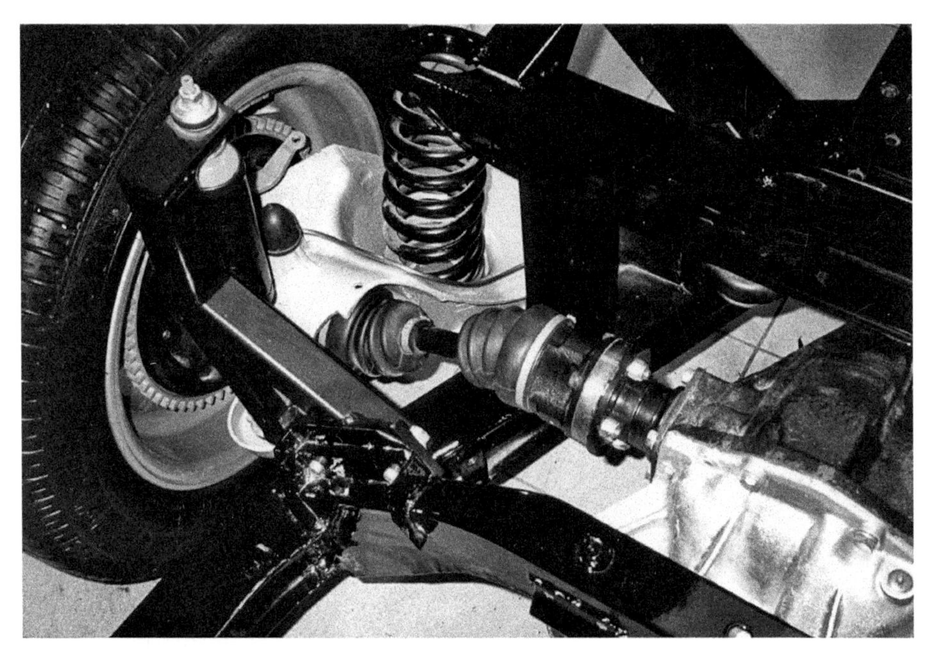

13-6. Another overview of the rear suspension but with TR Bitz's excellent telescopic shock absorber and mounting bracket. The shock is well located since it is attached to the trailing arm right at the point of maximum movement, thus giving maximum control. The bracket design is good, too, since it 'picks-up' on the original lever arm mountings and exerts no stress whatsoever on the body.

serious consequences if they are not fully absorbed.

So, to recap, the chassis section rots badly right at the point where the trailing arms mount to the chassis; to make matters worse, the structure rots from the inside out. The critical chassis mounting member is shown in photograph 13-7. Many a so-called repair has been carried out by the owner or his garage, which has involved merely plating over the top of the corroded section, often to get the car through its annual MoT examination. Often the car is still sub-standard since the new metal has been welded to very little. Worse, in the case of the trailing arm attachment members, there are still no solid internal spacing tubes (visible in photograph 13-7). Many a trailing arm mounting bracket has been re-assembled to a 'repaired' and apparently strong chassis section, but, without proper spacer tubes within the chassis members, the trailing arm bolts only squeeze the new plates together without ever tightening in the intended manner. This very unsatisfactory situation is compounded by inexperienced repairers re-using the original, but now too short, bolts to secure the trailing arm mounting brackets to the useless chassis leg!

This really is a very highly dangerous area to bodge, overlook or get wrong. Rather like the steering or brakes on a car, if the trailing arm mounting point(s) 'let go', the car could go straight into an oncoming vehicle or a ditch with no warning or time to react. If there is any swelling whatsoever of the chassis members in the area of the rear suspension mountings, replace them, and please do so with the proper pre-assembled spacered product specifically made for the job.

The differential bridges snap just inboard of the pocket where the rear coil spring sits in the suspension/ differential bridge, which would result in the car leaving the road as the rear springs come up through the body! Sadly, these weaknesses can be difficult to detect in a car, so look closely again at photographs 5-9 and 5-10 and for any sign of weakness, corrosion or repair to your car.

Home restorations rarely include having the four trailing arm mounting brackets crack tested, which really should be done if you plan to re-use the originals. For safety reasons I would advise that you buy new replacement brackets. There were different combinations of brackets for different models. Initially bracket 141399 (identified by 1 notch) was fitted as the inside mounting and 141398 (2 notch) on the outside. At commission numbers CP52867/CC61570 bracket 1555502 (3 notches) was fitted to the inside and 141399 (1 notch bracket) on the outside. It's possible to put these brackets in upside down, which will destroy rear geometry, so note that all notches are positioned upwards. Furthermore, it's not generally appreciated that at the same time as the mounting bracket 'notches' were changed, a change occurred to the trailing arm castings, and the angles varied from earlier castings. This can be another reason why some home restorers have difficulty re-establishing rear suspension geometry: the rear brackets and castings need to be compatible with each other. Fortunately, Triumph's casting sub-contractor was meticulous in incorporating the date of casting on the outside of each trailing arm. You might consider this another reason to purchase new brackets, for you should be able to tell your TR specialist the date of your castings, and get not only crack-free brackets, but ones that are definitely compatible with your castings. The mounting bolt positions shown in picture 13-8 also warrant attention.

13-8. Close-up of the trailing arm taken to show how the securing bolts must be fastened – if you ever need to remove them once the body is in place! Furthermore, the longer the bolts, the more impossible they are to remove at some future date if assembled the other way round – which precludes camber adjustments, etc. Always fit the heads of the bolts so that they are 'looking at' each other.

The original TR5 rubber bonded bushes fitted to mount the trailing arms were very hard. Those fitted to later IRS cars were a little softer, but the original bushes are likely to have deteriorated with time and will inevitably need replacing. The question is, what to fit now? The latest standard bonded rubber replacements are made from an even softer rubber compound than originally, which does reduce road noise but will not improve handling, and they are very likely to wear out (particularly those fitted to the inside of the car) quite quickly. On the other hand, there are special TR5 rear bushes that are actually harder than the original bushes, which should improve handling but are really difficult to fit. Nylatron bushes are harder still and, probably as a consequence, seem to transmit too much road noise. Superflex seem to have got the balance between ease of fitting, hardness, handling and road noise about right for most IRS owners. Superflex bushes will improve the entire drivability of the car, and should be a standard change for every new IRS TR owner.

The easiest way of removing the old bushes – which, after years in place can be reluctant to move by conventional means – is to heat the bush until the rubber catches fire. By this point the metal sleeve will offer little resistance to being pushed out, and the remaining rubber is then easily removed.

The polyurethane (Superflex) replacement bushes can be equally difficult to re-assemble if you are unaware of a couple of tips. The bushes are shaped like a cotton reel, with a small lip at each end that makes them reluctant to enter the trailing arms at all, never mind entering them square! Take a few minutes to make the angle iron/bolt aid sketched in drawing D13-1; it will be well worthwhile. Obviously, the bolt shank needs to be only marginally smaller than the bore of the new bushes. The aid holds the bush square to the trailing arm while you use your vice to squeeze the bush into its housing, aided by pre-warming the bush in hot (but not boiling) water, and liberal doses of washing-up liquid. Do put lots of copper slip on the bolts before fixing your refurbished trailing arms back on the car.

Look out, too, for cracks in the rear trailing arms caused by stress and age. They most frequently occur in the two places highlighted in photograph 13-9-1. The trailing arms are also vulnerable where the six hub mounting studs screw into the aluminium arm, as they can pull out or the thread can strip. You will need the expert help of your favourite TR specialist to heli-coil the casting with 5/16 in UNC threads. Note that the original thread into the casting is UNF, but coarse threads are best for soft material like aluminium. The outer end of the studs should ideally match the original UNF thread, but you may have to accept a UNC outer thread, too.

If the casting or threads are particularly badly damaged you might be better off getting a replacement trailing arm, which is still possible, although TR arms are becoming scarcer. Ex-Stag or ex-Saloon trailing arm castings will not fit, so you will have to specify which of the three TR types it is you need, remembering that TR4A arms do not have a bump stop fitted to the arm; they were fitted to the underside of the body. The '5 and '6 had provision for a bump stop on the trailing arm itself but, as we discussed a paragraph or so ago, there are early and late '6 castings (as seen in picture 13-9-2) and it's important to match casting to mounting brackets.

DIFFERENTIAL ALTERNATIVES

All TRs have a tendency to leak a little oil; it's almost a marque trademark! The

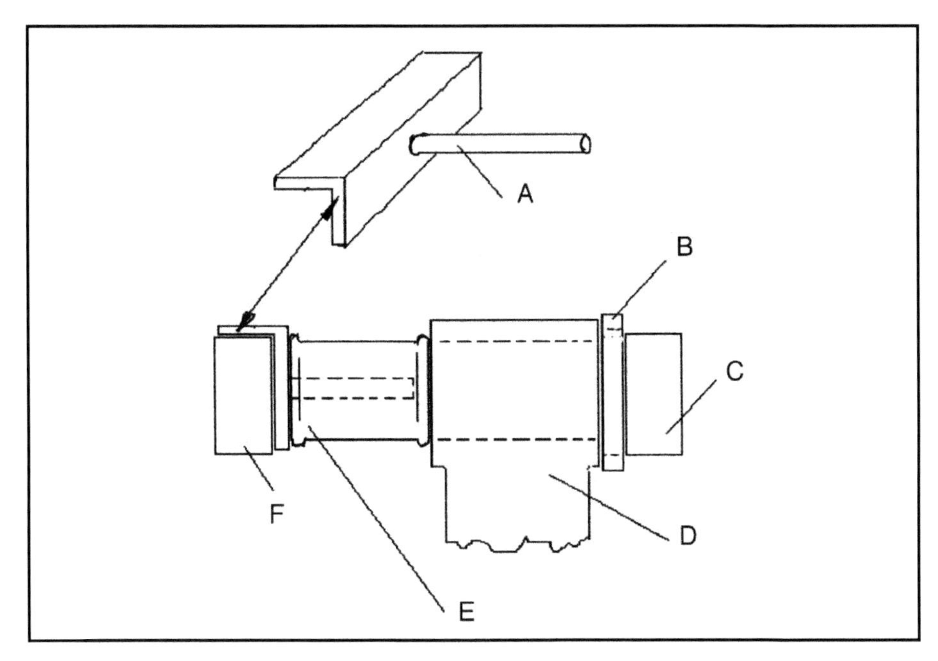

D13-1. Guiding a new trailing arm bush into position. A. Suitable bolt welded square to angle iron base. B. Spacer (short piece of pipe). C. Vice/vise jaw. D. Trailing arm. E. Polyurethane bush. F. Vice/vise jaw.

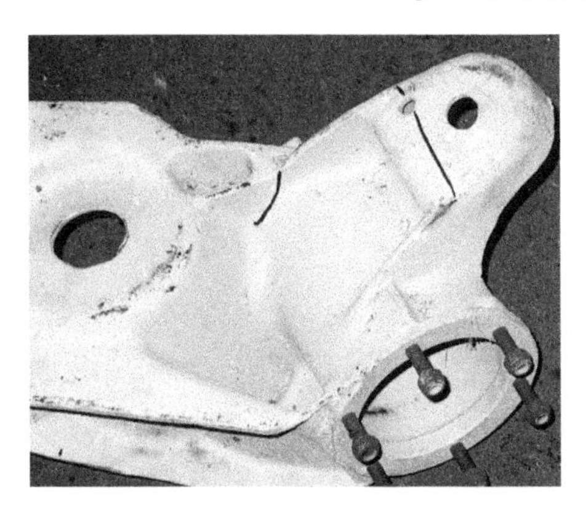

13-9-1. The IRS rear trailing arm with the two most common crack points highlighted. Don't confine your inspection to the areas shown, but be sure to inspect and even crack-test both arms where shown.

13-9-2. The post 1972 arms seen here on the right incorporate two changes adding material and thus strength to the arms. For comparison the original component is on the left but you will note the later casting has a smaller diameter hole for mounting the shock absorber and, as highlighted by the arrow, a deeper longitudinal web.

front bearing of the IRS differential is a case in point, and it's not unknown for TR owners to overdo the initial remedy. If your front differential oil seal leaks too much oil, check first that the breather hole in the rear cover is not bunged up. Over the years, oil and dirt collect and can close the breather, leaving the differential without a breathing opening as it warms up.

The only alternative for the trapped but expanding air is through one of the three oil seals. Either of the side seals are possibilities, but the most frequent route is via the leather front seal. All the seals are, of course, lower than the breather, so will allow oil to escape instead of the intended air, but often if the breather hole is cleared, it will relieve pressure on the oil seal that is leaking. Take care not to remove the split pin from the breather plug as it's quite a job to put it back. Also take care not to drop dirt into the differential as you clean the breather vent.

If you still have a leaking front oil seal after clearing the breather, then I am afraid it's a case of replacing the seal. As I mentioned, it is a leather seal which requires special preparation before fitting. The seal must be soaked in engine oil for at least 24 – and preferably 36 – hours before fitting. You can warm the oil a little from time to time during the soak to help its up-take, but do throw that oil away at the end of the softening/soaking period. Fill the differential with fresh oil as per the operations manual after you have fitted the new oil seal.

The IRS differential either works or it doesn't. If you are intent on the best repair possible, a service exchange is probably your best bet. A straight rebuild costs around £240 in the UK and, if you have ground to a standstill, then you'll almost certainly need a new crown wheel and pinion, too, at around a cost of £175. US readers should not have difficulty finding a complete secondhand replacement, though these are somewhat more difficult to find in the UK.

Fortunately, there are alternatives. Although you may have difficulty finding a complete TR unit with the higher 3.45 ratio fitted to UK 6-cylinder cars, all ex-US cars had units with a lower 3.7 axle ratio. These tend to be less popular than higher ratio differentials, so are slightly easier to find; in any event, they will be considerably cheaper. If you can locate one I would suggest you give it a try. You will quite likely find that fuel economy is not as good (by about 10 per cent), and noise might be slightly greater, but tractability and acceleration are improved. I run my '6 with a 3.7 diff as a matter of choice, although I have fitted a 28 per cent overdrive in place of the usual 22 per cent reduction.

I mentioned that complete replacement TR differentials are hard to find in the UK. Fortunately, with some ingenuity, they can be replaced or improved, and quite cost-effectively. Clearly, much depends upon the problem you have, but taking the worst case first, let's consider your options when you have a stripped or very noisy differential.

It's not always appreciated that Triumph was commendably frugal, using the same differential ratios (indeed, the same basic rear suspension) on a variety of cars. Other ratios were available but the two rear axle/differential ratios that interest us are the 3.7:1 and 3.45:1. The ratio used in your car can be identified from outside the differential by marking both hubs (wheels off the car first, of course) and the drive pinion flange with felt-tip pen or typewriter correction fluid. With the gearbox in neutral, get a friend to turn the drive flange 11 revolutions while you and another helper ensure that both hubs turn the same number of revolutions. If the hubs revolve (to all intents and purpose) 3 complete turns, chances are you have a 3.7 ratio in your differential. If both hubs turn past 3 revolutions (the actual figure is 3.2 revs), you almost certainly have a 3.45 ratio. From inside the differential it is easier to determine as you need do no more than count the crown wheel teeth with the following information in mind:

37 crown wheel teeth signal a 3.7 ratio, confirmed by 10 pinion teeth

38 crown wheel teeth signal a 3.45 ratio, confirmed by 11 pinion teeth

The model TR you have is a factor, and these ratios were originally used as follows:

TR4A 3.7
TR5.. 3.45
TR250.. 3.7
TR6 (US carb)...................................... 3.7
TR6 (PI)... 3.45

Fortunately, the 3.7 and 3.45 ratio differentials were also used in several other Triumph vehicles of the era as follows:

Stag.. 3.7
2500 Auto Saloon 3.7
2500 Manual Saloon......................... 3.45
2500PI Manual Saloon 3.45

Photograph 13-10 shows the usual rear casing for the TRs we are concerned with. If you have broken this rear casing, or your front mounting has broken, then the following proposal is no help, and you must search for replacements (possibly via the USA), or get the broken parts repaired. Note, too, that a TR4A (shown by photograph 13-11) rear casing is different to that of later TRs, so is no help.

However for the majority, the core differential, pinion, crown wheel, drive flanges and main casing in the above cars are identical to those parts in our IRS TRs, although you should note that pinion spacers can vary. The front mountings, drive flanges and rear cover plates may vary from car-to-car, but if you have a complete TR differential unit (of either ratio) and wish to fit a different ratio or a replacement differential, you can do so via the appropriate non-TR Triumph core unit. Clearly, you need to check that the ratio of the replacement core is suitable, and that it runs smoothly, but you will get the basis of your change for about £25-£50.

Personally I would be cautious about purchasing a replacement differential that exhibited signs of weeping oil seal(s), for, all too often, a weeping seal signifies a worn bearing. If the replacement is competitively priced, or your differential is otherwise in good shape, we can briefly explore changing oil seals, although there are very good reasons to ask an expert to tackle this task.

You should also be aware

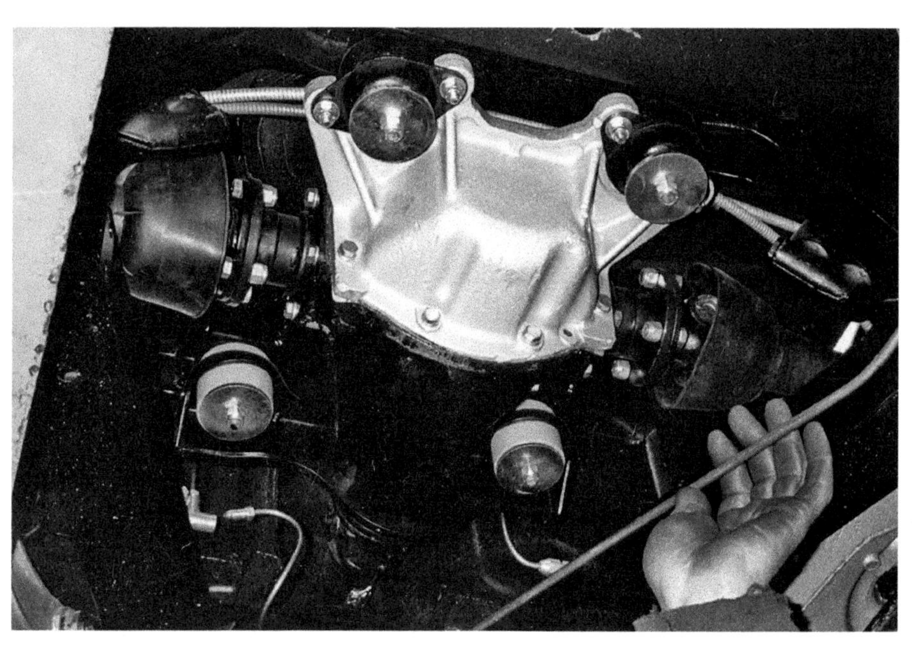

13-10. An excellent view of the IRS differential mounting arrangement showing the flexible mountings fixed under the differential. The large washers below are supporting the flexible mountings now that they are bearing the weight of the differential.

13-11. The slightly different rear mounting of a TR4A differential. This uses four pairs of the cone-type mountings normally associated with the front mounting of TR IRS differentials.

that Saloons and Stags had a front extension to the pinion and drive flange. The extension is certainly removable but, when changing the drive flange, you will take the pre-load off the collapsible spacer the Saloon cars used. This, therefore, will need to be swapped for the fixed/solid spacer used in the TR differential and pre-loaded. Unfortunately, this is not a job for most amateurs, so should be done by your local professional differential restorer or nearest TR specialist.

For those interested in finding out what is involved in changing the oil seals, there are 3 oil seals; one in the pinion and 2 in the driveshafts. How to remove the driveshafts from a car is obvious from the manual, or by looking at the car. If the differential is still in place in the car, don't be too quick to remove it for, as inconvenient as it sounds, it is often better to loosen and subsequently tighten the 11/8in AF nuts with the differential on the car. You can then use the car's chassis to hold things still while you tackle the 110ft/lb of torque required for the substantial nuts. However, you also have to hold still the drive pinion as you loosen or tighten its retaining nut. A special long, strong, locking bar that bolts to all four pinion drive bolts is essential: take a look at photograph 13-12 for an idea of what's needed and then leave the job to a specialist! There may come a moment when tackling the diff's two 11/8in flange nuts when you wonder if you are trying to turn the tommy bar in the correct direction; I can reassure you that the nuts do have a conventional right-hand thread.

Four bolts on the housing flange are all that prevent you taking the whole inner axle shaft assembly to the workbench. If you wish to replace either the axle shaft bearing or the oil seal, it is first necessary to remove the flange from the shaft. Use a three-legged puller and expect to have to exert considerable force, so a very strong puller would be a good idea. You would not be the first to find that the shaft/flange appear to have become 'welded' together. In this event, the power press at your local engineering works will rectify the situation, but don't wait until after you have damaged the shaft spline and/or flange, use the press first, before the damage occurs.

The oil seal and housing can now be pulled off and the oil seal carefully tapped out of the housing. Drive the axle shaft through the bearing and throw away the used bearing for it must not be re-used. Tap a new bearing onto the shaft by carefully striking the inner ('B') ring only in order to avoid damaging (or 'brinelling') the tracks inside the bearing. When re-assembling, do not forget to check the metal protector cup for damage, and to put it back against the flange before putting the housing and well-lubricated oil seal into position.

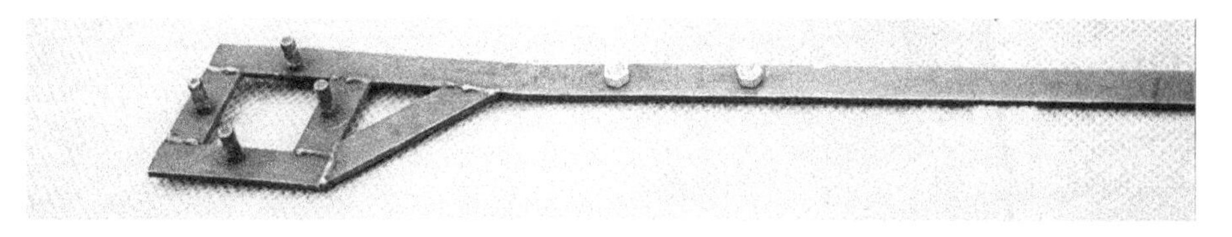

13-12. A means of locking the differential pinion and/or driveshaft flanges. The four fastening studs are, of course, for securely attaching the arm to the pinion in question, while you address the 1.125in AF nut. Note that there appears to be a slight 'set' in the arm, which is not an illusion, and serves to demonstrate the force often required to move these nuts.

When re-assembling the differential and differential mountings to the chassis, do ensure that the rubber differential mountings are in good shape. Frankly, for what they cost, you are better off replacing all four. The TR4A uses four rubber mounting cones as can be seen from photograph 13-10, which slightly reduces the importance of the rear pair. However, the condition of the special rear bonded mountings on the TR5/250/6 is particularly important. If those you have look new and you have half a mind to re-use them, at the very least test them by putting a large screwdriver into the main bolt hole and stressing the rubber with a substantial sideways pull. If any cracking of the rubber is evident around the edge of the through tube, replace the mounting, however good it looks.

Take care when re-assembling the mounting to the differential and the assembly to the car, and in particular double-check where the large mounting washers go. As strange as it looks at first, the rear mounting goes underneath the lugs that protrude from the rear of the differential rear cover (photograph 13-9), with the protruding end of the central tube pointing downwards. Yes, that does mean that the large washer then stands clear of the rubber, at least until the differential's weight is hung from the mounting, whereupon the protruding tube will be drawn up into the rubber and the washer will assume a more natural-looking 'flush' appearance with the rubber.

The front mounting assembly is generally more readily understood and rarely incorrectly assembled. Nevertheless, for the record, rubber cones go atop each side of the front mounting arm, which is then positioned over the front chassis pins. A second concave cone is offered up each side with the essential large washers in place before the two pin nuts are started. If, at the end of this operation, the differential nose points upwards, you have done something wrong (probably put the rear mountings above the rear lugs), and now run the risk of starving the front pinion bearing of oil, and some expensive remedial work if you do not fix it now.

Finally, the right side front and left side rear differential mounting pins are prone to break loose from the chassis. Whenever you are forced to remove the differential, you should take the opportunity to strengthen all four pins in the manner detailed in chapter 5.

One final recommendation for when working in the area of the differential: always check that the four bolts which secure the front end of the differential to its mounting/cross bracket are tight.

DRIVESHAFT DETAILS

There are a couple of common errors when re-assembling the rear driveshafts to the differential.

Firstly, it is best to feed the bolts through the flanges from the inside of the car so that the head of each bolt is 'looking at' the differential, and the nuts are on the outside of the universal joint flange. It's good engineering practice to always use new nyloc nuts (never re-use a nyloc nut), or new spring washers under all plain nuts; if you follow this principle it's very unlikely you'll have a problem. However, if you are unfortunate and a nut does come off one of these particular bolts, it will almost certainly drop away unnoticed. This is far from ideal, but the alternative is even less desirable: if you have put the bolt in from the outside, there is just a possibility that it will exit the coupling and jam in the universal joint.

Secondly, note that both left and right driveshafts carry the same part number (RKC454), so are interchangeable, and it may not a be bad idea to change the complete assemblies since you will then be driving on the other side of each spline. This does not excuse you from liberally greasing the splines with molybdenum grease before re-assembly! If this is on your mind I suggest you mark each shaft as you remove it to remind you which side of the car it came from originally.

Many restorers do choose to reposition parts of each shaft to the opposite side of the car in order to drive on the other side of the splines. I am not sure I would consider this good engineering practice, so think about it carefully. In any event, both shafts need to be checked for wear, particularly if you are to avoid – or are already experiencing – the infamous Triumph IRS rear end twitch. This is easily resolved provided the driveshaft sliding splines are not overly worn.

It is worth taking a moment to explain the reasons behind this alarming trait. If lubrication in the splines dries out, if dirt finds its way past a (split?) gaiter, or if spline wear has occurred, the torque from the engine can prevent the splines from sliding as suspension movement dictate. Obviously, the lower the gear you are using, or the larger capacity of your engine, the higher the torque transmitted to the rear driveshafts will be and the more pronounced the problem. When the torque causes the driveshaft splines to lock, rear suspension movement (which would normally cause the driveshafts to slide and change overall length) is either prevented completely or, at the very least, is severely restricted. In due course a sufficiently large road surface bump, or a reduction in torque when you lift off the accelerator, 'unsticks' the splines and the rear suspension takes up its unrestricted position, causing the rear of the car to 'twitch' quite suddenly, unexpectedly and very noticeably!

Fortunately, for those with driveshaft problems, just as with the differential we explored a short while ago, Triumph used many of the same components on TR4A, TR5, TR6, Stag, 2000, 2.5PI and 2500. The TR driveshaft assemblies are different from those of other Triumph models in that the lengths of the outer shafts vary (from the sliding spline covered by a rubber gaiter and the outmost universal joint yoke). However, these differences do not affect the interchangeability of the majority of components used in the sliding splines, or other parts within the rear suspension, for that matter.

To examine the splines, first pull back the gaiter. On the majority of IRS TRs this is achieved by cutting the two wire/cable tie clamps from the shaft. Original TR4As have a threaded collar, a steel washer and a cork washer on the female part of the sliding splines. The collar must be unscrewed and slid away. These parts were omitted from TR5s and TR6s. Thorough cleaning, examination for wear and testing of the sliding action is necessary. If all seems well a little torque should be applied, first one way and then the other, whilst the sliding check continues. Wear, or a feeling of roughness when doing a sliding test, suggests a replacement pair of splines are required.

Try using some parts from another sliding assembly (noting the respective differences outlined above), re-checking for smoothness and placing your new assembly on the opposite side of the car. This applies driving torque to the other side of the splines, but is not a recommended long-term solution.

The driveshafts rarely break. They can lock up if left ungreased (grease very regularly to prevent this), but it's more likely they will wear and 'clonk' when taking up drive or on overrun. One consequence of worn and clonking driveshafts is an abnormally short life of the universal joints. There are numerous rear driveshaft upgrades that space dictates are reviewed in a later book. However, I can tell you all of them are expensive.

Since it is hardly an upgrade I must mention that GKN Driveline (which made the original driveshafts) has employed modern technology and now uses a bonded resin 'Rilsan Fluidglide' coating on the splines. This is claimed to eliminate the infamous IRS 'twitch' by ensuring that the splines no longer stick, even under load. Furthermore, coating longevity is reputedly enhanced by grease nipples that lubricate the splines. The new driveshafts cost around £300 and come with new universal joints.

During re-assembly of the original driveshafts, do note that the halves engage in one position only, the importance of absolute cleanliness, of lubricating the splines all over, of using a heavy-duty molybdenum grease, of using new gaiters and of sealing the assembly to prevent the ingress of water and grit.

UNIVERSAL JOINT LONGEVITY

IRS cars unquestionably suffer from the consequences of using 6 universal joints to transmit power from the gearbox to the rear hubs. You only need a small amount of play in each universal joint for it to seem like an eternity before action at the clutch pedal translates into motion!

Whilst, strictly speaking, we are discussing the rear suspension universal joints, which are quite accessible, the universal joints at each end of the IRS propshaft can only be accessed with either the gearbox or differential out of the car. Consequently, they should also be on your list of "while the differential is out" jobs! It's worth mentioning that the universal joint circlips will be far easier to remove if you spend a few minutes with a 10mm wire brush and drill cleaning the 'mouth' of the yoke. As with the driveshaft splines, it goes without saying that everything must be scrupulously clean upon re-assembly.

There are two qualities of universal joint on the market. The original TR4A used sealed universal joints (without grease nipples), but a little later grease nipples were used on the inner universal joints, and later still grease nipples were incorporated into the outer universal joints. The TR5 retained grease-nippled universal joints but, no doubt for cost saving reasons, sealed universal joints were re-introduced for the TR6.

Whatever your car and whatever was originally fitted, the driveshaft universal joints should always be of the highest quality you can buy and always have a grease nipple. GKN items are the universal joints of choice in view of quality and size of the needle rollers used, but you may need to go to an 'AE' (Turner and Newell) distributor to get them.

A PVC cover can be fitted to any IRS car but was only originally fitted to the '6. I find mine flaps about and question its value but, in theory, it should act as a dirt shield.

All universal joint grease nipples should receive attention from a grease gun every 3000 miles or so. However, the outer pair of universal joints on each driveshaft require even greater care and attention. These two units experience the very minimum of flexing when in use, which you may think is an advantage and should extend their lives. In practice, the grease gets pushed out of the narrow angle through which the joints move, internal wear becomes focused on a narrow band and water gets in, making these universal joints the most prone to rust, stiffness and seizure.

You might be surprised to hear that universal joint stiffness also increases stress on the differential mounting pins. Therefore, annually – perhaps each spring when you are preparing your car for the season ahead – it would be a good idea to exercise the outer pair of universal joints through their full arcs. This can be accomplished by disconnecting the driveshafts at the inner ends, and rotating the outer universal joints through the widest arc you can, whilst thoroughly greasing them. If your universal joints are the sealed type, take the first opportunity to exchange them for high quality, greasable units.

REAR HUBS AND BEARINGS

IRS rear wheelhubs are further areas that require care, attention to detail and a certain amount of expenditure.

Going straight to the most important detail; on the end of each rear outer axle shaft there is a threaded portion that can snap off (as shown in picture 13-13), and allow the wheel to come off the car. The reason this happens is that a crack develops at the root of the threaded section, which eventually allows the end of the shaft and its wheel retaining nut to part company. Additionally, there is likely to be the inevitable wear (signalled by play) in the rear hub bearings.

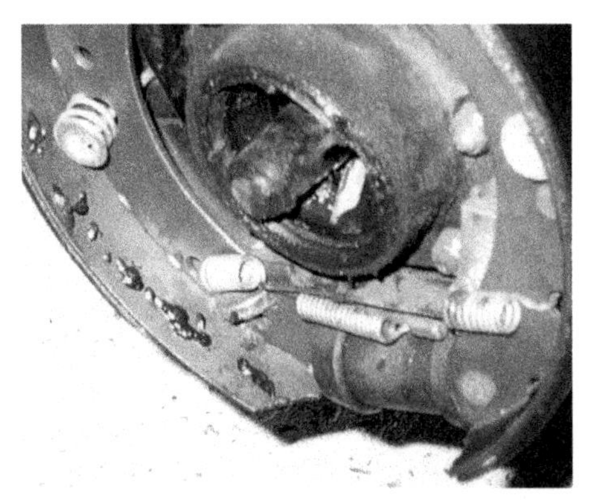

13-13. Cracked for some time, by the look of the fracture, this is the consequence of the threaded end of the rear hub shearing off – taking the retaining nut, washer, studded driving flange, brake drum, and all-important wheel with it. Some of the grease from the hub bearings can be seen splattered on the brake shoes, whilst, out of shot, is the chewed-up key and keyway. Strangely – and very fortuitously – I was only going about 10-15mph round a roundabout when there was a load bang, although just moments before I had been travelling a great deal faster.

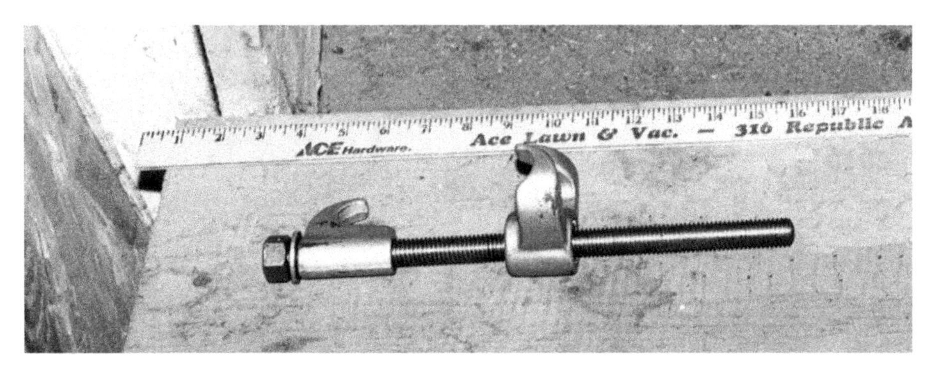

13-14. One of a pair of rear spring compressors you will need, particularly if you are assembling the rear suspension without the weight of the body to help compress the springs.

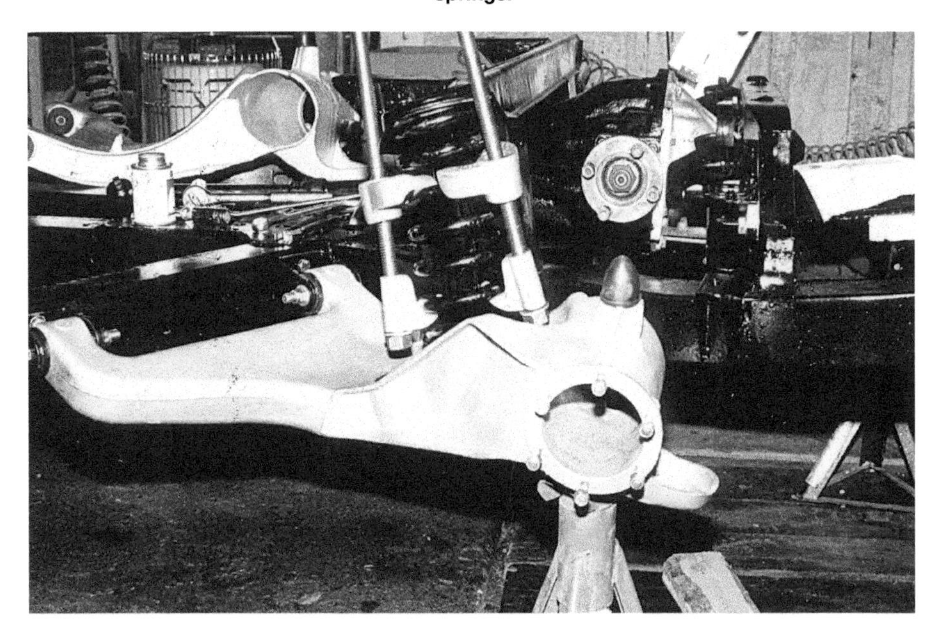

13-15. Here are the two rear spring compressors in action.

This problem can be resolved by having the existing stub axles crack-tested, or spending around £275 on a pair of re-machined shafts. The re-machined shafts are as good as new in that they are freshly machined, albeit from used driveshaft stubs. However, it will take a 100 tonne press to split the bearings off your morse-tapered stub axle. If you try to do this at home you are very likely to reduce the shafts to scrap by 'belling' or 'tuliping'

13-16. Another excellent telescopic shock absorber conversion arrangement. This is the CTM design, a company, perhaps, more widely known for its work repairing and/or building new TR chassis. These conversions are usually supplied with AVO adjustable shock absorbers.

the flange. Furthermore, whilst the workshop/operations manual explains it is essential to replace the collapsible spacers, bearings and oil seals as a set, it does not, in my view, put enough emphasis on the importance of also replacing the adjusting spacer (ULC2188) and tab washer (139057).

In short, there are so many opportunities for the home restorer to get this job wrong, and these parts are so important to the safety of you and your car, that I very strongly recommend you go the service exchange replacement route. All service exchange work on rear hubs for the industry is carried out by TR Bitz, and includes crack-testing of the shaft, new bearings and other such parts as are required. The cost – at around £75 each (exchange) – does not make it worth your while even contemplating anything else. Consequently, I do not propose to tempt anyone into a DIY attempt by an operational description here. In fact, I take the other view and emphasise the importance of not even trying to start the operation, for you'll more than likely damage the shafts and render them valueless.

You may be surprised to hear me suggest that, provided you do not try to get inside the hub, there is nothing to stop you using an ex-Stag, 2000/2500/2500PI hub assembly on your TR. Furthermore, if the hub is satisfactory but you have problems with the universal joints, there's nothing to prevent you from changing the universal joint just inboard of the hub.

SPRING COMPRESSORS

This is pretty routine stuff so, without going into detail, I'll just mention that you will need an external spring compressor for the rear springs, as shown in photographs 13-14 and 13-15.

SUMMARY OF UPGRADES

Whilst this book is primarily concerned with original restoration, I must mention, if only briefly, two upgrades.

First and foremost, fitting telescopic shocks to the rear. The standard lever arm shock absorber is shown in photograph 13-3, and the alternative, which will transform your car, can be seen in photograph 13-6. There are several makes to choose from but, if you are selecting adjustable shocks, do ask about the method and ease of making the adjustment. Some need to be removed from the car; some can be adjusted on the car. However, it is probably the design of the mounting bracketry that is the most important issue when it comes to fitting telescopics to a TR.

13-17. Revington TR's anti-roll/sway bar and rear shock absorber.

To the best of my knowledge I don't think there's one universal method of mounting the shocks, as each retailer/specialist has its own ideas and brackets. Avoid mounting brackets that in any way touch the body of your TR. The mounting bracket should be designed to contain all stresses within the chassis structure, not through the body. Ensure those that you are looking at are fit for the intended purpose: brackets designed for racing will be light but may not stand up to the rigours of thousands of miles of carefree motoring. The design must allow for the use of fuel injection pumps, filters and pipework. I have not looked at every design available but would suggest that the offerings from CTM and TR Bitz are worth exploring; they are shown in photographs 13-6 and 13-16.

Rear anti-roll/sway bars are also worth a quick mention, although, as you know, it is my intention to discuss upgrades/improvements in more detail in a follow-up book. However, I came across the demonstration exhibit shown in photograph 13-17 at Revington TR, and felt it offered such a simple assembly method that, at the last minute and in spite of a lack of space, I had to mention it in passing.

Chapter 14

Brakes

BACKGROUND

Overall the variety of brake variations is somewhat confusing, though this situation stabilised somewhat during TR4/5/250 and early TR6 production runs. It is not my intention to go into graphic detail about TRs produced prior to those we are interested in, but you may find a summary interesting, and possibly helpful when sourcing spares.

Put very simply, the vast majority of rear brakes are 9 inch drums, and are more-or-less identical. The position of the backplate on the axle or hub may change, but continuity runs right through the major part of classic TR production.

The front brakes are not so straightforward. Initially, the TR2 used drum brakes, but the TR3 introduced an 11 inch diameter front disc that was fairly quickly changed to a slightly smaller diameter (10 13/16in) disc on the TR3As. The respective calipers were also different and, needless to say, were changed simultaneously. A further change of caliper was introduced with the TR3Bs, but the period of stability then existed through to mid-TR6 production, when a metric version of the caliper was introduced.

The point I am working up to is to alert you to the possibility that the discs or calipers on your car may have been changed from original specification. This isn't a problem if both have been changed to compatible pairs of the earlier or later type. However, you may decide to fit new discs, and quite correctly choose those that are Original Equipment (OE) for your car. If they protrude well beyond the pads, you may have inadvertently fitted a larger disc than is appropriate for the calipers. However, the reverse scenario is very worrying as the pads may overlap the top of the disc, signifying that you need to fit the 11 inch diameter discs. If the brake pipe fitting seems very loose or very tight, check that you are not mixing metric and imperial threads. Forewarned is forearmed!

BRAKE LINES

We'll talk some more about metric and imperial brake hose couplings in a moment, but this section is to remind you about the advantages of copper brake pipes and braided flexible hoses.

The brake pipe kits made from copper material are available from your favourite specialist for all TRs and, whilst not absolutely original, would be suitable for all but a restorer striving for concours standard. In fact, the material is actually not copper but cupro-nickel, though it does have the same corrosion-resistant characteristics.

You don't need to worry about flaring tools, for the pipes come ready flared. They will arrive coiled (for ease of shipping) so you will need to straighten them. Put one end in the vice, suitably wrapped for protection. Hold a rag round the pipe adjacent to the vice and, starting with a 45 degree behind your rag/point of straightening, pull hard down the full length of the pipe. If you do not introduce a bend but try to pull the pipe straight, you will not achieve a truly straight pipe. There are then, of course, various raduised bends to position correctly if the pipes are to be both safe (i.e. not kinked) and look their best. Putting the radius in is best done using one of the large varieties of pliers available from most autofactors. You can buy 45 and 90 degree ones that are recommended since these little details will make a world of difference to the appearance of your finished re-piping exercise. Photographs 14-1 and 14-2 will give some additional help with the installation.

Moving on to the flexible pipes, you will not need me to remind you that these are a safety critical part of your car, and, for the small extra sum involved, the stainless steel braided

14-1. Some interesting detail on brake pipe installation. Clearly, you can fit many of the brake lines prior to fitting the body – if you are careful. As a rule-of-thumb, you should not fit any of the pipes on top of a chassis member, so the first point to note is that the lines have been securely attached to the sides of chassis members. These lines are made from 'copper' material and we can see that the front 'section' and the rear sections are not coupled together. This a dual-circuit car which awaits the body with its attendant dual-circuit master cylinder and pipework – which can only be attached once the body is securely in place, of course. Note the unusually long copper pipe connecting the front caliper to the flexible hose. The extra length is absolutely no problem provided it is routed so that the line will not foul on any part of the body or suspension when the steering is on any part of its 'lock'. Worth a double-check before the car goes out for the first time! Did you spot that this restorer has elected to supply the rear brakes via two lines down the length of the car?

14-2. Some detail for the rear brake line installation. Note how this caring owner has 'lost' some excess pipe length, mounted the pipes to the chassis, and how he routed and looped the flexible hose.

flexible hoses are better than the original hose on two counts. Firstly, they are more resistant to abrasion, not that I am condoning an installation where the hoses come into contact with the body or suspension components. However, it is added insurance! They are also more resistant to the inevitable, very slight, expansion that occurs when you brake, and will give the brakes a more 'solid' feel. This is particularly so with IRS cars where there are 4 flexible hoses within the system rather than the usual 3 on earlier TRs. By the way, I hope this next comment is totally unnecessary, but if any of your flexible hoses are swelling visibly when pressure is applied to the brake pedal, do not drive the car until you have replaced them. The amount of expansion I spoke about a sentence or two ago will be quite invisible!

IDENTIFYING BRAKE CALIPERS

One thing to be aware of is the change of caliper that occurred midway through the TR6 run, towards the end of 1972, at CP76094 for UK cars and CC81078 for US cars. Triumph went metric, which sounds of little consequence until I warn you to ensure you fit the appropriate short interim metal pipe to your caliper, since early cars use an imperial thread and later calipers (and pipes) a metric thread. One thread will, unfortunately, actually fit into the other but will be totally unsafe in use.

The metric calipers do, fortunately, have some identification in that the cross pins which hold the pads in place were reduced in diameter from 0.25in to 0.188in. The change also affects which pads you need, but once you know to measure the cross holes in your caliper (a twist drill is a handy way of gauging the size of the hole), you will know which pad to order.

Identifying the various calipers is a real headache, even for those with years of experience, as there are a baffling number of metric/imperial caliper options. Always be aware that a moment's inattention can have appalling consequences. It may appear quite clear cut from the commission number and year of manufacture of your car which type of brake caliper you should have fitted, but the various brake calipers originally fitted are not as easily identified as the manuals make it appear.

Not all small pin calipers were

made with metric threads, and not all large pin calipers have imperial threads. Thus, not all small pin calipers use a stepped mounting bolt. Furthermore, after 15, 25 or even 35 years, do not assume that every previous owner or garage has fitted exactly what the book says. It may be helpful, therefore, if I alert you to the various points to watch for when you change the front brake calipers, fit new lines, or both.

I think the first point to make is that you will be able to screw a metric male pipe into an imperial female threaded caliper, but it is highly likely that the union will jump a thread and you will lose the brakes under heavy pedal pressure. Fortunately, one – thankfully – consistent feature is that the depth of the threaded female pipe connection tapped into the inner top angle of the calipers does always tell us whether the thread is metric or imperial. Thus, if you are changing any components within your front brake set-up, you need to check whether you have a metric female tapped hole (about 12mm/0.5in deep) or an imperial threaded caliper hole, which will be some 3mm/0.125in deeper, and to order the correct mating pipes for your calipers. If you are changing calipers you need to check the size of the mounting bolt holes. It is probable, but by no means absolutely certain, that, if you have a metric (shallow) inlet pipe fitting, you will most likely have metric sized mounting bolt holes. Check your existing bolts in the holes of the new caliper: they should go through without a great deal of pressure but must not slop about.

If the bolts do slop about in the caliper mounting holes, you can be pretty sure you have metric sized mounting holes, but currently imperial bolts. The solution is, for once, easy: you need to buy four metric/stepped caliper mounting bolts, part number 158668. If, on the other hand, your existing bolts will not pass through the caliper mounting holes then (assuming everything is nice and clean) you probably have metric/stepped bolts with imperial holed calipers. This time the solution is to buy new imperial mounting bolts, part number BTB610.

Moving on to the different pad retaining pin sizes, up to autumn 1972, cars were fitted with 0.25in/6.3mm diameter (known as large) pins. From that point on the calipers that you should find on your car will originally have been 0.188in/4.8mm diameter, known, of course, as 'small' pins. Whatever pin size you have (originality aside), it is not material to braking efficiency. It is important that you order the correct size of brake pads (small hole/large hole) and, if you do not have suitable ones, the correct sized pins to slot through calipers and pads.

Never forget to fit the 4 tiny clips that go through the retaining pins.

BRAKE SERVO PERFORMANCE

If you feel your brakes could be improved, before seriously considering any of the checks outlined in the rest of this chapter, do take a couple of hours to ensure that your existing brakes generally – and the servo in particular – are functioning to maximum effectiveness.

First, check that the servo is operating by running the engine for a couple of minutes (to build up vacuum in the servo) and then switch off. Place your foot on the brake pedal and you should hear a 'chuff' noise. Leave your foot on the pedal and start the engine; you should feel the pedal depress if the servo is working satisfactorily. No 'chuff' and/or depression means you should question the effectiveness of the servo.

Carry out all the obvious maintenance checks. It might be a very good idea to replace the vacuum hose that leads to the servo before trying the foregoing test again. If the servo is definitely working but you still feel the on-road braking leaves something to be desired, take the relatively simple precaution of connecting a second manifold take-off point into the vacuum hose that feeds the servo. As unlikely as it sounds, particularly when you look at the balancing pipe installation, a second vacuum take-off feed to your brake servo can improve effectiveness of the servo and brakes.

For once this task is easier if you are using petrol injection since there is a pre-tapped take-off point on top of the central inlet manifold. For those using most carburettor inlet manifolds, unfortunately, the task will involve removing the manifold and welding a second pre-tapped take-off point to a convenient spot on the manifold. If your brakes are still substandard, read on ...

It's not unknown for the home restorer to forget the four washers from under the nuts that hold the brake servo to the pedal box/bulkhead/firewall. It is a difficult place to reach, but the force exerted on the servo is such that you must fit four largish, heavy-duty washers if you wish to prevent the servo from pulling through the bulkhead.

GENERAL BRAKE TIPS

- Remember that brakes are a safety-critical item, and treat components accordingly. If in any doubt whatsoever, ask a professional, and remember he will always err on the side of caution, as you should. Consequently, a completely new or factory reconditioned unit is usually the best option when it comes to brake parts.
- When removing the rear drums, do let off the handbrake and slacken off the adjuster located on the backplate.
- When bleeding the brakes, always start by bleeding the wheel furthest from the master cylinder – the left rear. I get on well with the 'Easy-Bleed' single-handed system, although as I recall, you may need to find an old master cylinder cap and drill a hole in it if you have a TR5 or 6 since the size of cap on the reservoir is outside those supplied with the Easy-Bleed kit.
- Many restorers feel it appropriate to replace the rear wheel cylinder seals, and certainly it is likely to be important that the rear wheel cylinders receive attention. However, on average, replacing the seals meets with no better than a 50 per cent operational success rate and I would suggest, for the cost involved and importance of the brakes, that you replace the cylinders completely with genuine Girling replacement parts. This is certainly particularly appropriate if the rear wheel cylinders need honing.
- Your best policy with doubtful calipers is to do a service exchange deal. If you really must explore the units further, do not split them. The experts feel that there is too much opportunity to get the subsequent re-torqueing incorrect, so not only will you find it difficult to buy the internal seals, but also the new bolts (with built-in plastic locking tabs) will be hard to find, too.

However, it is practical to change the piston seals once the calipers are off the car. Remove the circlip and dust boot on the later types of caliper and extract the pistons, making sure you know which piston came from which bore, and that they are not interchanged. Prise out seals carefully, ensuring no bore damage occurs, and thoroughly clean pistons and calipers in brake fluid or methylated spirit (wood alcohol). Assemble new seals to the caliper bores, lubricating with clean brake fluid. Assemble the dust boots on early types of pistons and carefully locate the lip in the groove in your caliper. Push the pistons home squarely. For the later types of caliper, fit new dust boots, and do not forget the circlip in each caliper. When refitting the caliper to the car, be sure not to forget lock tabs and the pad retaining pins (yes, they really are forgotten from time-to-time). Bleed the whole system.

- Stuck master cylinder pistons are not unknown, and some thoughts about remedies may be helpful. The master cylinder is a safety-critical component and may be best replaced. If you do so I suggest you use a genuine OE replacement part. Still want to recover your old unit? Okay, but do get the bore of the master cylinder honed before re-assembly, and try clearing the bore of dirt and rust and then soaking the assembly for several days in penetrating fluid like 'duck-oil' or WD40. If the usual tapping (best with wood, of course) and/or using the return spring to encourage the piston to come out fail to do the trick, try the 'Easy-Bleed' system. This uses brake fluid pressurised by tyre pressure (at no more than 20psi), although when attempting the latter, do not forget to close the pipe connection with an old union and folded pipe. If the piston is still stuck and you are really reluctant to scrap the master cylinder, try a grease gun coupled to the pipe connection. Remember that rebuilt cylinders only work satisfactorily about 50 per cent of the time, so it's best to buy new if you've got to this stage.
- Upgrading brakes will be a topic explored in some detail in a later book, but, very briefly, cross drilled TR discs (photograph 14-3) will help cooling, and are a first level improvement worth considering. Be aware that some brake pad materials only work at their best when warm, so are unsuitable for many applications outside competition. Nevertheless, you may care to look into alternative pad materials, but be aware that Mintex M1144 does take time to warm up, though is popular for TR competition applications. A significant upgrade in braking terms is to fit Wilwood Alloy calipers to the current discs at a cost of around £500. Available from TRGB, or approach Revington TR for information about its complete brake disc and caliper upgrades.
- Those keen to improve rear brakes should try Alfin alloy rear drums from TR Bitz, shown in photograph 14-4.

The handbrake on early TRs was good. It was mounted on the side of the propshaft tunnel, and had an

14-3. Cross drilled brake discs are an aid to cooling, although the effect is fairly minimal. I am not comfortable with cross drilled holes with significant countersinking at each side of the cross drilling as it reduces the pad-to-disc contact area. If you cannot buy discs with plain holes of about 0.125in (3mm) diameter, buy undrilled discs and have your local engineering shop drill plain holes to this pattern on a 'dividing head'.

14-4. Only some 30 per cent of braking effect comes from the rear, so it makes sense to spend the majority of your brake improvement cash on the front of the car. Nevertheless, these finned aluminium rear drums will increase rear drum cooling, provided you are sure the original drums are getting too hot. These are called 'Alfin' drums.

excellent mechanical advantage, or 'leverage'. From the TR4A, and for the TRs we are concerned with, the handbrake was moved to the top of the propshaft tunnel, its length was reduced and it used two cables instead of one, resulting in reduced mechanical efficiency.

We will look at compensatory upgrading in a later volume, but the most common causes of handbrake problems follow. Check that all are in first class order before going further:

- Worn linings
- Seized brake shoe adjusters (the square drive peg on the rear of the backplate)
- Seized and/or worn rear cylinders/ levers – particularly when the cylinder cannot slide fore and aft on the backplate
- Corroded and/or stretched cables
- Worn or distorted drums
- Worn clevis pins.

Before refilling your hydraulic systems (including the clutch), give some thought to whether you will use the traditional glycol-based 'oil', or whether the more recent 'silicone' fluid seems more suited. With all the hype for silicone hydraulic fluids, you could be forgiven for deciding, without much thought, to fill your refurbished hydraulic systems with it. After all, the original glycol-based fluids are definitely very damaging to paint work, and they are hygroscopic, absorbing water at every given opportunity. So, in with the silicone stuff?

I was considering – but anxious about – changing to silicone brake fluid. From time to time I had occasion to speak to AP Lockheed's technical department, and raised the question of silicone hydraulic fluid because several classic car owners were having trouble with their brakes – although no-one attributed the problems to the silicone brake fluid up until the phone call. The specific problems I raised were 'hanging-on' and spongy pedals. Lockheed explained that the basic problem with silicone fluid is that it does not have the lubricating properties of glycol-based products, hence the 'hanging-on' issue. Silicone fluids are also compressible at high temperatures, which is what brings about the spongy feeling to the brake pedal. Furthermore, the lack of hygroscopic characteristics could actually be a problem, because any water that did get into the system would be suspended as globules that could cause rusting/pitting of cylinders, and possibly brake failure. Bearing in mind that AP Lockheed does not make silicone fluids, it could be argued that it would promote its glycol-based products, but I think that too many classic car users have experienced problems with silicone fluids for that to be the case.

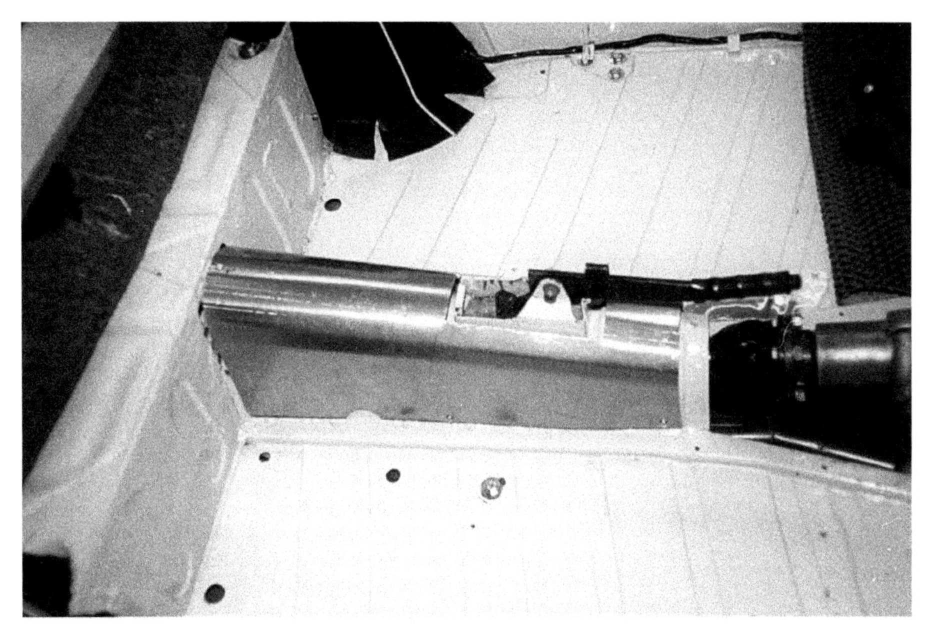

14-5. I thought this aluminium handbrake cable cover was likely to give better support when, like me, you lean on the propshaft tunnel. However, it may make carpeting the tunnel a shade more difficult.

Nevertheless, silicone fluids do have fans and very satisfied customers. Because they are not hygroscopic in theory, silicone fluids require changing less frequently than glycol. However, depending upon your climate, I recommend you change the fluid at least every three years in road-going cars regardless of the product you are using. This is in order to preserve the bores and seals of the components as well as, of course, to ensure the water content of your fluid doesn't boil under heavy braking, just when you need the brakes the most!

So it's glycol then? I think on balance I'd rather be using glycol fluid thoughtfully than wondering if my brakes were going to stick on. How do you use a glycol hydraulic fluid thoughtfully? By minimising the fluid's opportunity to absorb moisture. Do not remove the reservoir cap more often than is essential. When you change the fluid, flush and fill the system with brand-new fluid from an unopened container. Buy two small bottles rather than one mega-economy size. And seal the cap on any unused fluid as swiftly as is practical.

Chapter 15

Miscellaneous matters

ELECTRICAL SUGGESTIONS

Repair, renovation and maintenance of the car's electrical components needs to be addressed, if only because the majority of motoring breakdowns are electrically related. The problem is that one can fill a book detailing fault-finding and rectification of the numerous electrical components that make up even our relatively simple cars, and we just do not have the space.

However, a 181-page electrical maintenance handbook, especially written for the TR250/6, is available, and its content is very comprehensive. The book is of US origin so there are a few pages that will not apply to UK enthusiasts. Also, electrical issues relating to PI fuel injection have not been included. However, I have addressed these in Chapter 10, so the book will be a very useful addition to all TR250/5/6 owners and I commend it to you. Currently it is available direct from the author and details are included in Appendix 1 under 'Electrical'.

However, there are several mistakes that new owners of TRs do make, and also some pretty basic improvements that I will briefly outline.

15-1. Now that's what I call an uprated alternator, generating 65 amps. You will certainly need to install a much more substantial main feed cable, or duplicate the one you have. Furthermore, if your TR has an early alternator of the type with a separate control box, you will also need to slightly modify the wiring at the alternator (rear) plug, and fit a 'bridge' in place of the old control box. Why? Modern alternators are fitted with built-in rectifiers and control units.

- More power in the generating department seems essential if the larger headlamps – more in a minute – and/or a Bosch fuel pump are to be considered. In summary, an 16ACR alternator seems a very wise upgrade if you have an early car and are currently using a dynamo, or an 18 ACR alternator if you already use an alternator to generate the 'juice'. Even better is the 65amp monster shown in photograph 15-1.

You may find it necessary to make some fairly simple but vital wiring changes, too, and to introduce the wiring bridge shown in photograph 15-2.

- If you have a TR6, the rear (indicator and brake) bulb holders are a regular problem, easily resolved by fitting new/replacement bulb holders. The reasons for the problems are not easy to determine or rectify by any other route, since corrosion occurs inside the faces of the joint between the earth strips, and is impossible to remedy without destroying the holder. So, don't waste time; fit new ones. This problem is impossible to photograph but the potentially faulty part is shown in photograph 15-3, which may help.
- If your horn sounds when you go round bends, chances are that you have the steering column too far down inside the outer column.
- Do not buy a car without a set of dashboard dials/instruments, even if you recognise they are non-original to the car. The common, main problem of all TRs (with the exception of post-72 '6s) is non-availability of instruments. It is not recommended you swap the instruments on a TR2, 3, 4 and 4A. Not only is the clamping arrangement of the 5 inch instruments different, but the wiring to the smaller units could require some major surgery. Instruments from the Sidescreen and Michelotti TRs (4 and 4A) appear at first sight generally similar, but are very different in detail. The TR5 and early '6s instruments were identical and are generally in short supply, particularly the ammeter. Post-1972 TR6 (CR and CF) cars were not only slightly different again but are generally more easily available. The one exception is the oil pressure gauge which fits well with other instruments in the Triumph Stag, so has become sought-after as the Stag does not come with an oil pressure gauge.
- For maximum longevity, first crimp electrical terminations and then solder the crimped termination. US readers can buy a product called 'D5 De-Oxit', which I understand cleans and de-oxidizes old electrical terminations.
- A very light smear of copper-slip or petroleum jelly as you push your electrical connections together will aid conductivity, and ensure corrosion is kept in check. This is particularly true of connections within the engine bay.

15-2. This does not look too exciting, but it is essential to fit a bridge of one type or another to connect the wiring that once ran to and from the one-time remote rectifier necessary for the older type of alternators. You can simply connect a double-ended spade terminal across the two brown wires located near the fuse box (also in shot), or even solder the two browns together. However, this moulding was used on the original cars, it is well insulated, not unsightly, and not expensive.

15-3. Across the TR range these bulb-holders are unique to the TR6. They can look perfect, but give hours of endless 'fun' trying to get them to work. An invisible film of corrosion is often formed within the body. If you have any trouble with them at all, it's best to replace them with (freely available) new items.

- There is occasional confusion about what ballast resistors look like, which cars have them, and whether one is fitted to your car. Some aftermarket ignition coils are fitted with a ballast resistor attached to the outside of the coil, but in an OE sense a ballast resistor is a length of special resistance wire about 4.5 feet (1.5m) long. The covering insulation is coloured pink and the wire is laid within the front harness, running down towards the headlights, connected at each end to 'normal' (low resistance) cable, which in turn connects to the ignition circuits. OE ballast resistance wires were added as standard to all TRs from 1974. To verify your situation look at the wire going to the '+' terminal of your coil: white insulation means you have a non-resistant harness, and white with yellow tracer that you have a ballast resistant wire incorporated within the harness. Further verification is possible if you have a voltmeter (usually one scale on a multi-meter). Assuming all is well with your alternator, with the engine ticking over the electrical system generally should be generating circa 12/13 volts. If you measure the volts across the '+' terminal on the coil and the battery negative/earth/ground terminal and find it to be about 7 volts, you can be sure you have a low voltage coil and a (ballast) resister in the line feeding the coil. As an aside, an extra terminal on the starter motor will feed 12v direct to the coil when cranking in order to minimise the voltage drop the (7v) coil 'sees' when your starter motor is drawing lots of power from the battery.

IMPROVED LIGHTING

In today's traffic conditions it seems essential to improve the effectiveness of the car's lights. The following thoughts are not strictly original, but I think acceptable to most readers in the interests of your safety and that of the newly restored car.

Halogen Bulbs

The original 'sealed beam' headlamps should, at the very least, be replaced with bulb-carrying reflectors and halogen high intensity bulbs. Most Triumph specialists stock the basic replacement light units.

The potential problem with this upgrade, and certainly any subsequent improvements, is that the original switches and electrical wiring will not handle the current consumed by the upgraded light units. Thus it would be prudent to remove 95 per cent of the current passing through the switches by fitting a headlamp wiring upgrade kit, which introduces relays to the circuit and thicker wires to the lights. These combine to reduce the electrical resistance and voltage drop of both switches and wires. Not only does the original resistance drop the voltage available to loads such as the headlights, making them quite dim in many cases, the resistance also creates a lot of heat within the switch, which often results in switch failure and possibly a fire.

Advance Auto Wire's auxiliary/headlamp relay kit, seen at 15-4, uses relays and heavy-duty wiring to eliminate these problems. Most relays will operate with as little as 6 volts, so voltage drop caused by an old switch is no problem. With the extremely low current draw of relays, heat is not a problem either, so you protect the car's original equipment, improve your headlamps and increase the reliability of the car. Full instructions and a wiring diagram come with this upgrade, and consequently require no further expansion here.

Xenon bulbs

The next level of headlamp upgrade is Xenon bulbs – which you should not even think about if you have yet to improve the headlamp circuitry. Xenon headlights produce 300 per cent more light than standard halogens. The light produced is very white with a blue tint, and as yet rarely seen on anything other than recently made luxury cars.

HID/Xenon

Then we have high intensity discharge (HID) lights, of which there are a number of conversions available. These lights replace the original bulb with a HID assembly consisting of ballast and an arc unit, which gives the high intensity Xenon output. This is not the same as a straightforward Xenon bulb, but a true HID system, and can be found in a number of intensities – around 6000k is the figure to settle on. These units replace original bulbs, which in most cases will be the Halogen H4 with the P43 flange, and used to come with just the HID unit for dip and a separate halogen filament bulb for main beam. However, more recently two HID elements have become available. The output from the HID elements is stunning compared to halogen, making a halogen beam appear weak by comparison. The manufacturers claim that even the headlamp flash function is fast enough to give perfectly acceptable results.

Note that some of these conversions do not have approval marks for all the markets they are sold into, and you should check this detail before purchase. However, most units run at 35 watts, which is somewhat less than the standard bulbs and much less than uprated bulbs. In the UK, HIDS4U launched a Xenon HID conversion kit that requires no filters or auxiliary fixes and, it claims, works on any car.

Rear lights

In ideal conditions, the TR rear lights are adequate, but in adverse conditions such as spray/fog etc, the lights will not offer visibility equal to the vast majority of cars on the road, particularly when other drivers need to look past increasingly powerful headlights used by oncoming traffic. Low-level 21 watt fog lights fitted below the rear bumpers will have some value in adverse conditions, and are recommended.

It will not have escaped your attention that all modern cars are also fitted with high-level rear brake lights, as research has proven that they are infinitely more effective than additional brake lights fitted below eye-level. The task of fitting these self-contained units at the top of the rear window in a hardtop is simple and highly recommended, and picture 15-5 may give you some ideas – although I do concede a Roadster will provide more of a challenge, but the additional safety is worth the effort. Perhaps an enterprising reader will offer a fibreglass clip-on unit for the boot/truck on all classic cars.

15-4. These twin Advance Auto Wire units contain relays for the dipped circuit and one for full-beam function to take the majority of the current away from the headlamp switches. A set of substantial wires is included to carry the current to the lamps without any voltage drop.

15-5. An eye-level brake light offers additional safety – this is a Ford unit fixed to the glass by double-sided pads (use windscreen mirror pads). Most breakers yards will have dozens of similar units. Take the wires down the 'B' post, into the boot/truck and join the loom in front of the left-side tail lights.

IMPROVING SECURITY

The incidence of car crime has fallen in the UK, lulling many into a false sense of security. There are many different types of car theft; classic cars are usually stolen for breaking. Let's consider the most obvious types of car theft in an effort to help you decide on the most cost-effective way to safeguard your TR.

Theft from within the car first. High-security (circular) door locks are an absolute waste of time with an open car; you're better off removing temptation from sight and using a conspicuous steering lock. Never leave a locked box or other container on view for obvious reasons: the place for valuables is in the boot/trunk. However, don't forget that the Michelotti boot lid can be opened even when locked, although I am sure you will agree that it would be silly of me to tell the bad guys precisely how ... The Michelotti cars need relatively minor modification to guard against unauthorised entry. A spacer between the bootlid skins will make the handle much less pliable, and a large washer welded to the end of the latch pin makes it most unlikely that anyone will succeed in forcing the boot/trunk lid.

We're all aware of the opportunistic thief that strikes when your car is left on the drive, or outside a shop "for a moment". The moral of this is: be more careful with the keys. An immobiliser will help but only if the keys are not available; although the immobiliser I have seems easy to circumnavigate, it would take time to do and must deter this type of car thief. Although most cars stolen in this way are recovered, they have usually been vandalised and an attempt made to disguise identity. It's a good idea to hide something in the car (such as a business card), or have photographs of a couple of unique features on your car (such as a special carburettor/seat mounting etc.) in case this should happen to your car.

The next most easily deterred thief is the joy rider, who will be looking to make a fairly quick exit, albeit with your car. Security etching on the glass is not going to deter him, nor may a steering lock, since I understand that these thieves are able to remove all but a select few very quickly indeed. However, for those whose cars have electric fuel pumps, a well-hidden, fuel cut-off switch, particularly if the car also has an immobiliser, may delay the thief just enough to make him decide to try an easier target. My immobiliser cuts off the feed to the starter but I think immobilisers that also cut off either or both ignition and fuel supply are available. I would prefer to fit two separate devices to give potential thieves two obstacles to overcome, so suggest you fit a commercial immobiliser in addition to your own fuel/ignition cut-out. Unfortunately, the majority of carburettor TRs use a mechanical fuel pump, which precludes the use of a security device. Petrol injection cars use an electric pump, and will not start without an operational fuel pump, but any hidden fuel cut-off switches should be wired into the primary side of the fuel pump relay. If you feel you need to fit the switch straight into the feed cable to the pump, you must fit a switch with a capacity rating of at least 15 amps. These are not so readily available these days since the majority of cars use relays to reduce the load on switches.

It's very hard to put off the professional thief that wants your car for its parts, or in order to 'ring' it with another identity. I heard about a car that had been stolen from a double-locked garage, which had an immobiliser fitted and with the ignition keys safely indoors. In this case, the police established that the car had been broken within 24 hours of its theft. The thieves had winched the car onto a transporter, giving the impression that it was being taken away for repair.

Your only defence is not to advertise your address (area secretaries please note), and have the car fitted with a vehicle locating device such as Tracker, or Securicor's Trackback. These devices are the most expensive precautions that you can take and, although I don't have experience of them, believe they can pinpoint the location of your car very quickly indeed. Make a note of all the serial numbers and try to clearly but unobtrusively mark the chassis and body parts with some identification. A few centre-popped initials may enable you to prove to the police that a particular part is from your car.

Take particular care if offering your car for sale. Do not allow potential purchasers to see it in the garage, or tell them about its security protection until the deal is concluded and you have received the money.

It's a fact that because car anti-theft devices are now so effective, the only simple way a thief can steal a car is to break into the owner's house and steal the keys first. This is happening on an ever-increasing scale, which is partly why car theft statistics are coming down, since a car stolen in this way is officially categorised as a burglary. Such cars are often still broken within a few hours, but this trend extends your area of car-care to the house generally, and where you keep the keys to your car(s) in particular.

Finally, a few thoughts on immobilisers, the first of which is get one fitted! There are basically two ways to achieve this, and we'll discuss the professional installation company first.

Immobiliser prices range from between £100 and £200, and even up to £500, but this does include installation. Dual immobiliser/alarm systems are more expensive and I do not feel they are value for money. Whatever level of system and whatever fitting method you choose in the UK, ensure that the system is 'Thatcham', which means that it has been tested and approved by the Motor Insurance Research Centre located at Thatcham.

The professional installer will give you a certificate upon completion of the job that could get you an insurance discount, or put you on better terms with the insurance company should the worst happen. I believe you will only get a certificate if a professional installer fits your immobiliser, and is one reason why an accredited company should supply and install the system. You can add sensors to some systems (such as level sensors that detect the vehicle being lifted or hot wire ignition sensors), and you should discuss your options with a couple of Vehicle Security Installation Board (VSIB) accredited businesses.

Alternatively, you could install an approved system yourself, which is not without its advantages. Firstly, you can – and indeed should – 'bury' the wires within the car's main harness to make detection of the immobiliser very difficult. Secondly, you can take the precaution of using male and female connections at the points where the system interfaces directly with the car's wiring, to give you the option of bypassing the immobiliser circuits in the event of its failure. This feature is particularly useful in roadside breakdown situations. I remember one case where someone was not able to leave the cross-channel ferry as their car's alarm had been locked on, apparently by the ship's radar! If you are ever in a similar situation, you could quickly get mobile again. Obviously, you need to ensure that the bypass connections are not obvious (hidden under some black tape?) and are located so as not to compromise the security of the car.

BONNET RELEASE CABLES

New or prospective TR owners may already be wondering what's so special about the car's bonnet release cable? Look at the car's rear-mounted bonnet catch and imagine how difficult it would be to open the bonnet without a serviceable release cable. Be aware also that a broken or seized bonnet release cable is not an isolated problem; the consequences will be frustrating and time-consuming, and possibly expensive.

So do follow my advice, which starts with the obvious: lubricate both cable and latch assembly. It's worthwhile going to the trouble of separating the inner cable from the outer, and applying lots of 'copper-slip' before feeding the inner back inside the outer. Feeding a used inner into the outer can be frustrating, but I assure you that it's even worth buying a new inner/outer cable assembly and lubricating it properly before fitting to your car.

I also strongly recommend you arrange some form of secondary or emergency release mechanism; a discussion with your local TR group will provide a wide variety of ideas. My method was to fit the second bonnet release cable and bracket shown in picture 15-6, with a second bonnet release pull fixed to the right side of the car. It's important that you fix the second bracket to the right side 'A'-post very securely; you'd be surprised just how much strength is occasionally required to release the bonnet catch. Mind you, one significant benefit of a right side release catch is that you can (with an open window) press down on the bonnet (above the catch), simultaneously pulling on your new right side cable toggle: you will find that the bonnet releases very easily; so easily that I now rarely use the 'proper' release cable on the left side of my car.

Well, so much for secondary release measures; what do you do about the dreaded broken release cable? The good news is that, unless your car consumes vast quantities of oil and/or water, you should still be able to use it. The bad news is that I have heard of owners having to drill a hole in the bonnet to get to the release catch. Hopefully, one of the following suggestions will help you avoid such drastic action.

Assuming your cable has seized rather than broken, start by having a helper push down on the bonnet, perhaps an inch forward of the rear edge, as close to the catch as possible while you tug on the release handle. If this releases it, the most likely cause was a poorly lubricated cable or catch, though a poorly adjusted catch or bonnet stop can cause the same problem. If this initial step proves unsuccessful, but still assuming that the inner cable has not actually broken, next try unscrewing the outer cable securing nut. Ease the inner/outer cables out of the mounting bracket and, again with your helper pushing down on the bonnet, give both inner and outer cables a good pull. Still no luck? Time, then, to visit your nearest friendly garage with a car lift/hoist, for an 'underside attack' on the problem.

The first approach is a 'long-shot': see if a long rod will lever the catch arm towards the right side of the car. Take great care with this, for you will be very close to two power terminals; one on the rear of the starter motor and the other being the positive battery terminal. The latter terminal sits just a couple of inches (50mm) from the catch arm. Consequently, a wooden broom handle might be best.

It is possible to unbolt the hinges at the front of the bonnet from the underside of the car while you have it on the ramp, and this will most likely result in the rear catch unfastening. However – and particularly if you have a LHD car – you may prefer to first remove the glove/cubby pocket and try to reach the catch arm through one of the bulkhead/firewall holes. Remember the proximity of the battery termination, and that the catch arm needs to be levered towards the right side of the car (as viewed from the rear). You'll probably have to remove several air hoses from under the dashboard/fascia of later cars to stand any chance of satisfactory access. If you are able to get a screwdriver to the catch, have a helper take the pressure off the catch arm by pushing down on the appropriate point of the bonnet. Do ensure your helper pushes down quite close to the rear lip of the bonnet if you want to avoid a dent.

15-6. Bonnet/hood releases. The photograph shows a simple, effective secondary/emergency bonnet/hood release, which has been achieved via one extra hole in the side of the catch assembly and one extra hole in the arm, just behind the normal release arrangement. During manufacture, the 'skin' of the catch assembly metal becomes work-hardened, and each hole may first require a small ground spot to break through the 'skin' of the metal. The additional cable is a cut down original, routed through the bulkhead/firewall from an additional bracket fixed slightly in front of the right side 'A' post. It works so well that it has become my standard method of opening the bonnet/hood. The gauze-topped cylinder is an anti-run-on valve.

Chapter 16

Interior trim

RESTORING THE WOOD VENEER

The dashboard/fascia screws onto a metal dashback, and is made up of a plywood base with a thin hardwood veneer stuck to the visible face. The veneer is then finished with a coat of clear varnish (some satin finish, some gloss) in order to seal and protect the veneer and make it more aesthetically pleasing.

If your base plywood and/or veneer is damaged, there's nothing for it but to purchase a replacement dashboard/ fascia. These are readily available in original finishes as well as some, to my eye, very attractive burr walnut finishes.

However, the purpose of this section is to explore the possibility of salvaging a dash that looks cracked and split, but which may not be as bad as it first appears because the varnish can crack and dirt get into the cracks and onto the veneer. With such a dash, since you have nothing to lose, try careful, slow, gentle stripping of the varnish from the front of the veneer, once the dash is out of the car and all fixtures and fittings removed from it. At this point you could well find International Coatings' (free) booklet on yacht paint and varnishing very helpful; details in Appendix 1.

You will need a mild, varnish stripping chemical that will soften the varnish. A flat bladed wallpaper stripper, or similar tool (not too sharp or you endanger the veneer beneath the varnish), should remove most of the varnish if you have applied sufficient stripper and left it long enough to do its work. For any remaining stubborn varnish spots, apply a second coat of stripper and gently scrape again, before finally wiping down with white sprit to clean the dash and remove the worst of the dirt that will have become ingrained in the veneer.

Carefully inspect the veneer at this point; it may turn out that it is cracked and will need to be professionally re-veneered. If you have managed to remove most of the 'crazing', the probability is that your dash/fascia can be returned to its original glory by gentle cleaning (with white spirit) until all dirt is gone. Avoid sanding, if possible, although, if you feel the surface of the veneer will benefit from a light hand 'dusting', it is possible to use a very fine paper on a flat block to gently freshen-up the veneer, but it is essential that you proceed slowly and with care.

Most of those TRs that were fitted with a wood dash/fascia originally had a high gloss finish, with only the very late examples having the gentler 'satin' finish. This is your chance to apply whatever finish you feel appropriate for your restoration, so purchase the appropriate (gloss or satin) clear polyurethane exterior varnish. A chandlery would be one place you could buy this. Applying the varnish in several thin coats with a model maker's spray is likely to give the best result, but you could also experiment with a foam brush on a scrap piece of furniture before taking the plunge.

SOUND DEADENING FELT & CARPET FITTING

The first order of business is to equip yourself with a good quality, generously cut carpet and sound deadening kit, such as that from TR Bitz which has its own trim shop and knows the job backwards. I am, incidentally, indebted to Steve Wilson of TR Bitz for his expertise, and to Mark Price for his photographs which, together, helped me assemble this chapter. Mark's counsel and photographic support are in evidence right throughout the book, but he really surpassed himself when we got to this chapter!

You will need several sharp blades, or a Stanley knife, and two types of adhesive: an aerosol of spray-on adhesive (500ml) and a (half-litre) tin of contact adhesive. The former will allow you perhaps 5 to 10 minutes of adjustment time before it sets, whilst the latter, as its name implies, sets on contact. The second but equally important point before starting work is to appreciate that these adhesives can be both your best friend and a problem if, for example, you are careless with their application. Ensure you have plenty of working space around the car, and remember that the aerosol adhesive can be carried on the wind, or by a draught, with the result that you may end up with adhesive overspray all over your car. Never, therefore, spray adhesive outside. If practical, apply most of the adhesive inside the garage with the car parked and the actual fixing/working place well out of the way, outside the garage. When it is unavoidable that you spray directly onto the car, move the car inside and ensure that draughts are minimised. Ascertain which solvent will remove surplus adhesive without damaging the car, felt or carpet. My first choice is white spirit (known in the US as mineral spirit), but you may need to resort to petrol (gas) to remove some modern adhesives. Have some solvent(s) handy, for mistakes are best rectified as quickly as possible.

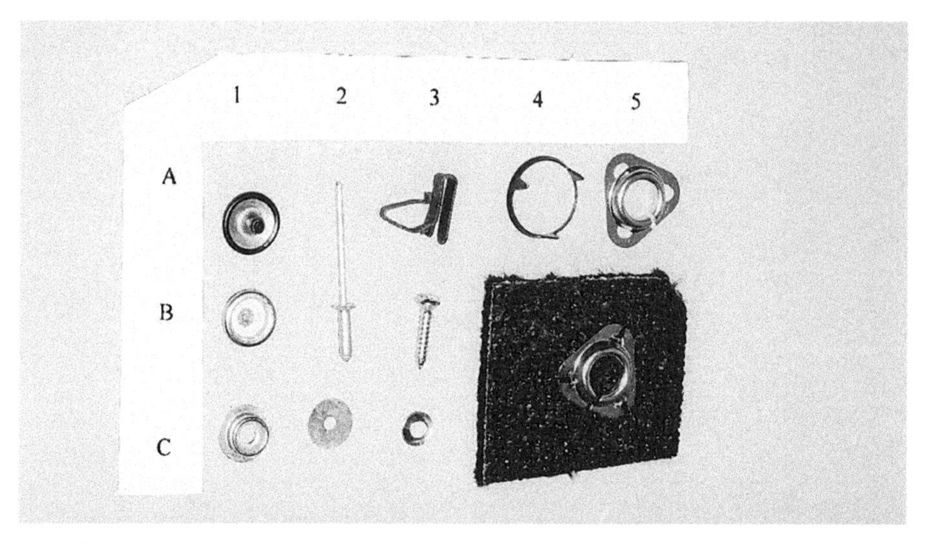

16-1. Trimming fastenings, shown here to ensure correct identification. Column 1 contains the three types of 'snap' fastenings used at the rear of the hood. Column 2 the pop rivet and its essential washer used extensively when fitting the hood. Column 3 has the panel fastening clips, chrome screws and conical, or 'cup' washers. Columns 4 and 5 show the two-piece carpet clips which, after assembly either side of the carpet, clip to the base fastening in the bottom right of the picture.

Assuming you are trimming right through the car, remove (or do not fit in the first place) the steering wheel and 'H' dashboard/fascia support, and undo the door-check strap bolts (one

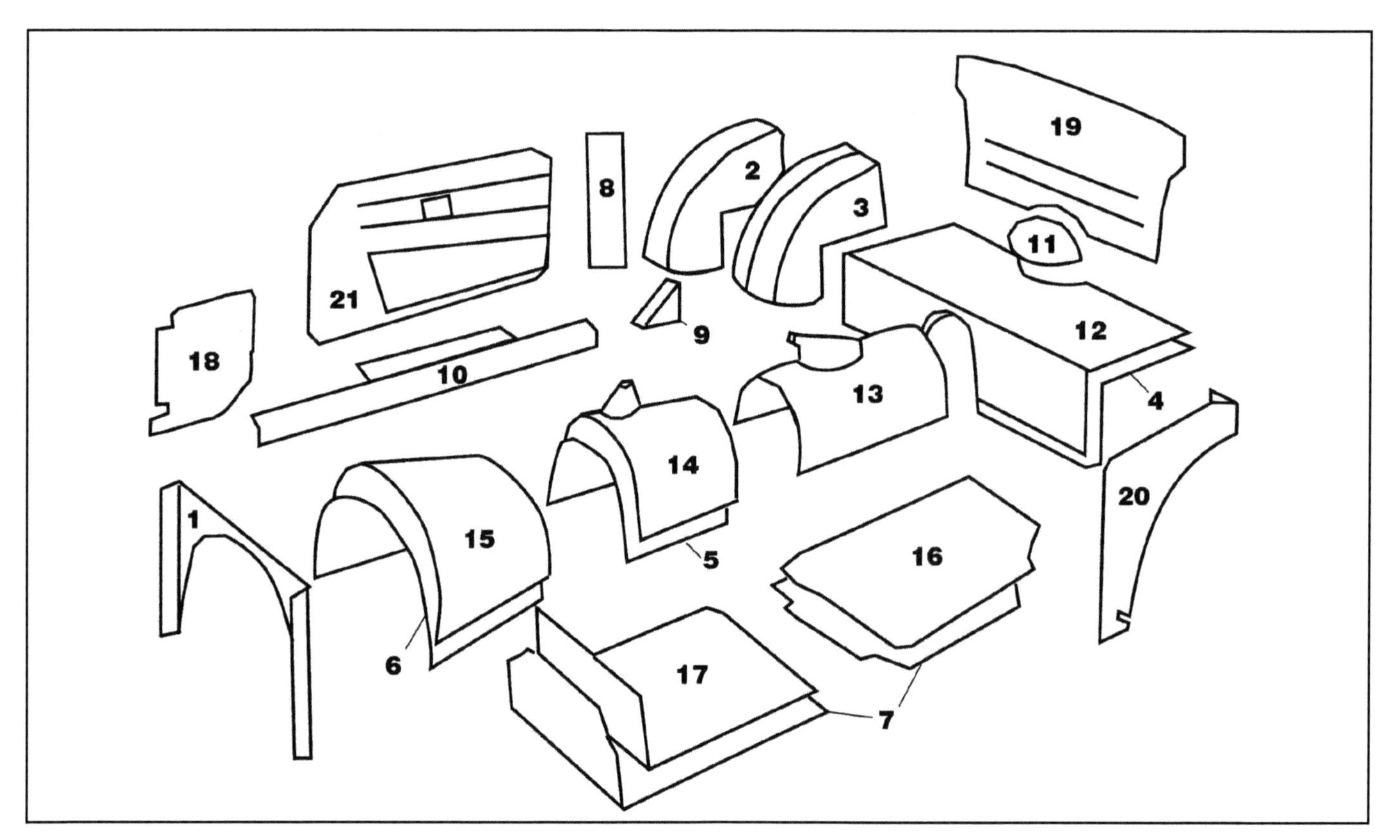

D16-1. Fitting sequence for felt, carpet, and trim panels.

per door), but leave them in place until you need the door wide open. Prior to fitting it really is a good idea to lay out overnight all carpets and felt. Try to minimise having to drill or screw through carpet but, where essential, take great care as the chuck or the screw can 'pick up' a carpet thread and pull out a very unsightly line of weave. Always use a very small pilot drill to drill through a piece of carpet (about 1/32in diameter or less than 1mm), and roll back the carpet and open the pilot hole first. Always stick several layers of masking tape on top of the weave to protect the carpet from the chuck when the drill breaks through. This will have the added advantage of telling you where your drill hole is! Don't laugh, it really is all too easy to lose the hole in the carpet and no fun at all when you are working at the bottom of a deep footwell!

Always use a (usually conical) washer under the screw head when fixing the carpet with a screw. During the course of this chapter I refer to a number of small washers, fittings and clips that not every reader will be familiar with. Therefore, it seemed a good idea to correctly identify most of your requirements, so I hope photograph 16-1 will help, while drawing D16-1 identifies the respective felt, carpet and trim panels, and should help you quickly sort out the sequence of fitting I recommend. With one or two exceptions, the sequence is not absolutely crucial, and you'll note that, on several occasions, Mark's photographs show that he used a slightly different sequence to the one I am proposing. Photographs 16-2 and 16-3 show a couple of the preparatory steps necessary with some carpet sets.

Finally, in general, pull surplus carpet and smooth wrinkles from the centre of the car outwards.

Start the trimming with carpet 1, as per the sequence shown in D16-1, the one with the substantial 'U'-shaped cut-out. This needs to be stuck to the bulkhead/firewall using spray adhesive, with holes for speedometer and accelerator/gas cables neatly cut before smoothing the carpet flat. This will ease you into the trimming as it is one of the simpler carpets to fit which does not require underfelt.

The rear wheelarches are your first serious trimming task. Your wheelarch kit will include some generous pieces of 1/8in (3mm) foam, which should be stuck to the wheelarches (2) to deaden road noise and give the subsequent cover a more luxurious feel. Apply the spray glue to the wheelarch, wait for it to become very tacky (almost finish 'flashing off'), and then apply the foam to the wheelarch. If you spray the glue directly onto the foam, or position the foam too early, you run the risk of the foam 'taking up' the glue and going rock-hard as the glue dries.

These foam pieces need to be undersized so that they do not reach (by about 0.5in or 12mm) the rear and floor panels. Cut away any surplus at the corners as one thickness of foam is sufficient. Let the adhesive set first, of course, and then draw a line with felt-tip or marker pen on the outside of the foam 4 inches (100mm) from and parallel to each inner wing, starting at the top/rear and finishing where the wheelarch meets the floor. Turn your vinyl wheelarch cover (3) inside out, and offer it to the wheelarch to check that the seat belt mounting point usually

16-2. Some sound-deadening felt has been fixed to the top of the tank – which is not the good idea it may at first seem. Fuel has been known to leak from the filler cap hole in the rear deck, and has been soaked up by the felt, creating a fire hazard. The place for that piece of sound-deadening felt is along the front of the tank. This picture is interesting for a second reason – affixing the carpet to the propshaft cover fixed behind the handbrake. Obviously, every car can be completed to individual preference, but my main text recommends you do not fix the propshaft cover until you have glued the carpet to it. Third point of interest; I wonder how many noted that the triangular gussets on this car have been cut away adjacent to the centre of the wheelarch? It really is a good idea to completely replace, or at least make good, these load-carrying stiffeners when you are doing your panel work and welding. Some restorers even weld an 18swg sheet across the body tub, fully closing the space in front of the tank to stiffen the bodyshell and act as a firewall.

16-3. This carpet set provides separate handbrake and gearbox gaiters. Most carpet sets have these, in leather or vinyl, already sewn into the relevant carpets. You will hardly need me to tell you that you must fix the gearbox cover before starting trimming, but you may be interested to know that this is a modern plastic moulded version, which, though tiresome to fit, offers excellent longevity compared to the original moulded board.

welded to each wheelarch does not cause the loose inside edge of the piping/beading to bulge. You may need to move the cover just slightly, and/or trim a fraction off the inside edge of the piping. With the cover still inside out, very lightly spray the whole of the inside of the cover, concentrating most of the adhesive round the edges of the cover. Allow this to 'flash off' for as long as you dare to minimise take up of glue by the foam but allowing some adjustment time. Do not spray the adhesive onto the foam.

Apply the vinyl arch cover, concentrating initially on getting the piping/bead to lay along your felt-tip line, starting at the top and working towards the bottom/floor of the car. Your next job must be to get the outside (4in wide) strip of the cover to lay smoothly and straight over the upper part of the arch from top to bottom. With this almost to your satisfaction, flip the inside out lower part of the cover over the lower inside of the wheelarch and ensure that this, too, lays pretty smoothly over the foam. Some final adjustment should still be possible before the glue sets, enabling you to remove any minor wrinkles from the cover.

Allow about 1in (25mm) of vinyl to carry over and stick to each of the adjoining panels, but the most important part of this job is the long radius over the top of the wheelarch. You will need to snip the 1 inch overlap every inch or so around its periphery (see photograph 16-6) so that it lays tight to the inner wing. Take great care not to make these cuts too deep initially, for they could show when the job is done. Better to first try a short snip, even if you have to extend it a bit later. The corners of the cover (where they meet each adjoining panel) will also require trimming. A single cut at 45 degrees to the angle is usually sufficient to allow the surplus vinyl to overlap, while still laying flat. These cuts are not too critical since carpets or panels will cover them as trimming proceeds.

Photographs 16-4 to 16-6 illustrate the sequence for completing the

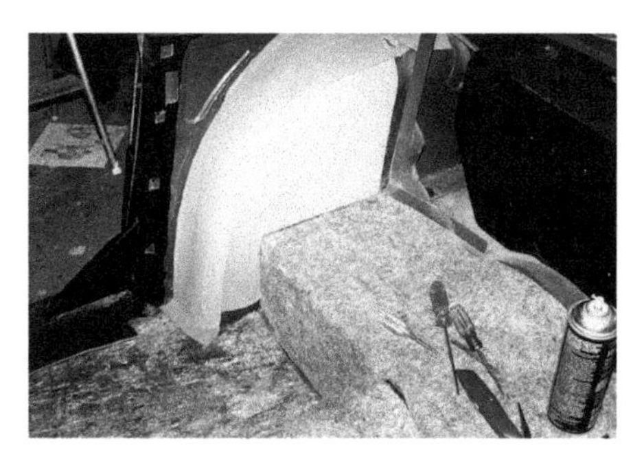

16-4. Back to the rear of the car and the first of certainly two – and in this case three – stages of covering the rear wheelarch: the thin foam stage. Note how the foam has been carried backwards almost as far as the back of the tank, which will also help deaden road noise. I would guess that the excess foam lying along the floor is about to be trimmed away, as, too, will be the last 0.5in (12mm) of foam where it abuts the rear shelf pan. This apparently minor adjustment is actually quite important since it allows the wheelarch vinyl cover to be glued to the actual wheelarch. Note the advice in the main text about keeping the foam as free of glue as possible to stop it losing its soft 'luxury' feel.

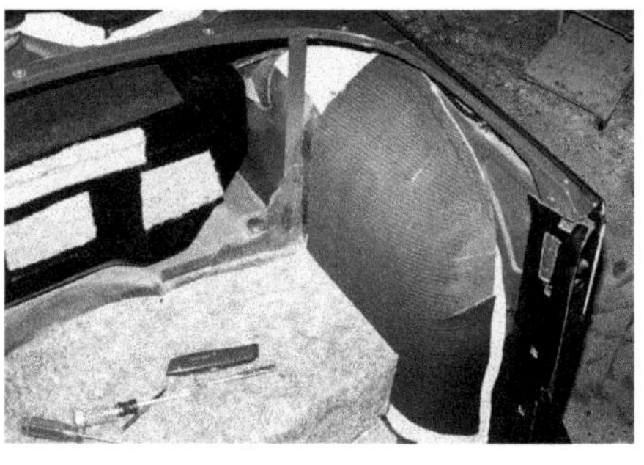

16-5. Step two. However, this non-original layer of sound-deadening will make the final (and very visual) wheelarch cover more difficult to apply smoothly. If it is included in your trim kit, I would think carefully and perhaps talk to someone who has fitted it, before proceeding yourself. Incidentally, the picture prompts an additional thought: you should wrap the 'B' post vinyl around the front lip of the 'B'post so that it cannot wrinkle when, in due course, you push the furflex door seal into place.

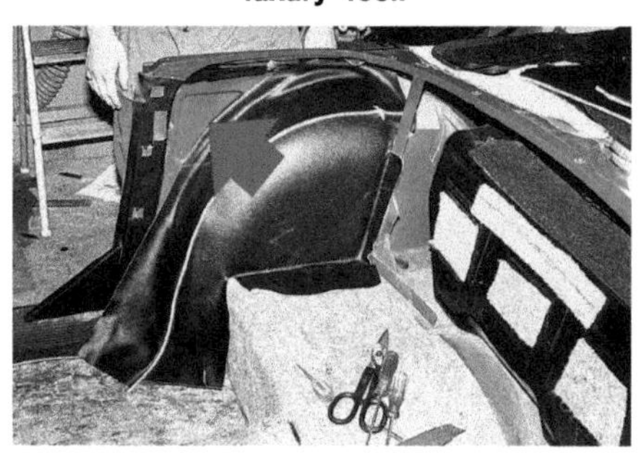

16-6. Finally, step three. The white – in this case – bead is carefully aligned to the inner wing, with as much vinyl as possible turned round the edges of the corner.

16-7. The sound-deadening felt going in on the rear shelf using the spray adhesive shown in the picture. The scissors need to be newly-sharpened if you are going to cut neat edges in the felt, and for any carpet trimming that may follow.

16-8. Cracking on now. The gearbox cover and prop tunnel felts will have been glued in place, but the floor felts will have been only laid in place and cut to size, but not glued, as explained in the main text.

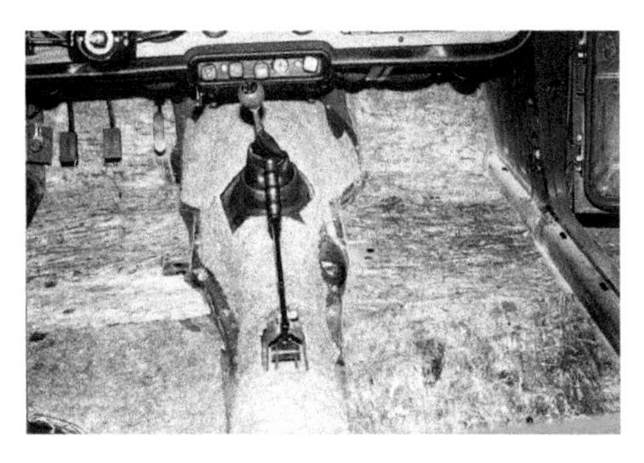

16-9 and 16-10 (right). Two views of the felt in place. Have you noted the fairly generous cut-outs for access to the mounting holes for the 'H' support bracket, and provision for the seat mounting bolts? Note, too, that the steering wheel has been removed for (slightly) easier access to the footwells.

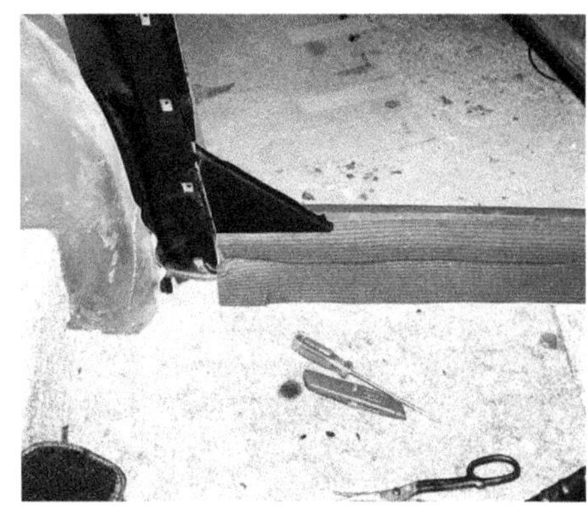

16-12.

16-11. The 'B' post cover in place. Note the careful cut-away to allow each quarter panel to be easily screwed/clipped in place in due course, but, as I mentioned before, the covering is better applied perhaps 1in (25mm) further forward. This allows it to be wrapped around the lip onto which the furflex seal will eventually be pushed.

16-12 and 16-13 (top right). This carpet set includes a special underlay for the sill carpet and wheelarch covers which, at least as far as the sills are concerned, is a definite plus and will give the sill a more luxurious feel, and deaden noise, too. Here we see that the 'B' post corner gusset has been trimmed and the underlay glued in place over most of the sills. Can you spot the rear wiring harness – correctly fixed in place before the carpet was laid, and now sneaking under the carpet for added protection?

16-14. The sill carpet looks great and gives the feeling that the trimming really is starting to take shape. However, we can see that, regrettably, this carpet set omits the original vinyl strip from the tops of the sill carpets. As a consequence, you have nothing to turn over the bottom door lip, so the furflex door seal cannot restrain the carpet, as was originally the case.

wheelarches which are one of the most difficult and obvious parts of trimming work. You would be well advised to carefully cut provision for the seat belt mountings, as shown in photographs 16-20 and 16-21, and to smooth the vinyl cover down before the glue sets completely.

Next up are some of the sound-deadening felt pieces, that need to be stuck in place with spray adhesive. The first pieces to position are those above the differential and on the adjacent vertical heel board. I will explain why shortly, but I would not glue the felt that goes over the fibreboard propshaft tunnel cover and move forward to glue the smaller gearbox cover felt (mostly behind the gear lever, but numbered 5 in D16-1, in place. The final piece of glued sound-deadening felt you should attach at this stage covers the front or main part of the gearbox cover (6).

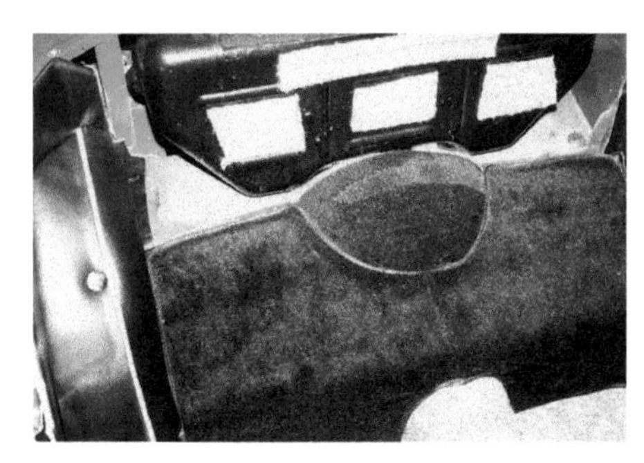

16-15. The hump of the differential was first covered with the oval carpet from the set, with the rear panel/heel board carpet swiftly glued in place shortly afterwards.

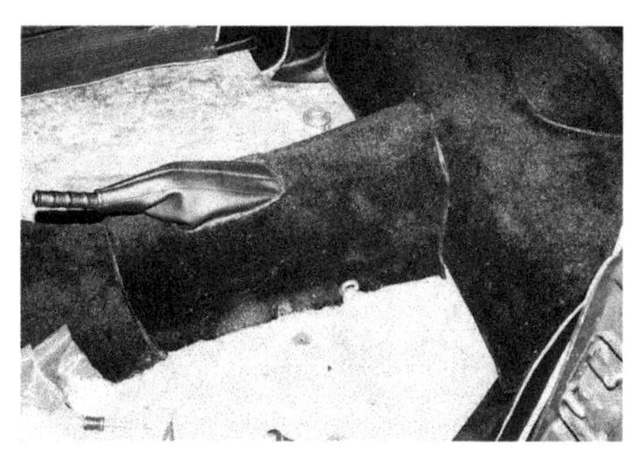

16-16. The propshaft carpet glued to the underboard (shown by the absence of wrinkles).

Photographs 16-7 to 16-9 will help you understand what has to be done.

Fit the rubber gaiters around the gear lever/shift. You can cut to shape the four footwell pieces of felt (7) as the photographs 16-9 and 16-10 show, but do not glue these in place; in fact, once cut to size they should be removed and placed to one side for the time being. Make sure that the felt is cut away (photograph 16-10) to allow the 'H' dash/fascia support casting mounting feet to be bolted in place in due course.

Spray glue and fit the two 'B'-post vinyl strips (8) so that they wrap around the door frame lip (they will eventually be covered by the furflex door seal), yet extend back behind where the quarter panels fit. It is a good idea to cut away where the fastenings will come a little later (as photograph 16-11 shows), or possibly to make a small hole with an awl.

We can now finish off the preparation at the rear of the car by covering the two triangular gussets at the base of each 'B'-post. These have a board stiffener inside each vinyl covering which needs to be bent over to lay along the top edge of the triangular gusset. Use impact adhesive for strength when fixing these trims.

Sill carpets (10) are next. With these, there are usually no sound-deadening foams or felt to worry about, just where to position the carpets relative to the sill, although the trimming kit shown in photographs 16-12 to 16-14 did actually include an underlay, just to confuse you! Underlay or no underlay, it is not difficult to position the sill carpets as the front of the triangular gusset at the base of each 'B'-post provides the longitudinal reference. Once you have established which carpet goes which side of the car (this will quickly become apparent if you lay them roughly in place), you will note that part of the top of each sill carpet is finished off with a piece of vinyl or tape. This is intended to go over the sill lip beneath the furflex door seal that you will eventually position around the door opening.

Your first point of reference is to marry the rear of this tape to the front of the 'B'-post gusset we spoke of earlier. You will also need to think about positioning each carpet relative to its sill. The plan here is to butt the carpet as tight to the top/outside edge of the sill as possible without allowing the carpet to 'turn up' towards the door – which, in due course, would prevent the furflex from fitting as intended. The vinyl strip or tape will, of course, 'turn up' towards the door, but that is as intended for it is thin enough for the furflex to slip over without a problem.

These carpets need to be glued with contact adhesive, which is stronger than spray-on adhesive, but allows very little (if any) time to adjust the fit. I always use 'Trim-Tack' adhesive, which does allow a few moments' grace to sort out any wrinkles. A tip you may find helpful is to put the sill carpets in place in 'step' sequence. By this I mean stick down the top of the carpet the full length of the sill, before turning your attention to the full length of the vertical face, and finally the surplus inch or so that sits on the floor. The end result is shown in photograph 16-14.

Now to carpet the heel board and rear shelf, both of which are situated behind the seats and numbered 11 and 12 on drawing D16-1. Spray glue over the small, oval-shaped piece of carpet, and locate it on the existing felt over the differential bulge. Position is not critical, but if you want to double-check all is well, you could offer up the (not glued as yet) main rear shelf carpet to reassure yourself and make any adjustments. Spray glue over the main rear shelf carpet and pop that in place. The vertical heel board carpet (using spray adhesive) is next; expect/allow about 1 inch of surplus carpet to run forward onto the rear of the floorpans. Photograph 16-15 shows where we have got to so far.

Forward a few feet is the propshaft tunnel (13) where there is a choice to be made. Many a TR has been successfully trimmed by having sound-deadening felt glued to the fibreboard cover, and the carpet pulled over the handbrake and stuck in place. However, this 'traditional' method tends to result in the carpet sitting loosely on the prop tunnel, so I am going to recommend an alternative that I think our premier TR restorers now use pretty much as standard. It does mean that the sound-deadening felt is done away with, but the neater appearance of the resulting installation more than makes up for this.

Unscrew the fibreboard prop tunnel cover from the car, lay it flat on the bench and find the rearmost centre point (across the width of the car). Lay the prop tunnel carpet face down on the bench and find and mark the centre point at the rear of the carpet. Match the centre of the fibreboard cover to the centre of the carpet, and roll the fibreboard cover from side-to-side, marking the carpet where the outside edges of the fibreboard cover come. Remove the fibreboard cover and re-mark the carpet with two lines, each about 1.5in (say 40mm) nearer the centre than the original outside edges of the fibreboard cover. Contact glue the area of the carpet inside the two lines and, ensuring that the centre points are aligned, stick the fibreboard cover to the carpet (with the carpet and board flat on the bench). The two outside lips of this cover/carpet assembly should not be stuck together, which means you should be able to lift both edges of the carpet to re-screw the fibreboard cover (and, now, propshaft carpet) back in place. If you are in any doubt, hopefully photograph 16-16 will help.

The gearbox cover now received our attention, starting with the smaller rearward piece (14) that should be positioned over the gear lever and

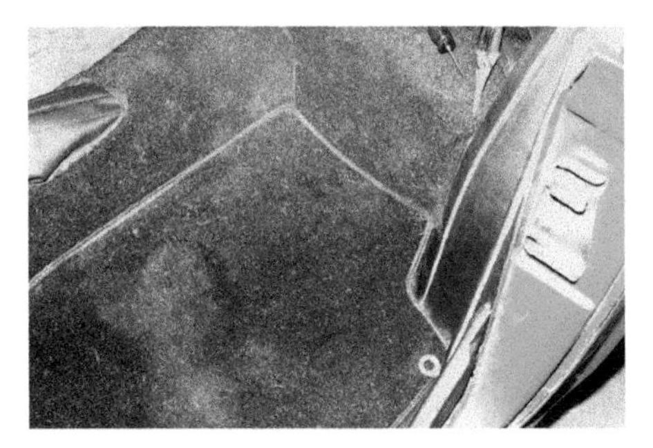

16-17.

16-17 and 16-18 (above). Two views of the floor carpets which are placed in the footwells but not glued, because of the possibility (a probability in the UK) of them getting wet and having to be removed and dried before corrosion sets in.

16-19. The central, almost 'U'-shaped carpet has been previously glued to the bulkhead at the front of the gearbox cover, and the side footwell carpets are obviously being temporarily located prior to spray gluing in place on the backing panel. Note how access to the footwell area has been improved by temporarily removing the door 'keeps' (just to the left of the top awl), which is a good idea provided you do not allow the door to open too wide and damage the paint. Might be even better to install the check straps simultaneously with the doors and just remove the (single) connecting bolt as and when you need extra access. Incidentally, Mark tells me the awl was indispensable for locating screw holes through felt, carpet and vinyl.

underfelt. The main gearbox cover carpet (15) has a reputation for being difficult, but you should find (or can easily rivet) four male press studs to your gearbox cover. Obviously, if you have added sound-deadening felt this will complicate the issue slightly, but only to the extent of cutting a 1in (25mm) (approximately) clearance around each stud – just as you would with the floor felts. Incidentally, if you are fitting the male studs and associated carpet clips, they are exactly the same as those you would fit to hold the floor carpets in place, although you can get slightly taller ones to help carpet fitting when underfelt is in place.

Moving on, we now need to position the four sound-deadening panels for the floors (item 7 on our sequence drawing) and, in particular, cut four 1 inch (25mm) diameter clearance holes in each felt; for the seat runner mounting bolts in the rear felts and the carpet clips in the front pair. I would use a sharp-pointed instrument such as a scriber or awl to find where the holes should go, and then cut them with a blade outside the car. You need to be more careful when it comes to cutting holes in the actual carpet, but with the rear pair of felts and carpets held in place by your seat runners, you will appreciate that gluing the rear pair is quite unnecessary.

The forward footwell felts (7) and carpets (17) are best spray-glued together, but I suggest you don't glue the felt to the floor in case you wish to remove the footwell covers for cleaning or drying out, or to vacuum the footwells. Instead, you should rivet 4 press-stud bases to the floor each side (do not forget the washer on the other side of the rivet!), and secure the carpet/felt with carpet clips. These special auto carpet clips are virtually invisible rings (see photograph 16-1) that are pushed through the carpet and turned over, whereupon they clip to the bases you riveted to the floor. Take care getting the alignment of bases and clips right. Photographs 16-17 and 16-18 illustrate this step.

Last but not least, the two front carpeted kick-panels (18) on the outside of each footwell need to be screwed in place. Pretty straightforward, really, except it's a good idea to have an inch or two of furflex door seal handy, as you do need to get the carpeted panels just right in a fore and aft sense. Too far back and you will not get the furflex properly in place, too far forward and you will have a (possibly difficult to see) gap between carpet and seal. Photograph 16-19 is interesting ...

16-20. The rear panel/tank board screwed into position. Read the main text to appreciate the importance of centralising the first hole/screw and the sequence thereafter. If you align the central panel line, you will see that the central screw does indeed fall dead centre to the differential.

INTERNAL TRIM PANELS

I have not described how to remove the trim, as it's pretty self-explanatory. However, one detail that may help is how to remove the old door panels, and the internal door and window handles in particular.

Window handles are actually pinned in place, but the pins are usually hidden by the trim panel, which is pushed outward by two coil springs positioned behind the door trim panels. To release the two pins per door, push the trim panel inwards at each handle, revealing a cross hole in each square drive, and push out each pin with a small screwdriver. Try not to lose the pins, although when we get to re-assembly, I will tell you a little trick of the trade, so all is not lost if you do.

Moving swiftly on, the rear or tank panel numbered 19 on our sequence drawing is the first trim panel to fix, starting by carefully measuring it to find the top dead centre point. You need to use equal care to find top dead centre of the rear deck of the car. If there's already a hole, determine whether it is indeed dead centre. If the rear deck hole is dead centre, drill the tank panel (judge the height carefully) to allow you to screw through with a chromed trim screw and special chromed trim washer available from any of our premier TR restoration specialists. If the rear deck centre hole is not dead centre, ignore it and drill both panel and deck at dead centre. In the majority of cases, the holes in the rear deck will be non-symmetrical, and you will almost certainly have to drill new holes in both panel and deck.

Whichever you have to do, make sure you position one screw at a time. Work outwards from the centre, alternating fixing screws from side-to-side to keep the panel position as even, flat and symmetrical as possible. You'll have only have 2 screws to fix each side of the centre fastening since the outside screw each side is best left off the car for a few moments. Photograph 16-20 shows the end result but, incidentally, when drilling through this – and the other vinyl panels – it really is a good idea to protect the vinyl from the drill chuck with a couple of layers of masking tape. It's a very skilled trimmer that can exert enough pressure on the drill to get through the rear deck metal, and stop the chuck from touching the vinyl when the drill does break through. It will help if you ensure that your drill tip is sharp and, for the cost involved, new ones are really worthwhile. Be aware of and guard against the potential to snag the drill tip in the layer of foam beneath the vinyl. The result is the drill caught in a ball of foam and a torn/ruined vinyl panel. It's also worthwhile, though a complicated process requiring three hands, to use a piece of metal to compress the vinyl/foam while drilling each hole. A 1 inch (25mm) 'washer' with about a 2-3mm diameter hole

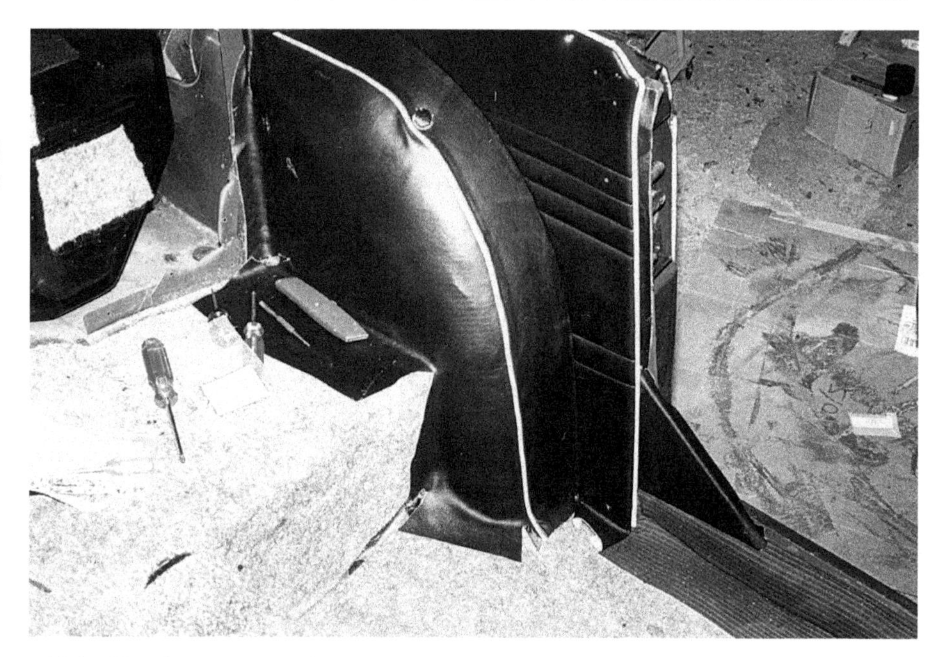

16-21. The vinyl wheelarch is glued down, the hidden/turned up edges around the inner wing will have been slit to allow them to lay flat without crinkling, and the quarter panel has been screwed in place. Note how the seat belt mounting has been opened up and a seat belt bolt temporarily screwed in place.

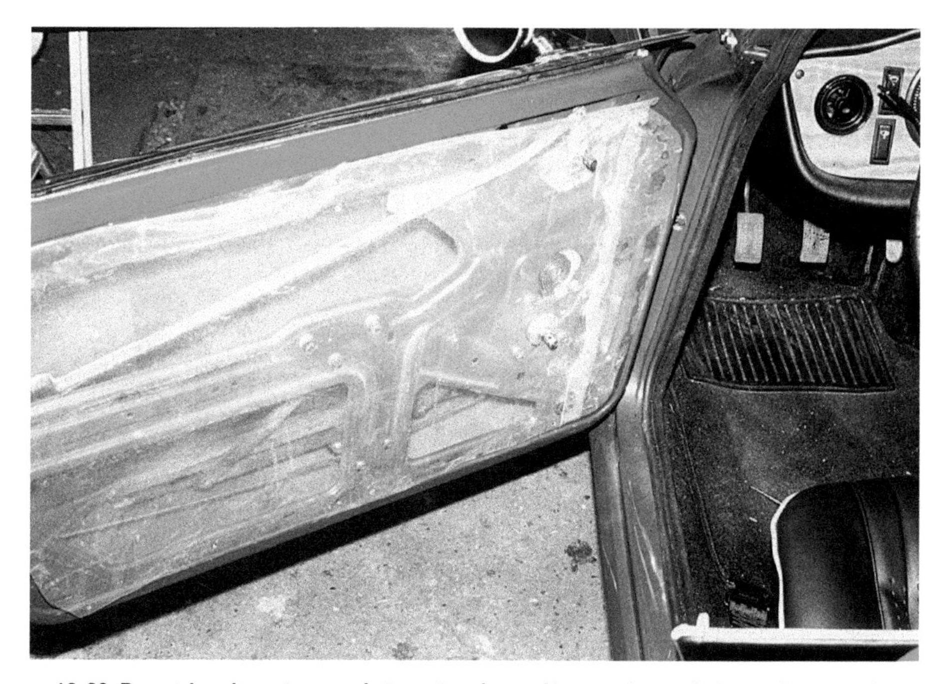

16-22. Door trimming: step one is to get a piece of heavy-duty polythene sheet on the inside of the door to protect the door trim panels from water.

16-23. A preparatory step you may need to carry out with some door panel sets is to cut the holes in the backing board for the window-winder and door handles. Some panels do have the holes pre-cut, which is a definite advantage in view of the risk you run of accidentally cutting the vinyl trim when getting through the board backing. If you do need to carry out this task, use the old trim panels (now you see why I told you not to throw anything away until the car is completely finished) to locate the board holes you need to cut. You will need to cut the vinyl eventually (for the square drives for window and door handles), but these holes should be more accurately located with the door trim in place on the door, and will definitely be in a different spot to any hole you accidentally cut when preparing the board. Moral: favour trim sets that have pre-cut door trim felt board holes.

should help with this, and also ensure that the chuck does not touch the vinyl. The masking tape is still important, however, to prevent the washer marking the vinyl.

Offer up the two rear quarter panels (numbered 20 on the drawing and also shown in photograph 16-21) to the inner rear wings above the wheelarches. Align the vertical edges; again, it could be useful to have a couple of inches of furflex door seal to hand to ensure that the vertical edge is the correct distance back from each 'B'-post. The furflex will need (eventually) to go fully onto the 'B'-post before it touches the front of each quarter trim panel.

The trim panels are held by 3 chromed screws each side, conical or 'cup' washers around the top, and a couple of hidden clips lower down. The key detail that may not be clear if your car has already had the trim panels removed is that each rear/inner vinyl 'tail' from each of these panels needs to be tucked behind the outside ends of the tank panel we put in place a few minutes ago. Consequently, it is only at this point that you should fit the outside screws to the tank panel. If your quarter panels are fitted using the clips I mentioned, you should never hit them with anything hard; a clenched fist should be sufficient and will not damage the panels. Care is required when fitting the front top screws to each panel. If these two screws (one each side) are too long, you will mark the deck panel by an outward 'bubble' in the paintwork, which would somewhat spoil that day's work!

The first step with the door trim panels is to buy and cover the inside of the door with heavy-duty polythene sheet, as shown by photograph 16-22, to protect against moisture. You will be better served in the long term if you find a heavy gauge material; I found just what I wanted at my local builder's merchant. Incidentally, the same heavy-duty polythene could be used to cover the seatbacks (discussed later in this chapter). I hold my polythene sheets in place with masking or 'duck' tape around the edge, but many restorers use a spray-on adhesive. Either way, fix it to the inner faces of your doors and then make two holes to allow the door handle and window-winder drives to poke through.

Assuming your door trim panels have their holes pre-cut (photograph 16-23), your next task is to position the twenty or so right-angled trim clips in the pre-punched holes in the hardboard backing to the door trim panels (21). These can really try your patience when you fix and position them for the first time, and there is much to be said for the "one-clip-at-a-time" approach. There is another complication in that you will need to choose your moment when to 'screw' (by about half-a-turn) the two cone-shaped springs into the hardboard trim panels. The top or small ends fit into the door handle and window-winder holes to hold them in place while you fix the trim panels to the doors.

If you have positioned all the panel retaining clips in one go, your next step is probably the worst part of trimming a TR – getting all the clips to align with their respective holes in the door. Obviously, you need to offer up the panel to its door and try, and get, say, the top pair of clips into their door holes. This at least holds the panel in place while you work progressively away from each top corner, manipulating each clip with some thin-pointed pliers until it can be persuaded to enter the relevant hole in the inner door panel. The rear edge of the door can be worked on in reasonable comfort, but to tackle the bottom row there's no alternative but to lay on your back. Perhaps a mirror will help with aligning the front clips?

Although it may seem that I am making a simple job sound more complicated than it is, give passing thought to positioning the clips one at a time, and then fitting the two coil springs to the door and window openings.

After that it's plain sailing, starting with the holes for the two square drives, which will not yet have come through the vinyl since the holes are rarely pre-cut. However, four light taps with a hammer on each edge of each square drive will effectively 'cut' the hole in the vinyl that you need, and each square

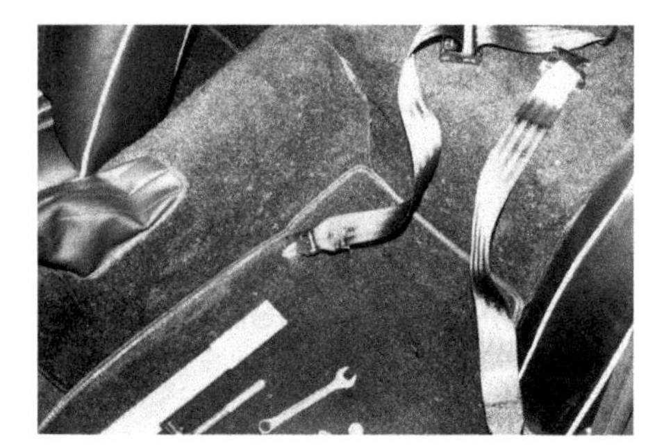

16-24.

16-24 and 16-25 (above). Towards the end of trimming you will need to tackle the seat belts. Photograph 16-24 shows the static/lap variety on the left side of this TR250, while photograph 16-25 may be of help to those considering inertia/reel belts, but are wondering how to fit them (I did). The solution, which works very well, was a pair of Securon Pn500/30 belts mounted as shown in this picture.

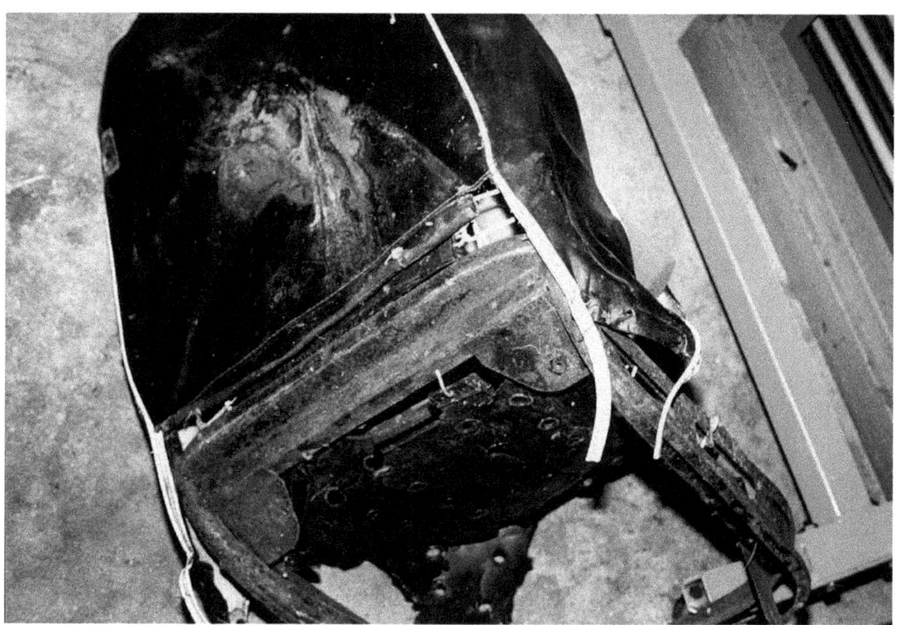

16-26. A seat that looks like a lost cause can be completely rejuvenated and can, I promise you, be made to look like new. This seat is a non-reclining one typical of those in the TR4/5/250, and has a rear cover that will eventually need to be 'pulled over' the seat back. There is some advice within the main text about how this operation can be carried out with minimal hassle.

will come through the panel. If you cannot bring yourself to hit the square drive hard enough to cut the vinyl, no problem, a light tap is all that's needed to mark the vinyl enough to enable you to cut the hole with a blade.

The large plain washers go on each square drive next, whereupon we come to the final 'pinning' of the door catch/ window winder arms. Take a length of welding wire of the correct diameter and mark the length of the pin you'll need to cover the complete diameter of

16-27. A not dissimilar view of a TR6 seat, where you will note that the seat back is not only adjustable, but the rear of the cover is actually a separate (clip-on) panel.

the relevant arm. Half cut through the wire with a junior hacksaw, push the half cut end into the arm (once you have aligned it with its square drive holes), and wiggle the welding wire until it breaks off at the cut.

This is the earliest that you should fit the Furflex door seals. Bearing in mind that there is still some climbing in and out of the car to do, you may be tempted to postpone fitting the seals. However, if you do not fit the Furflex door seals now, do at least check that your door glass is aligned with the windscreen/windshield end of the seal, and that the doors will shut nicely with the seals in place.

SEAT BELTS

The static seat belts shown in photograph 16-24 are available from many sources, but you may be interested to hear that Securon makes a range of aftermarket inertia seat belts which are available from premier TR outlets in the UK. Two models are applicable to the TR range.

Without the complication of a folding hood frame, the TR2, 3 and 3A will accept Securon's model 514/30, mounted atop the wheelarches and aligned across the width of the car. In practice, a slight inclination of the reel towards the outside shoulder is ideal. The TR4, 4A, 5, 250 and 6 have a folding hood frame to complicate matters. In these cases, you could

16-28. The TR6 backing board in more detail. Note the three holes each side of the board that takes the push clips shown in photograph 16-1, making it an identical mounting arrangement to the door panels. This panel has been removed by unclipping the lower six clips from the seat frame with a screwdriver, and sliding the board downwards to release it from the top three retaining clips. Do not take a screwdriver to the top edge of this type of seat back. For the benefit of this picture, Mark has turned the seat back over to show the rear of the seat and the rear of the board in one shot.

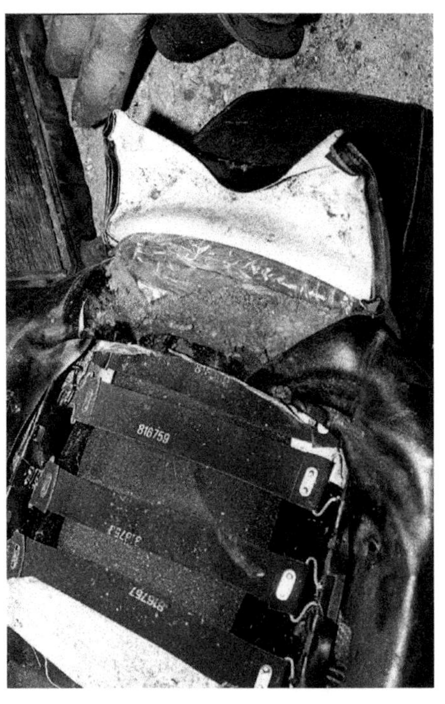

16-29. This picture is interesting in that it shows the make-up of the US-style, built-in head rest used in later TRs that may be relevant to some readers. It also shows one of the (three per side) holes in the back frame for the board backing used on the '6s. Finally, can you see the thin polythene 'bag' that was used during assembly to help pull the cover over the foam, and then left there?

consider Securon's model 500/30, which aligns fore/aft at the bottom of the inside wheelarch, with a pillar loop located at the top of the wheelarch, as per photograph 16-25.

REBUILDING THE SEATS

Obviously your attitude to this subject will be very influenced by the degree to which you want to retain the originality of your car. Since I have found the original TR seats to be far from comfortable on all but the shortest of journeys, I will suggest how to completely replace the TR seats in a later offering. This book, however, is concerned with restoring the cars to original specification, and this section will help you refurbish any of the original TR seats to pristine condition.

TR seats came as a variety of types. Probably the majority had no provision for a headrest (photograph 16-26), although there are the high-backed types with a built-in headrest shown in photograph 16-27, and also seats with a separate headrest. Then there are the seats with adjustable backrests and those manufactured in one piece, as can be seen in the majority of photographs in this section.

Some TR seats have a soft foam/vinyl back, others have a hardboard insert – as depicted by photograph 16-28 – to provide a slightly firmer back on the seat. This section will give information about refurbishment, regardless of the specific details of your type of TR seat.

Some intrepid DIY-ers make their own seat covers – which I do not intend to explore on the basis that experts have spent years developing and proving patterns and manufacturing techniques, and it is my recommendation that you buy a good set of specialist manufactured seat covers. How do you decide what is a good set? I would shortlist only those suppliers which are both specialist TR restorers and manufacture their own covers, such as TR Bitz.

At the same time as you buy your new seat covers, you may as well order a car set of underseat hessian, and diaphragms for the bases, for it is rare that these are re-usable.

It's likely that other parts will be needed; you'll get an idea of what from how your seats look before you start. If yours are as bad as those shown in photograph 16-26, then you might as well order foam mouldings at the same time.

It is tempting to start your refurbishment by separating the backs of both seats from their bases (in the case of seats with adjustable backs). In fact, I think you will find the pushing and pulling that comes later a little easier if you leave the seats in one piece. Furthermore, start your restoration by stripping one seat only, keeping the other as a reference until you are completely satisfied with the first seat rebuild.

A final thought: the left and right seats are not quite identical. For those with a pair of seats, the differences are minimal, but if you have to buy one seat because of damage, loss or whatever, it's important to remember this fact.

So, get started by taking a look at the series of flat and round metal retaining clips, and one long spring on both base and rear frames. These have to be removed, probably revealing the foam for the first time in many years.

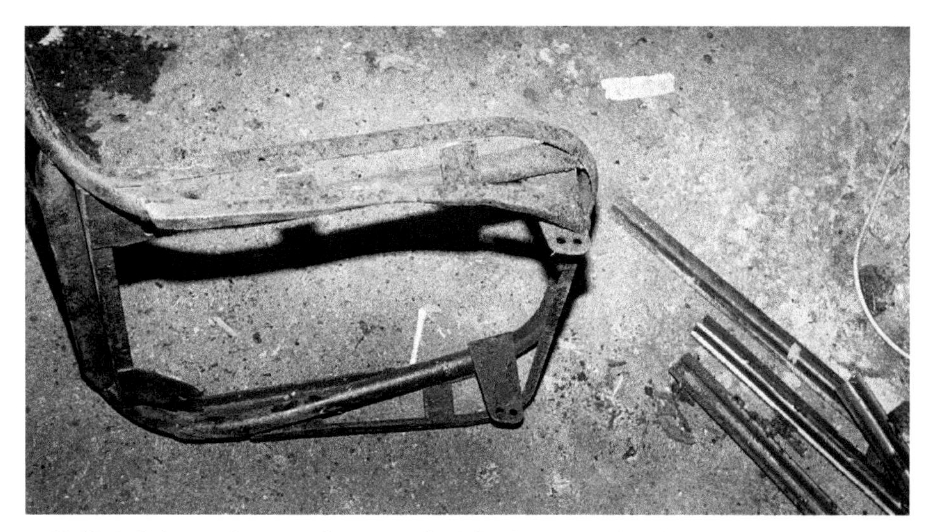

16-30-1. Strip one frame as far as required. Obviously, the first job is to carry out such repairs as are necessary ...

16-30-2. ... clean the frame and repaint as seen, in primer, here. it. This frame is from a TR250 and is a one-piece, non-reclining type. The other TR seats are different in detail but the rebuilding procedure is identical.

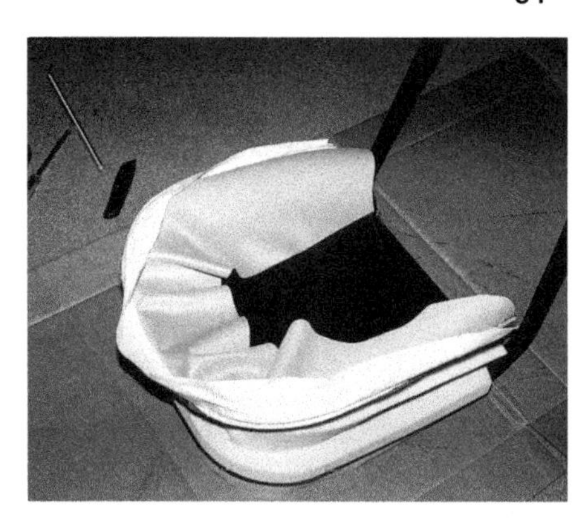

16-31. It is usual but not compulsory to start re-assembly with the base. This one already has the diaphragm in place and the picture shows the thin foam being placed in position prior to gluing to the seat frame. A view of the diaphragm is shown in photograph 16-34. Just slightly out of sequence, but this picture also shows (top left and arrowed) the long silver spring that will be fed through a sleeve sewn into the bottom/back of the new base cover. It clips into holes in the base frame and provides some 'give' in the seat bottom when in use.

Before you actually spring the clips, please take five minutes to note which clip/spring goes where, and make every effort to remove the clips with minimal damage. The foam mouldings could already be crumbling to dust, or may start to do so as you remove them from the frames. In a way, it's better if they are disintegrating as it takes all the decision-making out of the next step – you obviously need to buy new foams! If they look reasonable, you have the difficult choice of whether to buy replacements, or re-use what you have. Foams do, eventually, disintegrate, so it's a short-term solution to re-use the old ones. One compromise is to use a new piece of flat foam for the main seat bases, and pad out any slightly flat foam mouldings with a covering, or additional layer of

16-32. We have the 0.5in (12mm) flat seat base foam in place, and are weighing-up gluing the moulded base cushion to the base foam.

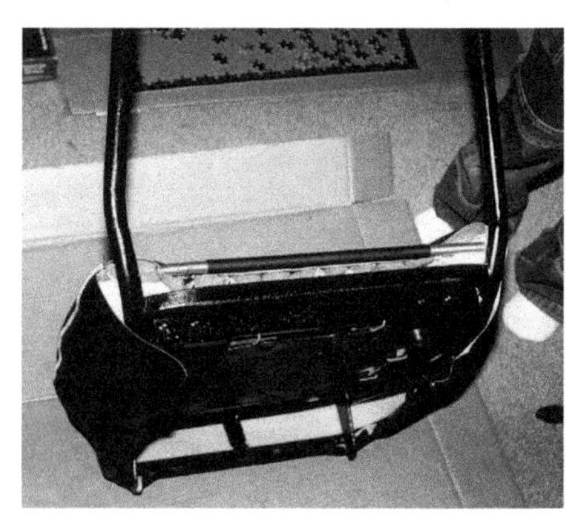

16-33. The base cover is on but we can see that the final clips around the bottom of the frame have yet to be fixed. Note the tension spring running across the rear of the base cover of this TR250/5 seat. The spring runs through a pre-sewn vinyl tube formed at the back of the seat base, thus securing the base to the seat back.

16-34. And now the clips are in place and the surplus base material has been trimmed away.

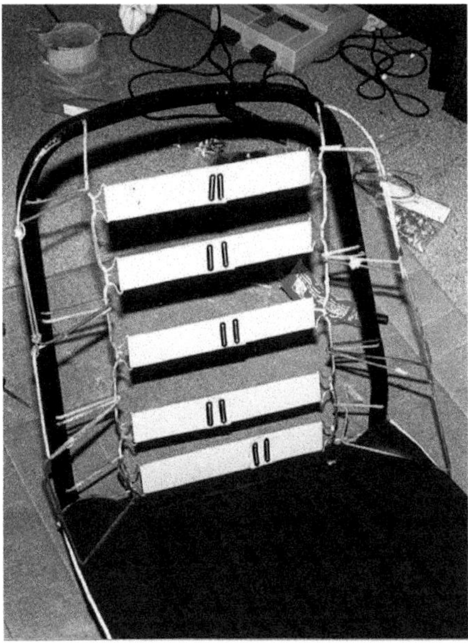

16-35. The first assembly operation on the back is to locate the new webbing and the side support wire frames.

new foam. This would be a low-cost compromise that would extend the life of your existing foam mouldings. Generally, though, I recommend you buy replacement foam mouldings.

Complete the disassembly by removing the hardboard back, where applicable, to the seat and the front/ centre section of the seat back. The hardback design (photographs 16-28 and 16-29) was used on TR6s only, and if you treat the sides (note, not the top) like a door trim panel, you should find that the clips pop from their frame holes. The TR6 back then slides vertically downwards to disengage the top clips.

You'll find that the more usual TR softbacked rear cover incorporates an internal piece of flat foam that will probably have to be placed to one side, along with its tired cover, although it may be prudent not to bin the discarded parts until the job is finished. The rear cover should come off in one piece, but it hardly matters if it disintegrates! However, take great care not to allow the two wire side support frames (seen best in photograph 16-35) to become bent or disintegrate, however fragile they are, as you will need to use them as patterns in the not-too-distant future. It may be better to cut off the rear cover in preference to bending the squab side supports.

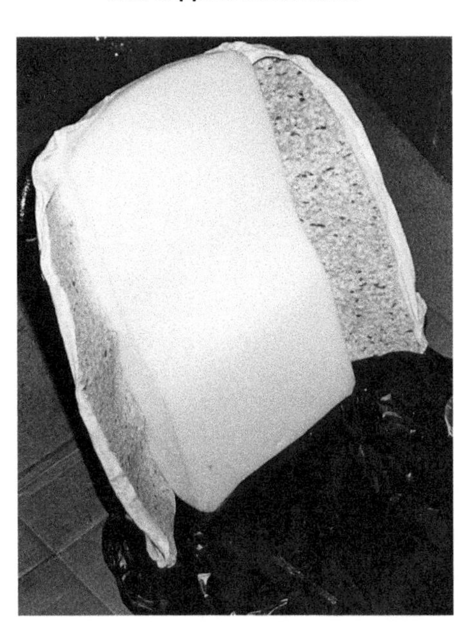

16-36. This picture shows two steps in one: the side foam mouldings are in place and the flat central foam has also been positioned.

The rear frame will now be revealed, along with the side foam mouldings and six cross webbings that provide most of the rear 'cushion' feeling. It is rare for the webbings to be beyond further use, but if yours are perished, get replacements. The repair kit comes as a roll of webbing, together with all the various fastenings needed.

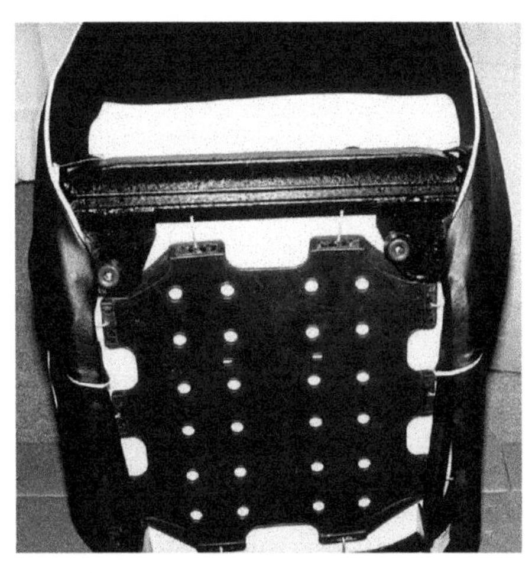

16-37. Finally, and just as I promised, looking as good as new. Well, that's not quite true, as the rear of the seat back has yet to be tucked under the frame and clipped in place.

It is very obvious but, for the sake of completeness, I will mention that you need to get the frame into good shape before too much else can happen. Photographs 16-30-1 and 16-30-2 show a not untypical frame that requires much preparation before you can get to the main focus of this section.

Re-assembly starts with the seat base, and replacing the diaphragm in particular. This is illustrated by photographs 16-31 to 16-34. Use a loop of string to stretch each of the wire hooks over the frame to secure the new diaphragm to the base. Lightly spray-glue the new hessian to the diaphragm and, if your kit includes it (not all do), fit the thin sheet of foam around the front and sides of the lower part of the frame. This 'skirting' (shown in photograph 16-31) provides additional cushioning between the metal frame and the cover.

I imagine that most covers fitted at home will be made from vinyl, since restorers investing in leather covers will probably have the seats professionally recovered.

Spray-glue a new, flat piece of 0.5in (12mm) firm foam atop the hessian as your base cushion. Well before you have got to this point, you will have noted that the sunken seat base shape is actually achieved by a 'U'-shaped foam moulding, which now needs to be repositioned to the front, and both sides of the base cushion, with spray glue, as shown in photograph 16-32. Allow the glue time to 'flash-off'. If you are following the compromise route of supplementing original foam with additional thin outer foam, this is the time to apply, with spray glue, the extra layer of, probably, 0.25in (5-6mm) thick foam sheet.

Slip the long spring through its tube/sheath at the back of the base cover (photograph 16-33). Now to the exciting part. Apply spray glue to around the edge of the frame and offer up the cushion cover to the base. This is best achieved by clipping the long spring at the rear of the base cover to the frame, followed by pulling/rolling the front of the cover over the frame before tackling each side. Spend a few minutes pulling and smoothing before taking the penultimate step of applying the retaining clips to the base frame. Finally, trim the excess material from the base and move on to rebuilding the rear of the seat.

Earlier we touched upon the wire seat squab supports beneath the two side foam mouldings. They are rarely damaged, which is just as well, as new replacements are no longer available, and I am told that they are a nightmare to repair/replace at home. If yours are damaged or badly corroded, try to find secondhand alternatives via your usual sources. If you feel the damaged sections are few enough to replace, you will need to obtain some heavy gauge, stiff – and preferably galvanised – steel wire. It's then a matter of hand bending each piece you need to replace.

Although they differ in size, each piece has a corner bend at each end that fits into holes in the seat frame. They are interlocked to a separate vertical piece of the frame on the TR250/5 seat, as can be seen in photograph 16-35. Many of the individual pieces will have been held together by small metal bands, most of which are unlikely to be re-usable; substitute with several turns of MIG welding wire. It will be very-time consuming and you will need to ensure that the various angles are correctly bent – but a repair is possible.

Finally, the webbing is hooked across two vertical pieces of the framework to maintain tension on the wire assemblies and hold them in place in the frame.

Moving on, for the cost involved, it would be best to replace the flat front foam piece in the centre of the seat back (photograph 16-36) with a new piece. If you have an early seat with a soft back, now is the time to lightly glue the rear foam to the webbing.

Whichever seat design you are re-covering, before pulling the one-piece rear cover over the seat back, take a moment to plan your approach for what could be considered the most difficult step. Start by letting the seat cover warm up in the sun for a few minutes; it will be softer and more compliant. Note that most covers incorporate foam sewn into the pleats and sides of the rear cover, so if you try to pull the cover over the seat back, you will encounter considerable resistance due to foam-to-foam friction, and risk stretching or even damaging it. A piece of heavy-gauge plastic sheeting placed front and rear (and curled around the sides) of the seat back should reduce the amount of effort needed to pull the (warmed) cover over the seat back. Get the pleats mostly aligned before withdrawing each plastic sheet. Incidentally, the same heavy-gauge plastic sheet will be absolutely ideal as a waterproof membrane behind your door trim panels, so buy a bit extra for the doors.

You will also find that your new cover is fitted with the flaps at the bottom edge of each of the front/side panels seen in photograph 16-37. These need to be spray glued and tucked beneath the bottom of each side foam. The rear cover clips need to be fitted and any surplus material at the bottom trimmed off. Hopefully, you now have a seat that looks very similar to that in photograph 16-38.

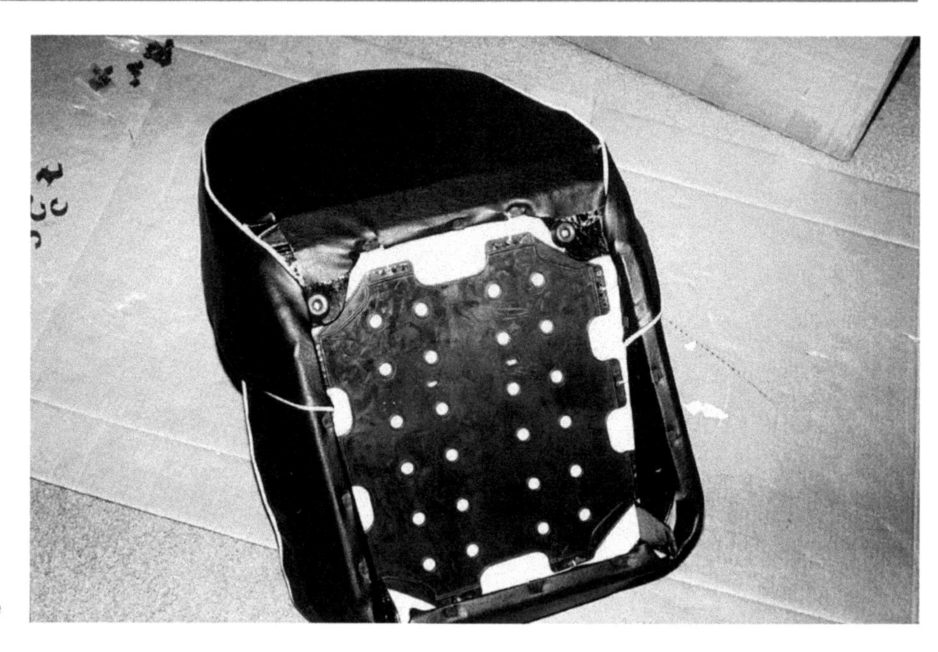

16-38. Done! – now for seat number two ...

Only TR6 seats have an extra board panel fitted to the rear, but this makes it easier to fit the rear cover as it is less bag-shaped! Fitting this rear panel is the last operation for, after positioning its securing clips, you need to slide the back panel up inside the top rear clips, and only then to engage the press clips (similar to the door panel clips) into the seat frame.

BOOT/TRUNK PANELS

Apart from saying that there are two types of boot/trunk panelling, there is not too much else to say about the four board boot/trunk panels, for their fitting is largely self-explanatory. The two types are for, respectively, carburettor cars (with no provision for fuel pumps and filters), and PI cars that have extra space behind the front panel for petrol injection gear.

However I might save you a few minutes if I suggest you install the two side panels first, regardless of which panel set you have purchased. The front and rear ones then follow without too much difficulty, although I found it helpful to have handy about a dozen large, chemically-blacked washers.

Chapter 17

Hood/soft top

REBUILDING THE HOOD FRAME

Whether you are replacing just the hood, as part of an upgrade or repair, or carrying out a full car restoration, allow enough time and cash to check and improve the hood frame at the same time. In fact, it is important that you sort out the frame first, since it is the foundation upon which a good hood fit depends.

Believe it or not, some owners lose the hood and hood frame. Typically, the assembly gets put up in the garage loft – maybe to fit a hardtop or because the climate makes a hood superfluous – and is forgotten. In due course the car is sold, sometimes without the hood that the owner has not seen for years! In this situation the frame could have remained unused for years, and become seized/frozen. Furthermore, it's not unknown for a frame to be bent, perhaps as a consequence of trying to free it from the car. Whatever the reason, hood frames are in short supply, so make sure that your prospective purchase has a frame, if not a hood.

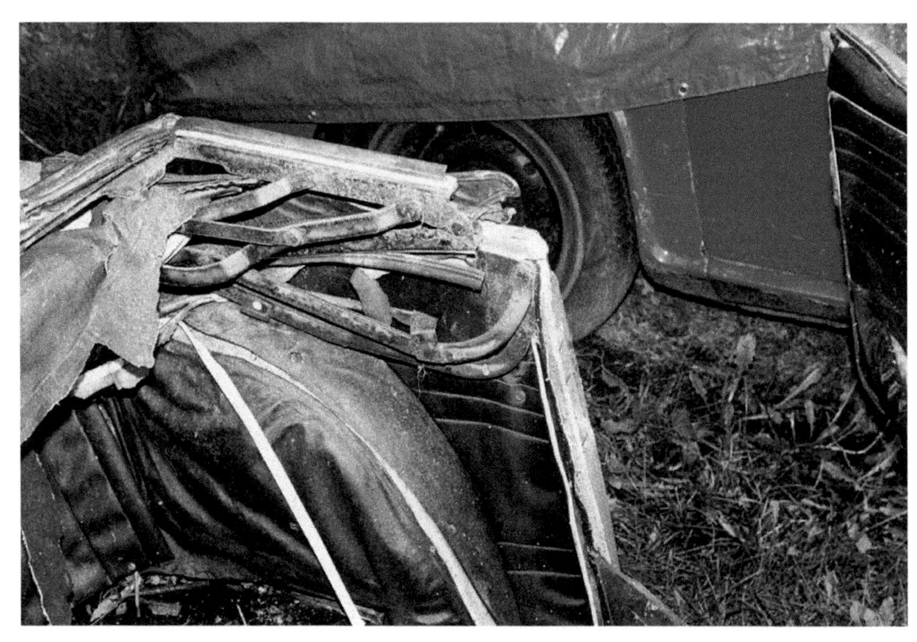

17-1. This could be the starting point of your frame restoration. Take lots of photos; it's amazing how many small details can cause concern if you don't have them to refer to.

The frame should be complete, too, and should look something like that shown in photograph 17-1. The ideal solution is to go for a reconditioned/exchange hood frame from one of our premier TR restorers. The exchange frame will come back with bent or mis-shaped sections corrected, beautifully and durably refinished (powder-coated if you wish), and with the correct length webbing straps riveted in place. The cost for this will be in the region of £125, which may stretch the budget on top of the cost of a new hood, so let's look at the cheaper DIY alternative.

The first – but most important – step is to check frame straightness. This must be done with the frame erected, on the car, and is easier to do if you can borrow a frame to erect alongside for comparison. Although it will be hard to think this far ahead, it makes sense to fit and correct the hood frame before the body is painted. I emphasise the importance of correcting the frame on the car since more frames are distorted off the car than on. The frame loses its two vital reference points when the rear mounting plates are unbolted from the 'B'-posts (photograph 17-1), so, if your frame is stiff or damaged, correct it in place on the car.

Assuming you are, like most restorers, fitting the frame to the car after trimming, proceed cautiously. The clearance holes for the 6 countersunk frame mounting screws make it seem obvious where the frame mounting bolts go, but take a thin, pointed tool of some sort (a heavy-duty needle would be good), and poke through the vinyl on the quarter panels until you can locate and align the 6 holes. Open up with an awl, or mark each with your chinagraph/ crayon, and cut a small clearance hole in the vinyl for each screw.

The next job is to (gently – probably with the help of a little heat) un-seize any stiff joints. If a joint is seized, do not force it. You can also, of course, apply lots of penetrating oil, twice a day if possible, for some days. The joints usually come unstuck eventually, but a forced frame distorts very easily and is very difficult to completely rectify. If yours already has some bent sections you need to straighten them, and generally ensure that the frame sits square and folds easily. Next I would suggest you rub down the frame; I found this is best achieved with the frame securely tied to a washing-line.

You will probably want/need to replace the frame webbings. You can remove the two original webbings, paint the frame and then fit new webbings, but I suggest you replace the webbings before painting, and do so one strip at a time (photograph 17-2). Each webbing needs to duplicate the original, and to tie together the header rail, the three frame hoops and the rear retaining bar. Sew or rivet a loop at what will be the rear of the first piece of webbing. A short rod through the loop beneath the rear bar retains the tails of each piece of webbing as illustrated by photographs 17-3 and 17-4.

The idea behind replacing only one web at a time is, of course, that the distance between each bar is retained. The webbing should be pop-riveted to each bar of the frame, and to the header rail (photographs 17-5 and 17-6). This is probably best achieved with the frame (and the rear bar) temporarily re-attached

17-2. One longitudinal webbing has come adrift from the front header rail of this hood frame, so should be first in line for replacement before the reference of the second web is lost. Note the extra white covering that this frame cross bar had. Not standard, but presumably added to reduce the possibility of the bar chafing the hood.

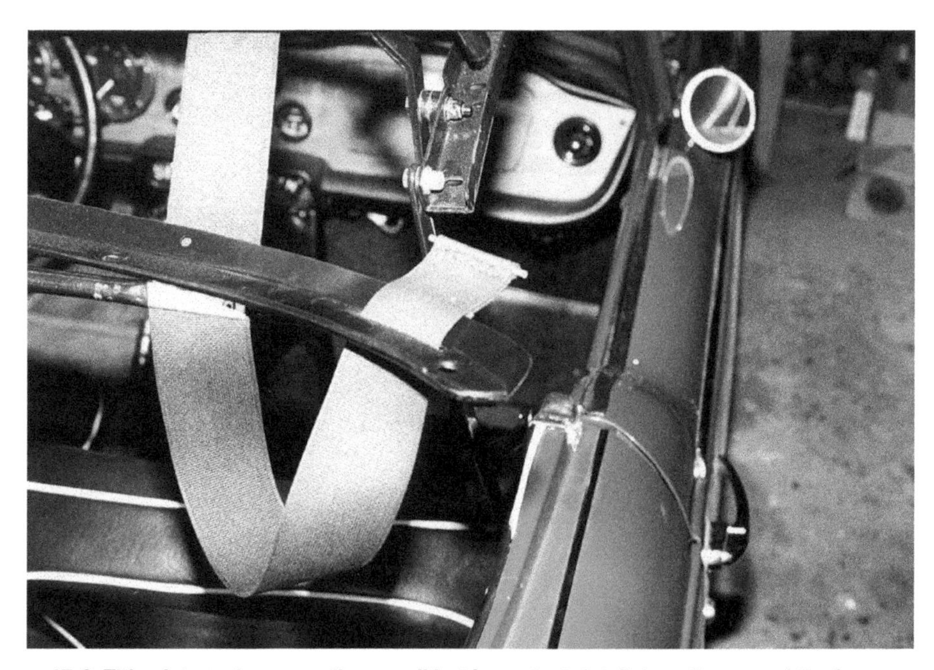

17-3. This picture shows another small but important detail: how the rear of the frame webbing is coupled to the rear retaining bar. This picture was taken during a trial assembly of frame to car, and shows the underside of the rear retaining bar. With the hood in place, the loop at the end of the webbing passes through a slot in the hood flap. Note, too, that one rivet hole is visible, as is the end mounting hole in the rear bar. Other relevant details are the one rivet that is just visible securing the webbing to the nearest hood bar, and the slotted holes for the (side) aluminium weatherproofing strip.

17-4. The rear bar and webbing correctly test mounted to the rear deck. You can just see the end rivet hole in the rear bar awaiting the hood rivets.

17-5. The frame front rail with one of the latches nicely framed. The webbing rivets have been removed from the header rail, but the picture shows how far the latch pins protrude, which prompts me to remind you to cushion this when folding the hood and frame to avoid the latch poking into the hood material. The old hood material has, of course, been cut away; nevertheless, you can still see where it was folded over the front of the header/front rail. You may even spot some of the original, well-rusted front seal mounting strip still in place. I would recommend you actually slightly extend the length of the hood material left in place and glued to the header, and also that you cut slots in the hood material for the latch to poke through.

to the car, as shown in photograph 17-4. This not only holds the frame securely, but also enables you to double-check that the webbings are even and taut when the frame is erected.

As a precaution, and in order to avoid chaffing the hood, carefully check that all plates, washers or rivets are

17-6. Now cleaned up and painted, this view of the header rail shows the new webbing riveted in place.

free from sharp edges/points. Hammer these flat if found – again, best done with the frame on the car, as shown in photograph 17-7. While not standard practice, you can glue a small piece of webbing over any really troublesome sharp pieces, but be aware that this may cause a bump in the line of the hood.

Some hoods have a hoop of material sewn round the hood frame bars, but the latest ones use a flat strip of velcro (photograph 17-7). This is the time to attend to such details. Once satisfied with the frame and webbings, remove the assembly from the car and re-attach it to the washing-line for painting.

FITTING THE HOOD

Never be tempted to buy a TR hood (or tonneau, for that matter) with fixing studs or holes pre-positioned, as they will be wrong for your frame. This doesn't mean that the hood manufacturer has been careless, for I have no doubt the hood fitted his pattern very well. It's just that every TR is different, and you cannot pre-determine where holes or studs are going to go.

Three types of material are used for TR hoods, and the following guide is relevant for all. Vinyl or PVC, which will top the vast majority of TRs, provide sterling service and probably are the easiest to fit. They are easy to clean, lightweight and less bulky, but tear easily and are prone to cracking in cold conditions. Rarely does one see a 'Double Duck' (canvas) hood these days, but there are some lovely hoods made from what is called mohair material. They do attract dirt and need additional cleaning, and the attention

17-7. The completed webbings, nicely taut the full length of the frame. Note the protection above the rivets at each cross bar and, in this case, a strip of Velcro across the central bar.

detailed later. They are also stronger and flexible in all conditions, but heavier and bulkier than vinyl hoods.

Fitting the hood well secures the top of the windscreen/windshield. This is important and is the last chance you will have to ensure good alignment between window (drop) glass and the 'screen frame. Therefore, check that your windows wind up and down easily, yet achieve a full-length seal against door and 'screen seals, etc. Once you fit the hood, you are stuck with what you have, unless you are prepared to buy another new hood, that is! Here we go, then, using the fastenings shown in photograph 16-1.

Drill the lower half of your rear retaining bar to take a central rivet of 0.125in or 3mm diameter, or, if it is pre-drilled, check that the centre rivet hole is indeed central to the bar. Check, too, that the five mounting holes in the rear bar are properly disposed about the centre, and do an alignment check with the holes along the front edge of your rear deck. It's worth mentioning that TR hoods have been fitted with the rear bar in place on the rear deck and, in error, the rivet holes have been drilled through into the rear deck. For the record there are no rivets into the rear deck, only into the rear bar, and the rear bar is only fitted to the car after all the rear riveting of the hood has been completed. That's got that potential mistake out of the way.

Moving forward, the next few steps really need an assistant to help swing and screw the frame in place. Start the hood fitting by finding the exact centre points of the rear and front edges of your new hood. This is best carried out off the car. While you can find the centre using a tape measure, the easiest and surest way is to fold the hood in half inside out. Take the two rear corners of the hood that will shortly sit above your 'B'-posts, and fold them together, whilst, of course, folding the two 'A'-post tops together at the same time. Now, clearly mark the centre points on the inside (which is currently outside) of the hood with a 1in (25mm) line at the rear and a 2in (50mm) line at the front. Use a chinagraph, chalk or soft crayon that marks the hood clearly but will rub off later. Fold the hood the correct way (outside out) and, from the inside of the hood, push the rear retaining bar down into the rear flap/corner of the hood. Carefully align the centre (rivet) hole in the bar with the centre line you marked on the hood, and, if there's any doubt in your mind about the assembly sequence you are about to commit to, take a look at drawing D17-1.

Check carefully that the centre hole is correctly positioned, that the rear bar is pushed tight into the hood flap, and that the turned-up edge of the hood is tight up against the rear bar. From the inside of the hood, pass an awl or similar through the inside edge of the hood (as illustrated by photograph 17-8), the centre hole in the bar and out through the back edge of the hood. Secure this position by locating a male press-stud base on the outside of the hood and, from the outside, passing a 0.5in (12mm) long $^1/_8$in pop-rivet through the hood, central hole in the bar and inner hood flap. Place a washer on the inside and rivet together the press-stud base, hood, bar and inside flap.

D17-1. Schematic assembly sequence for riveting hood to rear retaining bar, with an outline view on the left and a more detailed view on the right. A. Rear retaining bar. B. Hood. C. Rear deck. D. Rear retaining bar. E. Pop rivet. F. Press-stud base. G. Hood material. H. Hood binding. I. Washer.

17-8. Finding the position of the third rivet in the rear bar to hood sequence from inside the hood. The first rivet was dead centre (in front of the filler cap), and the second rivet was the first one to the right side. Now it's the left side's turn, before alternating back to the next rivet on the right side. You are seeing the tails of the rivets in this shot with the press-studs hidden from view on the other side of the hood. Note that the rear bar will be rolled forwards by some 180 degrees at the completion of all 7 rivets/press-studs.

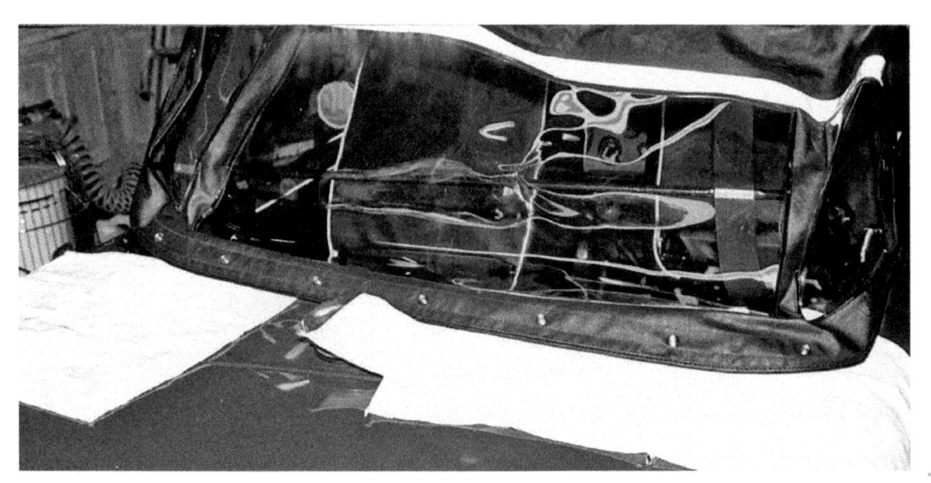

17-9.

This could be the moment to place the hood over the frame on the car (in which case, do use a dust sheet to protect the rear deck paintwork), or you might choose to complete the rear bar to hood riveting off the car. In either event, there are a total of 13 press-studs across the back of a TR hood, but we are currently only concerned with the central 7, or, more exactly, the 3 either side of the central stud/rivet completed a few moments ago. Each of the remaining 6 will have to be pop-riveted through the hood to the rear bar. It is vital that they are fitted one at a time, that you stretch the hood material outwards before pushing the awl through the next hole. Make sure that the hood material, particularly vinyl, is warm, that you work from the centre outwards, and that you complete each hole stud rivet on alternate sides of the bar. If you concentrate on one side of the rear bar and work straight across, you'll probably end up with an off-centre, unbalanced-looking hood.

Also make sure that you pull each section of hood between the rivets as tight as you can to avoid wrinkles in the rear window, and an ill-fitting hood generally. The end result of this stage should look like photograph 17-9.

The tail loops of both webbings should be fed through the rear bar and slots in the hood and, using something like welding wire, secured to what will be the underside of the rear retaining bar. To secure the rear bar to the car, use a large washer on the top of the bar and, starting with numbers 2 and 4, secure all 5 rear bar retaining bolts, thus completing the first stage of attaching the hood to the rear of the car.

Many hoods have two rows of three poppers in the roof of the hood; undo them, position the frame rails through them and re-fasten the poppers.

I mentioned earlier that the rear of the hood is attached to the rear

17-9 and 17-10 (left). All 7 rear rivets and press-studs are in place, and the rear bar has been rolled forward by 180 degrees. Photograph 17-10 also shows that the 3 male press-stud bases have been riveted in place along this side's deck extension pieces. Can you spot the join between rear deck and the extension, and that the press-stud is about ½in in front of the join?

17-11. The rear of this hood is compete with 13 press-studs satisfactorily located. We are just going to roll the front of the hood forward and enlist some strong help to stretch it forward.

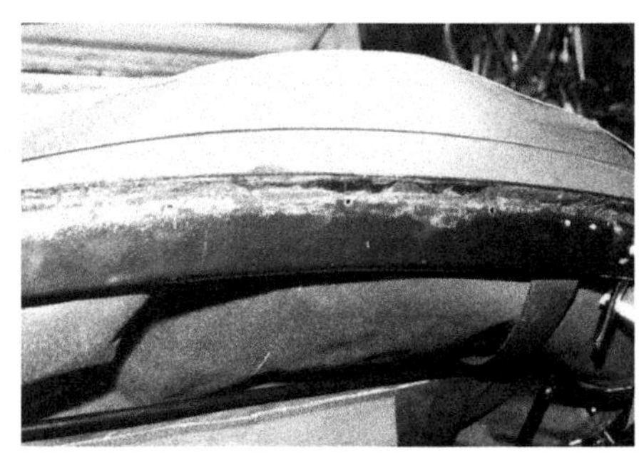

17-12. Note the central chalk mark and the frontal arc showing where the hood should be positioned fore and aft. This vinyl hood will probably benefit from being pulled a further ½in forward.

deck by 13 press-studs, but you will recall we were concerned with only the central 7. It is now time to deal with the outstanding 3 per side, located below the hood quarter windows. This time, the male press-stud base (item 1C in photograph 16-1) is located directly on the rear deck and, if not already in place, this is a good time to attach all 6 (3 per side). If you have fitted new rear deck extensions, there won't be any holes to rivet the press-stud bases to, and 6 holes will have to be drilled, preferably in the correct places.

Bearing in mind that the body will be painted at this point, I suggest you start by applying a couple of layers of masking tape to both rear deck side extensions, to guard against scratching the paintwork. The first deck hole to drill is nearest the door, and will actually be the last one you will use. However, you need to position the chrome 'B'-post finisher on top of the furflex door seal, and use the finisher's hole as your first rear deck 'spot'. This will give you the height of all three holes that side of the car.

Now go to the other/rear end of the side extensions and spot your second hole 0.5in (12mm) in front of the main deck panel joint with the forward extensions. The third and final hole this side of the car comes as near to the middle of holes 1 and 2 as you can get it. Drill your 6 holes, remove the masking tape and rivet the press-stud bases in place, as illustrated by photograph 17-10. Our next task is to fix 6 press-stud buttons to the outside of the hood and 6 press-stud sockets to the inside of the hood so that they correspond to the bases fixed to the rear deck. Since the hood is now securely centralised, there's no harm in concentrating on three press-studs on one side of the car, provided you work away from the central group of 7 towards the 'B'-posts. Complete each button/socket and secure the hood with it, stretching the hood material forward and downward, before moving onto the next. Never try to do all three in one go.

These 6 buttons require a small 'flaring' tool to spread the inside of the hollowed brass end to the button, and you may need to make, borrow or buy one. If you decide to make a tool, you need two parts: a support block of steel or brass, machined to accept the outside (usually black plastic-covered these days) head of the button, and a punch. The inside punch actually flares the brass button outwards over the socket, so the punch will need to start with a (blunt) point and taper progressively outwards. You might get away with a centre-punch to start the flaring, followed up with a piece of 0.25in (6mm) diameter bar to tap flat the opened end of the brass press-stud. Either way, your finished rear hood should bear some resemblance to that shown in photograph 17-11.

Moving forward, position the frame header rail on the top of the windscreen/windshield frame; it should just rest in place and not be secured. Again ensure that the centre of the header rail and the centre of the front of the hood are clearly marked and exactly positioned. Locate where the two header rail clamping pins are, and cut a couple of 10mm slots in the hood material, ensuring that they are not too long. If it's a vinyl hood you are fitting, it's a good idea at this point to re-affirm that it is pliable by letting the sun shine on it for half-an-hour or, in winter, running a fan heater inside the car for a similar time. Position the locking pins over the holes in the windscreen frame and, with the help of a strong assistant, stretch the hood forward as far as it will go. Mark a chalk line along the front edge of the hood material where it leaves the front of the header rail, and check that the two lines on your hood resemble those in photograph 17-12. Release the front clamps and, using spray glue, apply adhesive to the top of the header rail. If it is a vinyl hood you are fitting, pull the hood a further ½in (12mm) forward over the header rail and, working from the centre towards the outside of the hood, stick the hood to the top of the header rail – ensuring a smooth appearance. Using the locking pins, try clamping the hood/header rail to the windscreen frame. If the hood seems too tight (remembering that all materials, but particularly vinyl, will stretch a little, and that a loose hood flaps about in use), peel the hood off the header rail and reposition it slightly. Try the clamping again; if still too loose peel off and reposition, remembering that you have about 10 minutes before the sprayed adhesive sets.

Before you get to this point, you should be comfortable with the position and tightness of the hood. Now release the locking pins, fold the front of the hood under the header rail (photograph 17-13) and fix the underside with some really strong contact adhesive. Rivet the top channel to the underside of the header rail, and, with a nice sharp blade, remove surplus material from the inside, as shown in photograph 17-14.

Next up is to fit the new side seals and Velcro. The 4 aluminium pieces are screwed to the frame using the elongated holes to allow alignment, while the new rubber weather strips are usually best dealt with by sliding each into its aluminium channel from one end. Illustrations 17-15 and 17-16 show one of the four sections progressing. It's generally thought to be advantageous to leave the hood erected for a couple of weeks to aid the stretching process, and it helps if you can let the car/hood stand in the sun for as much time as possible during this period. The end result can be seen at 17-17 and is a credit to this TR250.

HOOD MAINTENANCE

We'll take a look at some hood maintenance details in a moment, but first, on the premise that prevention is better than cure, let's discuss a couple of details that will reduce hood maintenance to a minimum.

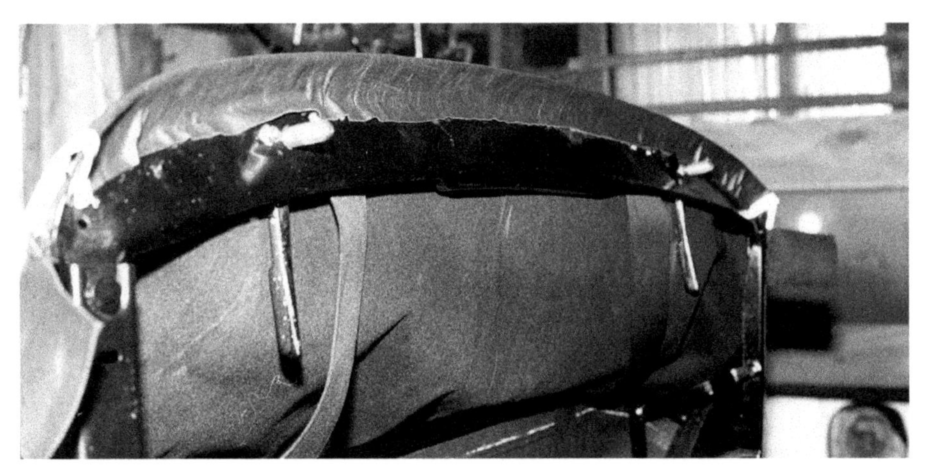

17-13. Second phase of gluing the front in place uses strong impact adhesive. I would extend the bottom glued area back towards the rear of the header rail.

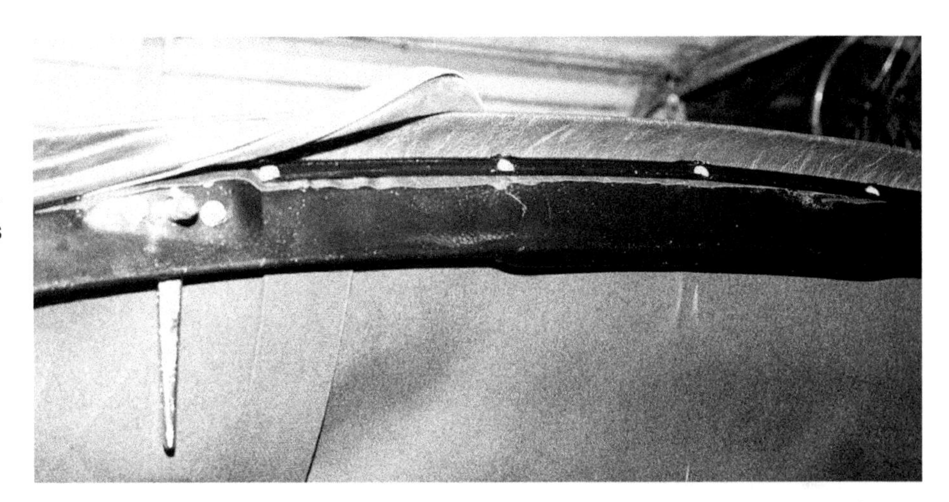

17-14. The front weather strip has now been riveted in place and we are partway into fitting the seal.

17-15 and 17-16 (right). There are four in all, but these pictures show one side Velcro strip stuck in place, a side weather strip aluminium channel screwed in place, and (photograph 17-16) the end result with the rubber seal slid in from one end.

17-16.

The way you fold your hood is a case in point; you will prolong the life of the windows if you fold the hood properly. Unclip the two locking latches from the windscreen frame, unpop the (3 per side) fastenings from beneath the quarter windows, and separate the velcro strips that run most of the length of the side windows. The 6 key poppers that do get overlooked are tucked up into the roof of the hood, 3 each holding the hood to the frame. Undo these or you risk tearing the corners of the hood.

As you fold the hood frame back from the windscreen, be sure to pull the top of the rear window out backwards until the hood frame is fully folded, and the rear window and most of the hood lays flat along the rear deck/boot/trunk with the two quarter windows sticking out sideways. You'll prevent window chaffing by placing a couple of old towels each side of the rear window. Place two pieces of foam or soft cloth padding (or part of one of your towels) over the windscreen locking pins, which will be pointing skywards at this stage in the proceedings. Fold the two quarter windows towards the middle of the hood without creasing the quarter windows, and, finally fold the very tail of the hood forward into the rear of the cockpit, making sure you do not fold or crease the rear window. Fit your tonneau or hood cover.

There's little you can do to repair a worn, torn or cut hood. If, however, your hood is not quite that bad, I trust some of the following tips will help you with 'rag-top' maintenance.

I am disappointed to report that hoods made from mohair material (shown in photograph 17-18) attract dirt, even with everyday driving. I have owned cars with vinyl and mohair hoods, but, in future, will stick to vinyl. The colour co-ordinated mohair certainly looked magnificent when first fitted, but very quickly got very grimy in today's traffic conditions, and needed an application of a specialist shampoo. (I can speak very highly of Renovo's products in this respect.) The vinyl hood has gone four times the mileage and time and still requires no more than the occasional soapy water wash and rinse, so has proved the most practical by far.

A garage once put an oily component on the mohair hood of my car, which left a dark stain the size of a dinner plate, outside and inside, which I thought initially had totally ruined the £300 hood. Not so, you will be delighted to hear, the problem was rectified after several applications of carb and choke cleaner! This product is made by Auto Chemicals, which was very helpful when I explained I was ideally seeking something that was virtually 100 per cent carbon tetra-chloride to act as an *in-situ* dry cleaning agent. The carb cleaner is mostly toluene but worked in the same way as carbon-tet. I had to rub the surfaces with clean towels to absorb as much of the dirty oil as possible after each application, and I would not recommend this treatment in anything other than a crisis. However, you can imagine what a relief it was to find a solution.

17-17. A US-style vinyl hood commendably home fitted.

17-18. This is a TR6 hood made from Mohair material and is included to allow you to compare it with the vinyl hood seen in the previous picture.

I mentioned earlier that prevention is better than cure, and suggested you protect your (folded) rear window with a couple of towels. However, if your rear window is already cloudy, microscopically scratched, or just covered in film, I recommend another Renovo product – this time the Plastic Window Polish. If your rear window has been folded repeatedly, it may be opaque, cracked or so distorted that it makes rearward vision difficult. Naturally, this condition won't be helped by polishing and may get you thinking about a completely new hood.

Certainly, if you are at the viewing stage of a prospective purchase, this is a prudent assumption. However, when you have the car home, take a close look at the rest of the hood first for, if it is in reasonable shape, you could at least postpone the purchase of a completely new hood for a while by having a new rear window fitted. This work can be undertaken in the UK by Perfect Rear Vision (020 8777 6764) for about half the cost of what you would have to pay for a new vinyl hood. The work can be completed within 24 hours, and you will be saved the task of fitting a new hood.

A final couple of tips. A candle lightly rubbed over the hood or tonneau zip a couple of times each year should ensure that they operate smoothly. A faded canvas, double-duck or mohair hood can be re-coloured to black, dark blue or brown, and the same materials can be re-proofed. Renovo can again supply appropriate products.

Chapter 18

Left-hand drive to right-hand drive conversion

PREDOMINANCE OF LHD CARS

Most Triumph TR production went to the USA, initially to help the UK's balance of payments, but, latterly, to satisfy the apparently ever-growing demand for UK sports cars. Sadly, this trend has not continued and the current reversal and repatriation of Triumph TRs is partly explained by the better weather conditions that prevail in large areas of the United States of America – particularly, of course, in California.

The sheer number of cars sold in the USA is another factor in the current west to east flow of classics. True, I can only count left-hand drive cars, but it's safe to assume that the majority went Stateside, as the table demonstrates.

So, not only did America have the lion's share of production but, by virtue of better weather conditions, were able to preserve more examples. In fact, it's quite extraordinary to see what detail is preserved in a 25 or 30 year old car that has been repatriated from a hot/dry climate such as southern California's. Fastenings undo with the minimum of effort, leaving the plating still in evidence, while, the chassis and panels can be totally free of rust, to the extent that UK restorers have to actually see an ex-US 'dry state' car to believe it.

Take care, nevertheless, when offered such a car, for it's all too easy to presume that a state is dryer than it actually is. Arizona is a case in point: parts of Arizona are unquestionably very dry indeed, but over half of Arizona is 6000 feet above sea level, with snow still in evidence in late March. Furthermore, it's not unknown for a car that has spent most of its life in a less ideal climate to be shipped to, say, California and sold as an ex-Californian car. So examine all dry state cars initially with a jaundiced eye; it should quickly become evident whether you are really looking at a dry state car or one that has, let's say, moved about a bit ... You also need to check out the mechanical condition, and weigh up what is involved in converting it to right-hand drive.

PREPARATION/STRIPPING

Disconnect the battery and remove the left-hand drive steering column, dashboard, instruments, metal dashback, steering rack, and carburettors and linkages. The latter can come out as one unit, but leave

Model	Total made	Number exported	LHD sales
TR2	8636	5182	2332
TR3	13,377	10,032	6019
TR3A	58,309	52,478	41,982
TR3B	3334	3334	3334
TR4	40,253	36,803	31,285
TR4A	28,465	22,826	19,400
TR5	2947	1171	995
TR250	8484	8484	8484
TR6PI	13,912	5542	5347*
TR6	77,938	77,938	77,938
TR7/8	114,463	88,007	74,800

*includes 3577 shipped CKD for assembly in Belgium

Production figures.

the manifold *in situ*. As surprising as it sounds, it's best to take out the heater to allow unrestricted access to the area where the new right-hand drive bottom steering column bracket will go. You'll also need to at least remove the right-hand seat, and if you plan to weld closed the original LHD holes, it's probably best to remove the left-hand seat, too.

It's sensible to remove any trim/carpet that is in vulnerable areas, and, if a new right-hand drive electrical harness is your choice, I would remove the left-hand drive front harness at this point.

Speaking of removing things, most UK owners of US cars remove the pollution control equipment straight away and, soon after, have the cylinder head compression ratio raised. However, these are details that can be done at your leisure once the car has been converted to drive on the proper (just kidding, no letters please!) side of the road, so let's get down to what's involved in the main event.

18-1. Alterations 'under' the dashboard involve removing the left side lower steering column support that's welded to the inside of the plenum chamber, and filling the hole in the bulkhead/firewall where the column went through. Here they are – gone! Needless to say, on the right side of the car you need to make or buy and fit the different RHD lower steering bracket shown in photograph 18-2, and cut the new steering column hole as described in the main text.

STEERING

The steering rack is probably the first item that most left-hand to right-hand drive converters consider. Of course, a RHD rack is unavoidable, and will have to be purchased on a part-exchange basis.

Externally, Triumph 2000, 2500, Spitfire and Herald steering racks are the same as the TR range of racks we are talking about, and are relatively easily obtainable, but unfortunately have different ratios. TR 'U' bolt mounts are the same for RHD and LHD, and can be re-used, but you'd be well advised to fit new rubber cushions. Solid aluminium rack mounting blocks are available as an upgrade which replaces the 'U' bolts and the rubbers – and this secures the rack far more rigidly than does the original design, but transmits more road noise and vibration. If you spread your 'U' bolts with a suitable tool (see photograph 12-10), or use a weld clamp to pull each end of the rack hard into its rubber cushion before tightening, you'll probably get the optimum rack mountings for a road-going car.

When finalising the fitting of your RHD rack, do be careful to ensure that a good earth/ground is looped from the new rack to the chassis. If you have pre-installed, or plan to later install a RHD electrical harness, you will find a suitable earth/ground wire (lengthy black wire with a ring terminal) is pre-supplied. If you are using a LHD harness, then a little ingenuity, or an earth/ground extension, may be required to ensure the horns eventually work. To get the horns functioning you must earth the rack and, incidentally, also loop an earth lead across the rubber doughnut top joint in the steering column.

The TR's outer steering column is secured in two positions 'under the dashboard'. The top mounting (nearest the driver) is actually bolted to the dashboard, and basically transfers from the left to right side of the car – provided you have secured a RHD metal dashback. These are available but you may have search a bit before locating one. Most converters get round this by having one of our premier TR Specialists modify the LHD metal dashback. The experts make 4 vertical cuts down the original dashback so that the large instrument 'panel' and the glovebox panel can be transposed, leaving the central and two outside parts of the dash in the same place.

The lower steering column mounting is a platform formed by a bracket welded to the inside of the bulkhead/firewall/plenum chamber between engine and passenger compartments. You don't have to remove the old bracket from the left side, but it does make for a more professional job if the original bracket and the old hole are made good, as per photograph 18-1.

Before you start welding brackets in place, however, first use a 2.25in (55-57mm) hole-saw to cut a new steering column bulkhead hole some 1in (25mm) from the corner of the bulkhead top panel. My car was a late TR6 and I used a 2in saw quite satisfactorily. A late TR6 (CR +CF model) will have a quite heavy pressed cowl added to the engine side of this hole, with the steering column hole closed off with a foam ring. The cowl is handed, so your right side hole will have to be to the earlier design, which used a substantial rubber ring/grommet to close the hole while allowing the steering column through, of course.

The bracket that goes just in front of the new hole is handed, as seen in photograph 18-2. Consequently, the LHD bracket is not suitable for

18-2. The new steering column hole has been formed, followed by securely welding a RH lower steering column support to the corner of the plenum box. Take this opportunity to effect any necessary repairs to the plenum chamber; a water test via the air inlet (just in front of the windscreen) is worthwhile.

18-3. The 'new' right side with, in this case, an originally-shaped hole for the servo mounting. Note the holes in the top of the footwell for the pedal box and (arrowed) the slot for the clutch pedal 'extension' to come up through in due course. One omission that ideally needs to be added before painting is the small hole for the throttle cable outer, which goes outboard of the pedal box holes.

RHD conversions, but the bracket is available as a separate unit – part number 815834SB – from Revington TR and, I expect, most of our TR specialists. The original bracket was a single-piece pressing, but it is quite practical to manufacture a two-piece RHD bracket using the original bracket for the dimensions. The advantage of making your own two-piece bracket is that you can, and should, if yours is a non-concours car, use thicker material. I used 0.125in (3mm) thick material to make my 'bridge' support for the steering column. Obviously, the completed bracket was originally spot-welded to the bulkhead, although it is more practical to plug weld the replacement.

In view of the importance of the component, it's worth emphasising the benefit of cleaning up the bulkhead beforehand, and ensuring the highest quality plug welds.

I suggest you replace the two internal steering column bushes before refitting the steering column to the car. There's a neat trick that makes this otherwise difficult job much easier, and it really is too easy to ruin the steering column if you try to remove the old bushes. Instead, use a blade to cut off the original rubber locating 'pips' as near flush with the inside of the column as you can. Now, push the new bushes in from, respectively, the top and bottom, simultaneously pushing the old bushes further into the column.

Another important component that needs your attention is the top steering column rubber 'doughnut' flexible joint of the TR250/5/6. In the interests of safety, it's best to replace this as a matter of course, along with the thin, braided earth strap which is important to the operation of the horn.

PEDALS

There's no need to buy a new pedal box, just remove and completely strip the LHD unit, noting the position of various spacers and washers, and offer up the box in the right-hand footwell as far forward as it will go, and as far to the left as practical. Wear all-enveloping protective goggles when you do this. Hold the pedal box in position by hand, or with a bottle jack or piece of wood, and scribe 4 hole positions – 3 horizontal and 1 vertical. Put down the pedal box, drill 4 small holes and fix the pedal box in position from inside the car, using 4 self-tapping screws.

Check that the pedal box is as forward and to the left as it will go, and drill as many of the remaining holes as possible, but not the hole for the servo for the moment. This operation needs extra care as you will mostly have to drill through the captive nuts fixed to the pedal box, and won't want to damage the treads. It's probably worth pre-drilling a short bolt with a 2mm central hole and using this to protect your pedal box threads.

Remove the self-tapping screws, lower the pedal box and open out 14 holes to the respective correct screw clearance sizes (2 sizes involved). Bolt the pedal box in position with, perhaps, 6 bolts, and carefully find the centre of the four mounting holes for the servo hole. Drill one small (3mm max) 'pilot' hole from inside the footwell, check it is central and lower the pedal box again. You have the option of forming a hole, as shown in photograph 18-3, or, using a 2 inch diameter hole saw and the pilot hole, cutting a hole for the servo body from inside the engine compartment.

The final operation is to cut a rectangular slot in the top of the footwell to allow the clutch pedal to be coupled (eventually) to the clutch master cylinder. It is a good idea to have the clutch master bracket and rubber boot-shaped cover (seen in photograph 18-4) to hand when cutting this slot, and to initially cut it on the small side. There remains the task of closing the LHD holes as shown in photograph 18-5. Jumping ahead for a moment, it will be helpful when finally bolting the pedal box assembly in place to initially omit the 4 bolts that surround the slot just mentioned until you are ready to incorporate a UK clutch mounting bracket, part number 146413.

LHD and RHD pedals seem to me surprisingly different. Nevertheless, I'm sure it's possible to re-use the original LHD brake and clutch pedals. I must confess to a compromise in that I re-used my LHD brake pedal as a RHD clutch pedal with an extension piece cut off, some resetting and an extended top, as shown in drawing D18-1. The result is shown in photograph 18-6. I am sure those with super welding skills could reset the original LHD clutch pedal, and weld a side extension to convert the pedal to a RHD brake pedal, but, in the interests of safety, I elected to buy a new RHD brake pedal, shown in close-up in photograph 18-7. Subsequently, I recall that the quality of the new part made me wish I had not bothered, but at least you have a choice. The RHD throttle pedal is difficult to reproduce, and you're better off buying the long, bent RHD pedal and RHD bowden cable to operate the throttle, regardless of the car's induction. You'll need a couple of nylon

18-4. This is the sort of clutch master cylinder mounting arrangement you should be aiming for if you are converting to RHD, particularly if you anticipate fitting petrol injection sooner or later. The mounting bracket is the standard used on all UK RHD cars.

18-5. Again, the procedure involves closing most of the holes on the left side of the footwell. You need to leave the large hole (just visible) for the harness that will still feed through into the engine compartment at this point, whether LH or RH drive. You will also need to leave one hole for the bonnet/hood release cable. This conversion is clearly well advanced. The main LHD holes have been beautifully closed, and the converter has added the holes for the wash bottle cradle mounting to the base of the footwell. This must be a TR250, since TR250/5 cars had the wiper motor bolted to the firewall/bulkhead, but a TR6 would have a mounting bracket welded to the top of the footwell. It may be difficult to see, but the owner has also replaced the platform between the old master cylinders and the rear of the inner wing.

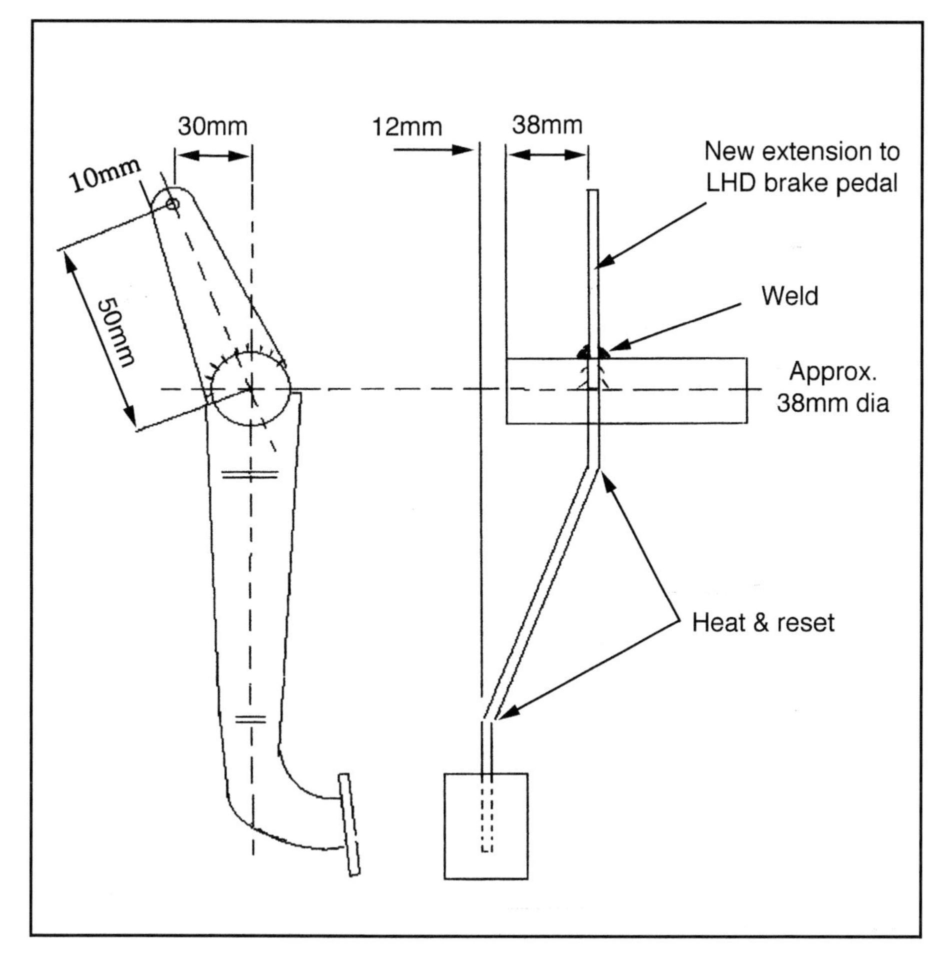

Pedal modifications for RHD clutch.

18-6. The general RHD pedal arrangement with the more complex throttle pedal clearly visible. As described in the main text, the clutch pedal is, in fact, the one-time LHD brake pedal.

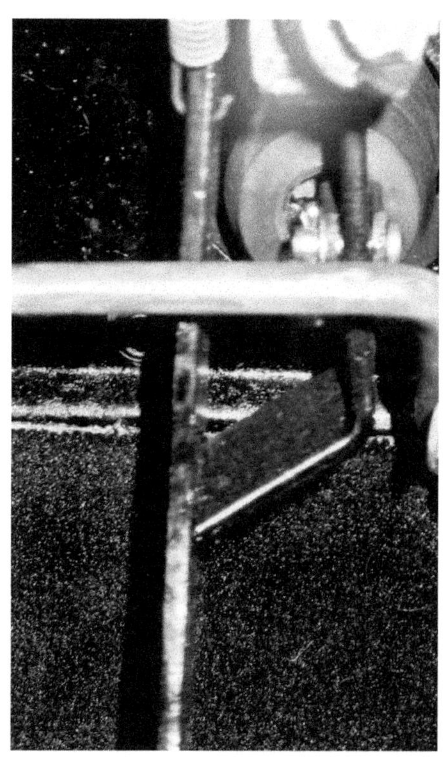

18-7. A close-up of the offset arm on a RHD brake pedal.

bushes and will have to drill a hole for the bowden cable outer. Photograph 18-6 will give you an idea of the layout of the clutch and brake pedals, and complexity of the throttle pedal.

HYDRAULICS

Brakes first. Provided you have pre-drilled the pedal box holes correctly, and opened the large hole centrally, the original LHD brake master cylinder/servo should just bolt into its new location, no problem. Do not forget to refit the original or replacement large washers to the four servo studs that come through the bulkhead/firewall and pedal box, and to use new nyloc nuts or the original nuts with new spring washers.

Which brake pipes you need to install depends to some extent on whether you plan to do the job properly, and replace all the brake pipes on the car, in which case, it's a question of removing the US brake balance valve and pipes and installing the UK (copper) pipe kit. However, if, for whatever reason, you have elected not to install a full brake pipe kit, you can still discard the US balance valve, but will need a couple of special conversion pipes. One will take advantage of the convenient US rear brake line coupling located roughly under the left side 'A'-post. This needs to be coupled to the front female opening in the master cylinder. The rear opening in the master cylinder has a slightly larger fitting and needs routing via a female line coupling to the original front brake line. As one TR friend put it: "Just remember that the 6-cylinder TR's brakes seem piped up back-to-front". These conversion pipes are available ex-stock from our premier TR specialists.

Bleed the brakes, starting with the furthest wheel from the master cylinder.

The clutch master cylinder is less straightforward than you might think. The mounting bracket needs to be positioned atop the bulkhead (photo 18-4) as we touched upon a moment ago. Many LH to RH drive conversions have had the clutch master cylinder in the original position on the almost vertical bulkhead/firewall. However, this will mean that the car cannot have petrol injection without first repositioning the clutch master cylinder. For the minor additional work involved I recommend you use the UK clutch master cylinder location, although it is an unfortunate fact that the US LHD clutch master cylinder pushrod will be 0.3in (7mm) too short for its RHD application, as demonstrated by photograph 18-8. The master cylinders are otherwise identical.

However, although the UK pushrod carries a Triumph part number (122296), you'll be lucky to find a TR stockist that can supply a separate 'OE' UK pushrod. Your local Lucas/Girling outlet may be able to help by placing on special order part number 64673863, otherwise you face some minor complications, or will have to buy a complete new master cylinder/pushrod assembly. If you do the latter, ensure it is a Girling unit. I would take the opportunity to buy a 0.75in assembly for the reasons explained in Chapter 7.

Let's explore the cost-saving option. First and foremost you can remove the pushrod from an old, unserviceable RHD master cylinder to replace the LHD one, provided that the hole in the fork is in absolutely perfect order. See the chapter on engine and clutches for more detail, but note the importance of ensuring there is no mechanical 'play' in the system generally, and the forks in

18-8. A comparison of original US LHD clutch master cylinder (at the bottom of this shot) with a somewhat non-standard replacement shown at the top. The shorter (about 0.3in or 7mm) US pushrod can be seen, but you will appreciate from my earlier comments that the top, longer pushrod is my low-cost solution. Very sharp eyes may spot that the top cylinder has '3/4' cast into it after the word 'Girling', while the lower cylinder has '.7', indicating that the top example has the larger bore which displaces more fluid when the clutch pedal is depressed. The larger bore will increase pedal pressure, but my wife did not comment when I changed to the larger bore, so the increase in pedal pressure cannot be unacceptable! The final detail that will not have escaped your attention is the hexagonal 'adapter' screwed into the outlet connection of the top cylinder. If you have a car built before about 1970, you may already have the larger master cylinder and a clutch feed pipe to suit. Later cars had a smaller master cylinder but a larger feed pipe. The late/larger pipe will not couple into the early master cylinder without this special adapter/coupling.

particular. One alternative that I used, partly because it was cheap but also because it gave me an adjustment that I felt might be helpful, was to cut a male thread on a slightly longer pushrod (from a slave cylinder) and weld a suitable nut to the rear of a new fork. The result is shown in photograph 18-9. Apart from initially setting up the clutch, I have in fact never used the adjustment – so far!

THE WIPER CHANGE

The LHD wiper arms, with their different crank angle, will not park as RHD arms are intended to. You could leave this detail for the moment, but sooner or later you will want to fit RHD wiper arms onto the existing wiper spindles, and change the parking position of the wiper motor. You hardly need me to describe the first operation, but a few words about the second may be helpful.

18-10. The park position of the wipers is changed by repositioning the white cam seen here on the main wiper-motor gearwheel.

TR250/5s need the top the parking cam within the wiper motor gearbox rotating by 180 degrees. Ease the top metal plate (and pin) away from the plastic gear wheel with a couple of thin screwdrivers and, once clear of the locking posts, turn the plate through 180 degrees. The two parts then just push back together (as shown in picture 18-10).

ELECTRICS, DASHBOARD AND INSTRUMENTS

I recommend you purchase a new RHD electrical harness, although you can get away with buying a new front section only, and making a few alterations to the original LHD rear half where the two join by the 'A'-post on the left side of the car. For those on a tight budget, it is possible to alter the original LHD front harness, but not worthwhile in my judgement. The original TR6 LHD rear harness is so wired to illuminate the two wrapped around US/red rear lenses let into the rear wings. Not only do these lenses need to be switched to amber for UK use, but the lamps have to be changed so that they do not act as side-lights. I switched mine to work simultaneously with the direction indicators.

All things considered, it may be just as well to buy new front and rear RHD harnesses to provide a professional basis for the electrics.

Be aware that almost all of the lights will need your attention in some

18-9. A ex-US clutch master cylinder with its pushrod altered as described in the main text. Note the nut welded to a (new) fork, and the extended rod screwed into the nut.

way. In the majority of cases, it will be a change of lens that is required, but you will need to change the complete headlamp sealed units to the opposite dip pattern to be legal on UK roads. This is your opportunity to upgrade them, too. If you are not sure what coloured lenses go where in the UK ask your preferred TR specialist for a UK lens set for your particular car, take a notebook to your next Triumph meeting, or ask the local group to help you out. All the UK lenses are available, but you may get quite a shock when you come to change the front side/direction lights to UK specification. There are many more differences to US lights than it appears at first sight, and they turn out to be far more expensive than ever I, for one, expected! (circa £200 the pair).

The 3-position dip switch/light switch on early '250/6 cars does present a problem, and it is probably best to buy a new RHD one. You can re-use the original switch, but it will either have to be mounted on the 'wrong' side of the column, or upside-down on the right side – with 'on' being in the highest position instead of the lowest. CF cars can have their two position switches reversed and the blue/white and blue/red wires reversed where the switch connects to the car's loom.

It is only an electrical item by virtue of its light bulb, but be aware that the cubby/pocket will, strictly speaking, need to be changed, for they were handed. No-one will, however, be surprised when I tell you that you will need a new dash-back for the right side instruments.

Right-hand drive CP and CR metal dash-backs are very rarely available and you can see the difficulty of changing a LHD dash-back from pictures 18-11-1 and 18-11-2. A professional specialist may be able to help alter your unit, but I contacted Revington TR who had a hand-made RHD dash-back in stock, which only required minor fettling on my part.

You have lots of choices when it comes to veneer finishes, and all are available ex-stock. The instruments themselves transfer across without difficulty; but you will need to buy new drive cables for speedo and tachometer, and also get the speedo recalibrated if you swop the differential ratio.

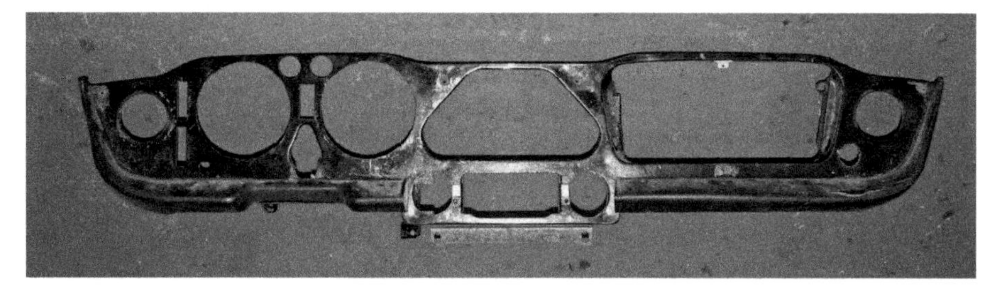

18-11-1. This is a LHD CC model metal dash-back which you may need to alter to ...

18-11-2. ... form this RHD CP dash-back. Perhaps this can be achieved by cutting two identical rectangles from the original unit but the task is made infinitely more difficult by the fact that the main instrument 'half' needs to be welded in place as a mirror-image (ie front face to the rear) if the light switch openings are to be correctly positioned but the cubby aperture requires that the pressed rear recesses remain behind the dash.

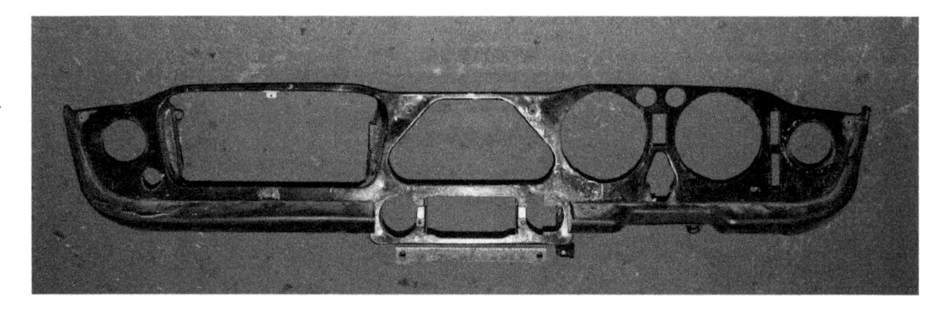

Chapter 19
Conclusions

All of the TR models that this book covers were built to be used, and you should – must – use them. Part of the pleasure of ownership for a good many enthusiasts is 'tinkering' with their toy. However, whilst many readers will enjoy the tinkering just as much as the driving, it should be done at an appropriate time.

I really enjoy the planned repair or overhaul of a particular part (even the whole) of the car – in my garage. But no-one dislikes a roadside investigation, repair or breakdown more than I do. I have gone to great lengths to establish the problems that TRs have, and how to guard against them during refurbishment, in an effort to maximise 'on-the-road' reliability. Triumph initially built very solid, dependable sports cars but, from the introduction of the TR4A, the IRS cars became less rugged. Still highly desirable, of course (remember, I have a TR6), but time has revealed more retrospective chassis strengthening than was required by all the pre-IRS cars combined. You now know not only the weaknesses, but also the corrective and preventative measures.

A final cautionary thought. We all want the best – the best looking car, the fastest, the quickest off the line, the most outstanding specification, etc. It's only human nature. So I implore would-be restorers to consider carefully what they can afford and to prepare their specification, budget and a game plan that is comfortably within their financial limitations. Allow for the apparently inescapable fact that, however carefully you think the project through, it will still end up costing more than you expect – perhaps as much as 20 per cent more. Accept that some specifications or cars are just outside of your budget.

One method to achieving the car of your dreams is to tackle it in stages. Obviously, the bodyshell and the integrity of the chassis and brakes must be the first consideration, but you could then pause and run with your original engine and gearbox, or a somewhat tired suspension for, say, a year or so before starting on the second phase. Top-notch suspension improvements might be the focus of phase two, possibly fitted during your car's winter lay-up, followed by a further period of financial recovery. Finally, maybe a further year or so later, it's time to put in that fast road engine and/or overdrive on the gearbox. You could even re-trim the car a year later still.

Think your own project through; please don't become one of the all-too-frequent "Abandoned Project" advertisements. It really is very easy to overstretch resources and enthusiasm as expectations and component costs rise, and the project takes longer than expected.

Enjoy the camaraderie of the TR fraternity; not impossible with your car off the road but much more practical with your car on the road, even if it is a 'rolling project'. No-one will think any the less of you or your car if you turn up with a roadworthy but obviously unfinished project. The relevant TR club will help you enjoy the car, generate some great friendships and solve some technical difficulties, too. The premier TR club in the UK is the TR Register, whilst Stateside there is the VTR (Vintage Triumph Register).

So do have fun on and off the road, and show off your TR as often as you possibly can.

Appendix 1

Clubs, suppliers & specialists

CLUBS

TR Register,
1B Hawksworth,
Southmead Industrial Park,
Didcot,
Oxon OX11 7HR,
UK.
Tel: 01235 818866
E-mail: tr.register@onyxnet.co.uk

Vintage Triumph Register,
15218 West Warren Avenue,
Dearborn, MI 48126,
USA.
Web: http://www.vtr.org
E-mail: vtr-www@www.vtr.org

Triumph Sports Six Club,
Sunderland Court,
Main Street,
Lubenham, Leic,
LE16 9TF UK.
Tel 01858 434424,
email info@tssc.org.uk

TR Drivers Club,
http://www.trdrivers.com/index.html

TR 6-Pack,
http://www.6-pack.org/sixpack/index

Club Triumph Ltd.
Durkan House,
5th Floor, 155 East Barnet Road, New Barnet, Herts
EN4 8QZ UK
http://www.club.triumph.org.uk

SPECIALIST TR RESTORERS, REPAIRERS, DEALERS & SPARES SUPPLIERS

Moss Europe Ltd (London branch)
Hampton Farm Industrial Estate
Hampton Road West
Hanworth Middlesex
TW13 6DB
Tel: 020-8867-2020
E-Mail: sales@moss-europe.co.uk

Revington TR,
Thorngrove Barns,
Middlezoy,
Somerset TA7 0PD,
UK.
Tel: 01823 698437
E-mail: info@revingtontr.com

Rimmer Bros
Triumph House
Sleaford Road
Bracebridge Heath
Lincoln LN4 2NA
UK
Tel: 01522 568000
E-mail: sales@rimmerbros.co.uk

TR Bitz,
Lyncastle Way,
Barley Castle Trading Estate,
Appleton Thorn,
Warrington,
Cheshire WA4 4ST,
UK.
Tel: 01925 861861
E-mail: triumph@trbitz.u-net.com

TR Workshop,
Unit 4 Bankside,
Love Lane,
Cirencester
Glos GL7 1YG,
UK.
Tel: 01285 659900

TR Enterprises,
Dale Lane,
Blidworth,
Mansfield,
Nottinghamshire NG21 0SA,
UK.
Tel: 01623 793807
E-mail: stevehall@trenterprises.com

TRGB Ltd,
Unit 1 Sycamore Farm Industrial Estate,
Long Drove,
Somersham,
Huntingdon,
Cambs PE17 3HJ,

UK.
Tel: 01487 842168
Web: www.trgb.co.uk

Overdrive repair specialists

Overdrive Repair Services,
Units C3/4 Ellisons Road,
Norwood Industrial Estate,
Killamarsh,
Sheffield,
Yorks S21 2JG,
UK.
Tel: 0114 2482632
Web: www.overdrive-repairs.co.uk/products

Overdrive Spares,
Unit A2 Wolston Business Park,
Main Street,
Wolston,
Nr Coventry,
West Mid CV8 3FU,
UK.
Tel: 02476 543686
E-mail: odspares@aol.com

Gearbox rebuilding

First Gear,
3 Church View,
Beckingham,
Doncaster,
South Yorks DN10 4PD,
UK.
Tel: 01427 848101

Chassis repair and manufacture

CTM Engineering,
Unit 3A, Bury Farm,
Curbridge,
Nr Botley,
Hants SO30 2HB,
UK.
Tel: 01489 782054
E-mail: colin@ctmeng.freeserve.co.uk

Paint manufacturers/specialists

International Coatings,
24-30, Canute Road,
Southampton,
Hants SO14 3PB,
UK.

Jmautocolour,
Unit 30, Luther Challis Business Centre,
Corrinium Avenue,
Barnwood,
GL4 3HX,
UK
Tel: 01452 310502
http://www.jmautocolour.com/information.htm

Hood renovation products

Renovo International,
PO Box 404,
Haywards Heath,
West Sussex RH17 5YN,
UK.
Tel: 01444 443277
E-mail; renovo@dial.pipex.com

Chrome restoration

Central Engineering Services,
Unit 1 Riverside Industrial Estate,
West Hythe,
Kent CT 21 4NT,
UK.
Tel: 01303 268969
Web: www.c-e-s.demon.co.uk

Electrical

TR250/6 Maintenance Handbook
Written by Dan Masters and available from DMP, PO Box 6430, Maryville, TN 37802-6430, USA.
Web: http://members.aol.com/danmas6/

Advance Auto Wire
(Headlamp upgrade relay/wire/fuses)
Web: http://www.advanceautowire.com

HIDS4U
(High Intensity Discharge/Xenon headlamp conversions)
Web: www.hids4u.co.uk/products.asp

PI specialists

Prestige Developments & Injection,
77 Box Lane,
Wrexham,
Clwyd LL12 8DA,
UK.
Tel: 01978 263449
Web: www.prestigeinjection.fsnet.co.uk
(petrol injection parts and service)

Cylinder head refurbishment/ unleaded computability

Bailey & Liddle,
Unit 16, Upper Brents Estate,
Faversham,
Kent ME13 7DZ,
UK.
Tel: 01795 535068
(Engine reconditioner and unleaded cylinder head conversions)

BODY REPAIR TOOLS & EQUIPMENT

Frost Auto Restoration Techniques,
Crawford Street,
Rochdale,
Lancashire, OL16 5NU,
UK.
Tel: 01706 658619
(Clamps, cutters, tools and equipment for auto body repairs)

SPARES SUPPLIERS (USA)

Moss Motors,
PO Box 847,
440 Rutherford Street,
Goleta, CA 93116,
USA.
Tel: (800) 667-7872
Web: http://www.mossmotors.com/

The Roadster Factory,
PO Box 332,
Killen Road,
Armagh, PA 15920,
USA.
Tel: (800) 234-1104
Web: http://www.the-roadster-factory.com

Victoria British Ltd,
Box 14991,
Lenexa, KS 66285-4991,
USA.
Tel: (800)255-0088
Web: http://www.longmotor.com

Paint manufacturers/specialists

Interlux,
2270 Morris Ave,
Union, NJ 07083,
USA.
(Marine industry paint suppliers)
Multi-language website: www.yachtpaint.com

Appendix 2

Welding, tools & techniques

WELDING

Your restoration plan will almost certainly necessitate some welding, as even the best-restored car will require some attention sooner or later.

A relatively simple job – like left-to right-hand drive conversion – will involve a small amount of welding almost as soon as you have the car in your garage, and a full body-off restoration will give you lots of welding practice! The prospect of welding can be daunting, especially if you don't own a welder, but don't be put off; and ladies, that includes you (ladies made up 25 per cent of my welding course). Your restoration plan, budget and timescale should allow for the purchase of a welder of one size or another, and possibly some tuition on how to use it. Mind you, preparation for welding is dirty, noisy and unsociable; give some thought to whether your family and neighbours will put up with the cutting, chiselling and grinding involved in removing all of the old rusty material before the quieter/cleaner/constructive work begins!

THE ALTERNATIVE METHODS

Although it is unlikely that your first task will involve welding, it is something you must prepare for, and buy not only the obvious welding equipment, but some important ancillary items, too. Without some form of welding facility you won't complete a body restoration at home, so let's look at two most relevant forms of welding: gas and MIG. You may see references to stick and/or TIG welding, both of which have their advantages, but we will gloss over these techniques simply because they are less suited to DIY bodywork repair.

The first principal to explore is the advantages and disadvantages of heat and distortion. I certainly do not want distortion, I hear you cry, which is true, but you may want to shrink an area of steel, which can be accomplished by (gas) heating the area you wish to shrink and then allowing it to cool. As it cools, it shrinks beyond the starting point. However, in general, heat generates distortion and, in the vast majority of cases, this is not welcome.

So, the welding technique that generates the least heat is the preferred method for most home bodywork restorations. MIG welding is the almost universally preferred welding method for home panel repair. Let's explore MIG welding and the variations that are available first, starting with a brief overview of what MIG welding is.

MIG stands for Metal Inert Gas. A MIG set consists of the welding set itself, an umbilical cord and a hand-held 'torch', which has a trigger set in the handle. The trigger, when squeezed on, activates three things: a welding current is switched on, welding or filler wire starts to feed forward from the welding set, and the inert gas starts to flow through the umbilical cord and out through the torch. The wire speed, or amount of wire fed through the torch is variable, as is the welding current. Very broadly speaking, the thicker the material you wish to weld, the faster the wire speed and higher the amperage settings required.

The main advantage of MIG welding is that it does generate less heat than gas in the panel(s) you are welding, but it has a downside, too. The filler wire has to be reasonably hard to withstand the feed mechanism without kinking; consequently, a weld applied by MIG is hard and has to be ground flat. A MIG weld cannot be dressed by hammer and dolly without splitting, whilst a gas weld is generally softer and more malleable.

If you plan to mostly plug or seam weld (as would be required to, say, fit a new floor or bulkhead), then MIG welding would be your first choice. If, however, you are carrying out a concours restoration and want an invisible wing repair, then butt welding by gas would be the preferred method. It will be much easier to plenish the weld flat and, with a very light grind, it would be virtually invisible with just a coat of zinc primer.

Please don't think that it's impossible to butt weld with MIG. To do so you'll need a 1mm gap between the two panels (that might come as a surprise). However, it is easier, quicker and just as aesthetically acceptable to most restorations to joggle the new panel and plug weld or even seam weld down the edge of the overlap. A short spell with an angle grinder is all that is subsequently required to form an equally almost invisible repair from the outside of the car.

You can, of course, have the best of both worlds by tackling the tub of the car with a MIG welder, and sub-contracting the gas welding repairs to your detachable panels, such as wings, etc. However, for the majority of home restorers aiming for a high quality usable restoration, MIG welding will be the choice right throughout the car. Plug welding is a very frequent restoration requirement, and MIG is infinitely superior for plug welds. MIG and gas are equally effective on thicker chassis welding, but most of us are only able to buy one piece of kit, so I would opt for MIG. A MIG set is easier to buy and store than a gas set, in any case. With gas equipment there are two large gas bottles to worry about. Of course, you will have a similar problem with MIG, unless you go for one of the much advertised no gas MIG welding sets.

Most manufacturers of MIG sets designed for domestic use offer no gas models, usually at the bottom end of the range. As the title implies, they are designed for use without the shield of inert gas the more professional MIG set employs. The no gas sets actually generate a local inert gas shield at the weld point, via a flux core that is incorporated into the centre of a no gas welding wire. If you have a no gas set you must use no gas wire, but you will not want to use no gas cored wire when you have a gas shielded set.

Why choose a gas shield set when you can avoid the problems and costs associated with a large gas bottle by using a no gas set and wire? Good question. Firstly, note that no gas cored

Mild steel MIG wire (no gas wire has fluxed core)

0.45kg of no gas wire 0.9mm diameter	£8
0.8kg of gas wire 0.8mm diameter	£4
4.5kg of no gas wire 0.9mm diameter	£50
5.0kg of gas wire 0.8mm diameter	£12

Gas cylinders

Gas charge	**Approx. bottle capacity**	**Bottle monthly rental**	**Gas**
Oxygen	10 cu. metres	£5	£15
Acetylene	6 cu. metres	£6.00	£35
CO2/argon mix	12 cu. metres	£5	£30
Disposable canister of CO2/Argon	about 30 minutes of gas usage	NIL	£8

US prices

The flux core (no-gas) wire is sold in 2lb and 10lb spools. Wire sizes typically vary in inches as 0.030, 0.032, 0.035 and 0.045 although with enough persistence one can find other sizes. A 2lb (.9kg) spool of 0.035 (about 0.9mm) will cost approximately $17. A 10lb (4.5kg) spool of the same wire will cost about $60.

Regular MIG wire comes in the same wire sizes as the flux core wire. A 2lb (.9kg) spool of 0.035 (about 0.9mm) will cost approximately $9. An 11lb (5kg) spool of the same wire will cost about $34.

Gas cylinders

These can be purchased outright or leased on an annual basis. The most recent lease rate quoted was $60 per year, regardless of bottle size. Information below is for the refill. Bottle sizes come in 40, 60, 80, and 390 cu. feet. (390 is a little larger than the 10 cu. meter bottle you refer to). The 60 cu. feet size is about right for most home hobbyist needs.

A refill of CO2/argon mix for the 60 cu. feet bottle will cost about $35. A refill of the 390 cu. feet bottle will cost $85. Costs of other sizes vary accordingly with bottle capacity.

Acetylene comes in a 150 cu. feet bottles and costs $40 per refill.

MIG welding consumables – price comparison.

wire is more expensive than the more usual plain MIG wire. Secondly, I have seen experienced MIG welders lay down some excellent welding using a no gas set, and have even declared the set to be first-class. However, I must confess that I found it much more difficult to weld with a no gas MIG set, although my gas shielded MIG welding is quite acceptable. Furthermore, I do not believe anyone would argue with the fact that no gas sets generate much more by way of spatter, slag and smoke. My advice would be to buy a conventional gas shielded MIG set. Get round the gas bottle problem by hiring a large bottle for your main welding stint (say, for 3 months) and using small disposable canisters for smaller, intermittent welding that comes up from time to time. The table should help you to decide.

Which gas you use is very important and, for MIG welding, always use carbon dioxide with 5 per cent argon – called Argoshield. This makes for maximum ease of welding. When ordering the large rented gas bottles, expect to pay for the gas when ordering your cylinder(s), along with a cylinder deposit equivalent to at least 3 months' rental and a delivery charge (about £15). Many suppliers do not charge for collecting the empty cylinder(s).

WELDING EQUIPMENT

When setting out on your restoration, decide whether or not you want to buy all of the equipment that will be required. For the cost of a welding set, extra gear and possibly some training, you might be better off paying a specialised/professional restorer to do most of the welding, or buying an ex-Californian shell/part-shell that should need very little welding. If you buy an ex-Californian shell it may be practical to hire a welding set for a week, or buy the smallest size of welding set and very minimum of ancillary gear.

Only you can decide but it may help if we spend a short while deciding what equipment you might have to budget for. The size (or capacity) of the welding set you buy is very important. A very small capacity set – say, 90 amps – may be a false economy for a full restoration, although perfectly suited for small jobs. In my judgement the best value for money – particularly if a full restoration is planned – is offered by a 150 amp capacity set, available from selected retailers for under £240. Choose one with a non-live torch; in other words, the torch will not transmit current until you pull the trigger. You should get various valves and gas pipes included for this sum, but not a gas bottle.

ADDITIONAL EQUIPMENT

Allow a minimum of £300 for tools and equipment specifically for the sheet metal work involved in a full restoration. Assume at least the following additional equipment:

- A substantial electric drill. Choose one with the highest torque at low rpm that you can find. It might be prudent to pay the extra for an industrial standard drill to get the torque and robustness you will need, so assume at least £85.
- Angle grinder at £60.
- Face mask/goggles for protection when grinding at £15.
- Pneumatic nibbler at £54 or pneumatic 3 inch cut-off saw at £30.
- Hand edge setter (or joggler) combined with hole punch at £35 or Sealey pneumatic version at £67.
- Combination welding head shield with flip-down front at £20.
- A small selection of cold chisels and bolsters – say £20.
- Best quality tin snips – about £15.
- Any angle magnet clamps at about £13 each x say 3 = £40.
- Locking welding clamps about £5 each x 4 = £20.
- 4 to 6 'Intergrip' butt welding clamps – about £14.
- Gloves – cut-resistant for handling very sharp sheet metal.
- You will need to separate numerous spot welds, and zip-cutters are okay for this although I get on better with a sharp 10mm drill reground to give a flatter than usual (say, about 130 degree) point.

Most welding and metalworking tools can be purchased from Frost Auto Restorations.

It will not have escaped your notice that the above pneumatic equipment will require a compressor. Whilst not something you need to buy – although it is possible to – you may need to have some way of rolling over the body tub to make it more accessible for underside welding. A pair of triangular angle iron frames clamped front and back of the shell is the usual method. If you need to completely roll the shell, ensure the pivot is sufficiently far off the ground to allow the full width of the shell to pivot through 180 degrees.

SAFETY

This is largely a question of common sense. In the garage, do not smoke when welding or around flammable gases. Do not have heaters near to compressed gas bottles. Chain gas bottles to a wall when not in use. Always, absolutely always, wear a good welding mask with the dark eye shield in place. Always wear thick welding gloves and protective overalls when welding, and thick leather or the latest no-cut gloves whenever handling rusted and new sheet metal. Never grind without proper eye protection. Remove the fuel tank from the car, drain or blow through fuel lines, and disconnect the alternator before welding.

LEARNING TO WELD

Although you will benefit from some expert tuition, which we will discuss in a short while, the following are a few of the absolute basics for seam (or continuous) MIG welding for you to experiment with first.

Hold the body of the torch about 5-10mm from the workpiece, but angled slightly (about 30 degrees) so you can see the end of the filler wire touching the workpiece/arc. If you are right-handed lean the top of the torch away from you while you move the arc towards your left side (called pushing the torch) for best penetration. Here are the main problems you will probably experience –

1 If your wire 'bounces' off the work, pushing the torch away from the workpiece, then, almost certainly, the wire feed is too fast. Rectify by turning down wire speed; a slightly higher amperage may also help.
2 Opposite to the above problem is if the torch virtually touches the workpiece. Speed up the wire feed.
3 Frequent holes burnt in the workpiece suggest too high an amperage, rectified by turning the amps down a little.
4 Welding is all about joining molten metal from both panels. A good strong weld is therefore only possible if the amperage is high enough to fully penetrate both panels. Full penetration (melting) is achieved when a bubble appears along the line of the weld on the lower panel. This is often difficult to see working on a full panel on the car, so do try numerous 'off the job' test pieces until confident you are getting proper penetration. A discolouration of the lower panel is not sufficient and, in this event, a higher amperage setting is almost certainly called for.

You can undoubtedly learn to weld from a specialised tuition book and lots of off-the-job practise. However, remembering the safety aspect, some expert guidance is very worthwhile. The obvious way to get this in the UK is via evening classes at your local Technical College. Most run Motor Vehicle Technician courses, and one year's classes (about 30 two-hour sessions) will get you to an adequate and safe welding standard. I cannot speak for all Colleges but my experience is that, in addition to learning to weld, you will be able to take smaller panels (wings, doors, boot lid and bonnet) into college and work on them under expert guidance. Furthermore, if you need to make up, say, a couple of curved false lips to weld to your tub's inner wings, then the requisite joggle or rolling equipment will be available at college, and all you will need pay for is the material you use. In the USA, via the American Welding Society (phone 800-443-9353) you can get in touch with your local chapter, which will have knowledge of local learning opportunities. Alternately, contact your local Community College or High School, either or both of which will quite likely know of, perhaps even run, evening vocational classes.

You can buy a full sheet of mild steel from your local steel stockist, but if you attend college you will be able to buy the steel in small, manageable pieces, guillotined to the precise size you need, when you need it. Most steel stockists are listed in the telephone directory and will be pleased to deliver one 2x1 cold reduced mild steel sheet to your home on a next-day basis. The gauge you should ask for is, in today's parlance, 1.2mm, which anyone of my vintage will know as 18swg or 0.048in. Cost will be around £30, including delivery. By the way, if you need the 2x1 bit translating too, it means roughly 6'x 3' .

In the next section we will look at the basics of forming steel panels from scratch, and, of course, adjusting pre-made panels to fit your car. It's unlikely you'll have both gas and MIG equipment at home, nor are you likely to have a wheel or rollers, which are indispensable for even minor tweaking of a pre-made but slightly errant new panel. Your local Technical College will have all of this equipment and will help you use it. You can always take the old panel, or maybe a template, with you. The cost of a year's evening classes is about the same as a MIG set – £200. Go for it; you can't afford not to!

TYPES OF WELDS

You will come across the following terms as the body restoration progresses. Tack welding usually precedes all of the following and involves a series of 1mm long 'tacks'. These can be easily broken to allow repositioning of the panel, but are sufficient to hold the panel in place while measurements and/or other checks are done, or even additional panels put in place.

SEAM/STITCH WELDING

Seam or continuous welding will be required very frequently, and is usually best done in 30mm long stitches, with a gap then a second stitch then another gap. Eventually, you go back over all the gaps and fill them in. This technique minimises distortion.

Butt welding – As the name implies, this is a weld that joins two flush-fitting panels. A typical application would be joining the front half of a curved wing/fender to the rear half, or letting-in a repair section. A small (1mm) gap between abutting sections helps the welding. After tacking, you will effectively seam welding the join.

Plug welds are perfectly acceptable and often used as a substitute for spot welding, but are best positioned closer together (more frequently) than the original spot welds. They must be carried out in such a manner as to achieve first-class penetration of the lower steel sheet. It's too easy to weld closed the hole you have made without properly penetrating the lower sheet. Consequently, what seems to be a good weld has no strength at all. Start with a good-sized hole in the top sheet, about 1/4in (6mm) diameter. This is slightly bigger than most factory spot welds, but gives you a sporting chance. Prepare the lower sheet by linishing it with a paper disc on an angle grinder. Start the welding in the centre of each hole by melting the lower material and spiralling outwards to puddle the hole, including the periphery of the top material layer. Do lots of trials on scrap material to ensure you are putting the maximum amps into the welds and that penetration is right through to the lower material. You will see a molten 'bubble' or spot on the lower material if you have got it right. A circular discolouration is NOT sufficient. Your life may depend on having good welds so take a hammer to some of your test pieces and see if they really have 'stuck'. It depends on the application, but most plug welding is carried out with the 'spots' pitched about 3/4in to 1in (20 to 25mm) apart. You can minimise distortion of plug welded materials by plugging only every 5th hole initially. Then go back and plug the 'middle' hole in each block of 5, again welding, in effect, every 5th hole. Now there will be a strong temptation to fill in all the remaining plug holes in one run, and in some circumstances this will be quite satisfactory (say, when there is a lipped flange within the joint). However, the safest approach is to go down the length of the seam repeatedly, again welding every 5th spot on each run. Finish with a paper linishing disc on an angle grinder.

'V'-notch welding – Not widely practised but does have advantages in certain circumstances. It can really only be used when two pieces of metal are to be joined at an angle of between 45 and 90 degrees. You can, of course, seam weld the join, but strength is added to the structure if you fold a lip of, say, 12mm to the edge of the joining piece, Your first inclination would probably be to seam weld or plug weld the panels together. However, you can snip a series of 'V'-shaped notches from the 12mm flange and then seam weld down the edge of each V. You will spread the stresses further by this method and put more weld length into the joint than via a straight seam weld run.

BASIC METAL FORMING & ADJUSTMENT

Assuming you are planning to repair an existing rusted panel, the first step is to establish the area to replace. There is an understandable initial temptation to replace only the most severely rusted part of the panel to minimise the size of the replacement, but you must plan to completely remove all of the old rusted metal, whilst bearing in mind what repair panels you have available. You do not have to use a full repair panel, but it's absolutely pointless choosing a repair panel that leaves rusted material in the original panel. If the largest repair panel available is insufficient to allow you to remove all of the damaged area from the old panel, consider either a full replacement panel or whether you should make your own repair section. Earlier TR replacement panels are not a wonderful fit so, with the exception of TR6 panels, you're probably better off repairing the original panel if this is feasible. There's always the good old standby of second-hand replacement panels, and wings in particular, from an ex-Californian car. The trouble is, ex-Californian cars are getting harder and harder to find, and those panels that are available are less and less suitable. In the not-too-distant future, the superb hot climate body (and chassis) parts we have been used to in recent years may be exhausted. You may still be better off repairing the bottom 4 inches (100mm) of a disappointing ex-Californian panel than, say, 20 inches (500mm) on original panels. Either way, a repair is a repair, which brings me to conclude that a few words of basic advice on fitting, even making, a replacement panel, are possibly appropriate.

Basically, panels have either a single curvature or double curvature, and although it may look daunting, remember that steel is inexpensive, and if you have to make a replacement several times, so what? The first thing to do is make a couple of cardboard templates of the original panel profile. Most single curvature panels, like, say, a wing or a door, curve from top to bottom, and this is the section, plane or curvature your first template should record. The second template is, in effect, the shape of the panel as you view it from the side of the car. It's a good idea to mark the original panel with a couple of blobs of paint, nail varnish or typing correction fluid to record where your first or curvature template has been prepared, obviously ensuring that you don't subsequently cut away your template positioning points. Paper really is not suitable for making templates and I would suggest a roll of pattern maker's card. This is usually green, about 0.5mm thick and will have the rigidity you need.

Do not be tempted to sand or shotblast the old panel before the repair is complete. You will more than likely have insufficient metal left to act as a pattern of the area that needs replacing, and could even distort the area you plan to keep. We might give the edges a light blast later, but only after we have completed our repair and satisfactorily offered up the repaired panel to the car. Purely from a economic point of view, you don't want to be taking one panel at a time to the blaster for even a light edge-only treatment, so it's best to complete all the panel repairs and if there is any light surface (and it should only be light surface) corrosion, have the lot lightly, gently and ever-so-carefully blasted around the edges and zinc primed straight away (within the hour is best if possible). By this route, if a repaired panel does gather a little surface rust from fingerprints or atmospheric dampness, there's no real need to panic. That said, obviously, repaired panels are best kept as dry and as rust-free as possible. If you think it might be 12 months before your collection of repaired panels are ready, a light spray of oil, a wipe with an oily rag, a coat of lanolin or even a coat of zinc primer, is a good idea. The oil or lanolin can be largely removed with thinners just before blasting, and the primer will last little more than a second under even the lightest blasting.

Returning to actually fitting a single curvature repair panel to the original, you'll appreciate that a pre-formed repaired panel does much of the hard work for you, but I hope that the following will be of help if you are forced by cost or availability into making a repair section yourself. Whatever you do, if using a pre-formed repaired panel remember to allow at least 1/2in (13mm) overlap between the repair panel and original. The first step for your homemade repair panel is to cut a blank piece of steel sheet that is initially at least 1inch larger in each direction than you think you will need. Do not try and be too accurate at first. Mark out the edge you plan to marry to the original panel, remembering that you will need a shade over 1/2in (13mm) of extra material above the join line for a joggled, overlapped, plug welded joint. Your next step is to form the basic single curvature. Slowly is the operative word here, but frequently checking progress with your curvature template, bend the blank panel over your knee or a (usually large diameter) tube until it conforms to the first template. If you attend college you should have access to a set of rollers, and these will do a wonderful job. If you make a mess of the bend, get a fresh piece of steel sheet and start again, but don't throw away your first effort, it could still be a very useful trial piece for subsequent steps.

Next turn your attention to the joint between the original panel and your repair panel; this is point of no return when you need to cut the original panel at your joint line, which should be absolutely straight.

We now have a choice about the order in which to complete the next two steps: joggle first or form the plug weld holes first. I will proceed on the basis of holes first and explain why a little later. Punch or drill a series of holes across the edge of the repair panel where it will overlap the original panel. These holes will eventually be used to plug weld the new repair panel (from the rear) to the original panel. The holes should be at 3/4in to 1in (20–25mm) intervals and be about 1/4in (6mm) in diameter. It is far better to have too large a plug weld hole than too small, so if you are not very experienced at plug welding, go to 7mm diameter with your test pieces. You can even consider 8mm diameter holes if you not absolutely certain of getting a weld that fully penetrates the lower (original in this case) panel. The centre of each hole should be 1/4in (say 7mm) in from the edge of the repair panel. Drilling is a perfectly good way of making these holes, but you must ensure that you support your new curved repair section with a piece of wood behind the drill line before you start drilling the holes. Nor must you distort the lip too much, so use a sharp drill.

There is, however, a better way to form these holes. A Sealey hand-held pneumatic punch does the job far quicker than drilling, and with no chance of distortion. It is my recommendation that you buy or hire one of these admirable tools. The Sealey also has the invaluable advantage of also enabling you to put the 1mm joggle along the overlapping edge of your repair panel (be it pre-punched or handmade). This is where the 'joggle first or hole first' question comes in. With the Sealey equipment you can put the holes in, then run along the overlapping edge and in 1/2in (12mm) 'bits', with no more effort than a squeeze of the Sealey's trigger, can effect the joggle or step in the overlap joint. This is important as it will subsequently minimise the amount of filler needed to make the repair joint invisible. Sealey equipment is available in most Technical College vehicle repair workshops.

If you don't have access to a Sealey punch/joggler, you need to joggle the 1/2in overlap joint on a folder. Even then it's not easy to achieve the joggle since it consists of two 90 degree bends virtually on top of each other. You can use a set of joggling rollers, or choose to ignore the joggle/overlap joint and go for a straightforward overlapped joint. In this case I would put the holes in after achieving the joggle/overlap.

So, we have a curved repair panel with the top 1/2in joggled and holed to provide for an overlapped plug welded joint. Now we get to the point when some of the surplus metal needs to be trimmed off.

Your second template comes into its own for the first time. Lay it on your repair panel, lining up the top cut or joint line and (soft) pencil round the edge of the template. It's best not to use 'tin snips' or 'Gilbows' as both are hand-held trimmers that cut by a scissor-like action which will distort the metal somewhat. Use a powered nibbler to cut all but a 1/2in (12mm) lip allowance down one side of your repair section. The basic thing you must remember before starting to form a lip on any curved, even slightly curved, panel is that it will try and straighten the panel curvature unless you take precautions. This paragraph is intended to help and we will cover initially an outside lip that is relatively unusual in external panels like wings, but could be required as part of an inner wing or bulkhead repair panel.

The outside lipped single curvature panels need the lip to be progressively stretched as the (90 degree) lip is turned out. As with all metal working, slow and progressive progress brings the best results. Tap the 1/2in lip over the end of a (flat) steel mandrel or steel dressing dolly. For slight single plane curves, the mandrel/dolly can be 1in (25mm) wide, but as panel curvature increases you will need to reduce the width of your mandrel and also the size of the 'bite' you take. Go right down the length of the panel lip and turn the lip out by 20/25 degrees. Check panel curvature with your first template and tap the outside edge of the lip (to stretch it) until the curvature again coincides with the curvature (first) template. Repeat this slow but sure approach, reaching the lip's 90 degree angle only after about 4 runs down the whole of the flange length. Never turn one end out completely and then tackle the other end.

There's a very useful shrinking and stretching tool that would be invaluable for the above and also in the following circumstances. Called a shrinker/stretcher, it costs about £200, but will enable you to fabricate repair panels for numerous situations. It can be bolted to your workbench or gripped in your vice and is available from Frost Auto Restorations.

Turning a 1/2in flange inwards requires the same approach as described in the foregoing, except that you will have a surplus of material in the flange to cope with, so the last thing you want is to stretch the flange. The technique, in this case, is to deliberately wrinkle the inside lip as you turn it in. A pair of pliers and a sort of twisting motion will create a series of ripples or puckers. Where the repair panel has only gentle radiuses you could shrink the lip by oxyacetylene heating it to red heat in the knowledge that it will shrink on cooling without further dressing.

As the severity of your main panel radius increases, so too will lip wrinkling. In fact, you may have to resort to snipping out a shallow V from the lip to compensate for a very tight main panel radius that can occur at the very bottom of, say, a wing. In these cases you will need to oxyacetylene weld the V back together and then plenish flat the resultant weld. Where lip wrinkling is considerable, but not so great that you have had to resort to V snipping, you will need the help of oxyacetylene to heat the lip to cherry-red. You then need to plenish the wrinkles, actually thickening the lip material in the process – so do not overdo the hammer/dolly plenishing or you will stretch the lip material!

If you seek re-assurance, just take a look at any pressed repair panels you have to hand. You will note a crinkled lip at, say, the bottom of a wing pressing where the internally formed lip has 'bunched-up' the surplus material. That is how a pressed component handles the surplus material. However, when you are hand-forming a panel, the 'support' of a press die is not there to make the metal flow in the same way' so we have to adjust the compensation – and that is what edge wrinkling and occasional snipped cuts will do.

Two other areas will require you to cut into the lip; firstly at each corner of your panel and, secondly, at any 'swage' you need to introduce into a panel. A swage is basically a large joggle, which is mostly introduced horizontally into wing or door panels for aesthetic reasons, presenting you with something of a challenge, but swages (small swages, anyway) are not without advantages, too. Dealing with possible advantages first, when joining a single curvature panel down its flat plane or length, it's all too easy to generate an unintentional curve in what was the flat plane of the panel. This occurs due to heating and cooling when welding and can look like a gentle ripple or wave when viewed end-on (say, from the front edge of a door, looking backwards down its length). A swage increases the resistance of the door to ripple if your repair section or panel is married to the original panel in close proximity to the swage. Indeed, if you look at some pre-made repair panels, they will often be pressed so as to come to a swage line in the original panel, and may have a lip pressed into the repair panel at that point. Many amateur body repairers do not appreciate that the lip on the repair panel is there to strengthen the repair panel, and, if left in place, will dramatically reduce rippling of the finished assembly. Sadly, these lips are often removed at home and their objective nullified.

Reverting back to the original swage line, any repair joint made immediately adjacent to a swage line is also less visible. So the area adjacent to a swage line is actually a good spot to consider for your cut line of a repair panel. Not that you should cut away large areas of good, un-corroded metal just to make your joint adjacent to a swage line some 10-15 inches (250/375mm) further up the panel!

There are bound to be occasions when it is necessary to make a repair panel incorporating a swage line, so we'd better spend a moment explaining how this can be done. Firstly, carefully measure the height and depth of the swage, and even cut a small cross-section template to allow you to check your efforts to replicate it. Secondly, form a swage in your blank panel before introducing the single curvature and, thirdly, you MUST snip the turn-in or turn-out panel lips at the top and bottom of the swage before you start turning the lip. Forming a swage is not difficult, but forming a specifically dimensioned swage will take a few attempts. Get it right on some off-cuts before you approach the repair panel blank.

At college you should have a folder available to make the job easier. At home, you'll need a little more ingenuity. To form the swage, first find a piece of bar or tube as long or a shade longer than the length of your swage. On a repair panel this may be too daunting, but if you were tackling a door, the bar would need to be at least the width of it. Not only must the length be adequate, but the diameter must also be correct – which is where that small section of original swage again comes into its own, for the tube you have in mind needs to nestle comfortably in the radius of the original swage. Now we are set to go, and start by clamping the flat repair panel blank onto the workbench under a stout piece of angle iron. A couple of large 'G' clamps will usually do the job. Now place the sectioned piece of correctly radiused bar or tube on top of the blank, and up against the clamped stop, and bend the blank up by the prescribed angle (rarely 90 degrees; often not even 45 degrees). When the front is correctly in place, unclamp the stop, turn the blank over, re-clamp it in the correct place and make the same degree of upward bend. Get your sample section and check right along the length of the repair panel that the swage is properly formed.

It will be obvious that the longer the swage the more difficult it is to get the full definition along the full length of both bends. You are unlikely to achieve the definition you need on anything but the shortest sections without dressing the central areas of each bend during each bending operation. This is best achieved by a polished square section of steel, but a square section of brass or hardwood will also do a pretty good job. I rarely use anything but wood, but whatever your choice the idea is to place the length of steel, brass or wood against the panel material and strike it fairly hard with a club hammer to ensure that the definition of the radiused bar is translated into the blank. If you use a rough-faced piece of hard material such a steel, you will reproduce the imperfections in the face of your component blank. Not so important when you are bashing the unseen side of the repair panel, but not a good idea when it comes to dressing the outside of, say, a door panel.

Throughout all of the above I have glossed over the vital 'preparation' that is essential to good welding. You must clean each of the mating surfaces until they look like new. All old welds must be ground off to allow panels to come together without gaps, and you must also remove all contaminates like paint, oil and grease. This preparation will take much longer than the actual welding, but is of the utmost importance.

Finally, remember that it is unwise to fully weld a series of panels until you are absolutely sure they not only correctly interface with each other, but that the fit of related panels and parts (such as boot/trunk lids, doors, bonnets, lights, etc.) are to your complete satisfaction. So, start by positioning each panel with self-tapping screws, pop-rivets and/or weld clamps. As your confidence grows, do some tack-welding of the whole assembly but continue to offer-up the mating panels and parts. Only when you are absolutely sure all is well should you start fully welding the assembly, removing the temporary fixings and filling the resultant holes right at the end of, say, that side of the car.

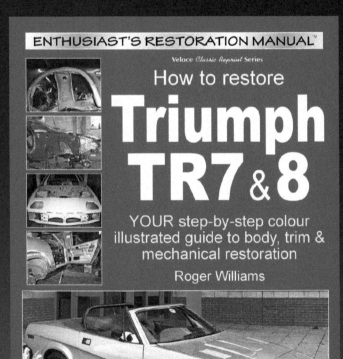

Index